Volume 67

FUNDAMENTALS OF ENHANCED OIL RECOVERY METHODS FOR UNCONVENTIONAL OIL RESERVOIRS

DEVELOPMENTS IN PETROLEUM SCIENCE

Volume 67

FUNDAMENTALS OF ENHANCED OIL RECOVERY METHODS FOR UNCONVENTIONAL OIL RESERVOIRS

DHEIAA ALFARGE

Iraqi Ministry of Oil, University of Karbala

MINGZHEN WEI

Missouri University of Science and Technology

BAOJUN BAI

Missouri University of Science and Technology

SERIES EDITORS

BAOJUN BAI

ZHANGXIN CHEN

ELSEVIER

Elsevier
Radarweg 29, PO Box 211, 1000 AE Amsterdam, Netherlands
The Boulevard, Langford Lane, Kidlington, Oxford OX5 1GB, United Kingdom
50 Hampshire Street, 5th Floor, Cambridge, MA 02139, United States

Library of Congress Cataloging-in-Publication Data
A catalog record for this book is available from the Library of Congress

British Library Cataloguing-in-Publication Data
A catalogue record for this book is available from the British Library

ISBN: 978-0-12-818343-4
ISSN: 0376-7361

For information on all Elsevier publications
visit our website at https://www.elsevier.com/books-and-journals

Publisher: Candice Janco
Acquisitions Editor: Amy Shapiro
Editorial Project Manager: Sara Valentino
Production Project Manager: Sruthi Satheesh
Designer: Miles Hitchen

Typeset by Thomson Digital

Contents

Author's bio xi

Acknowledgments xiii

1 Introduction to shale and tight oil reservoirs 1

1.1 Introduction 1

1.2 Concepts of unconventional resources 2

1.3 Geological characteristics of unconventional liquid-rich reservoirs 4

1.4 The necessity of IOR for unconventional oil reservoirs 6

1.5 Possible IOR methods in unconventional reservoirs 6

1.5.1 EOR methods and their applicability 6

1.5.2 Injectivity problems 10

1.5.3 Conformance problems for IOR methods in unconventional reservoirs 11

1.6 Summary 12

References 12

2 CO_2-EOR in shale-oil reservoirs based on a laboratory database 15

2.1 CO_2-EOR in unconventional reservoirs 16

2.2 CO_2-EOR mechanisms 18

2.3 Rock and fluid properties of unconventional plays 19

2.4 Dataset for laboratory CO_2-EOR projects 27

2.5 Data analysis results and discussions 28

2.5.1 Effect of formation permeability on oil recovery obtained by CO_2 injection 28

2.5.2 Effect of total organic carbon content on oil recovery obtained by CO_2 injection 30

2.5.3 Effect of matrix porosity on oil recovery obtained by CO_2 injection 31

2.5.4 Effect of mean pore throat radius on oil recovery obtained by CO_2 injection 33

2.5.5 Effect of oil saturation on oil recovery obtained by CO_2 injection 34

2.5.6 Effect of water saturation on oil recovery obtained by CO_2 injection 36

2.5.7 Effect of exposure time on oil recovery obtained by CO_2 injection 37

2.5.8 Effect of core-sample bulk volume on oil recovery obtained by CO_2 injection 37

2.5.9 Effect of pressure and temperature on oil recovery obtained by CO_2 injection 39

2.5.10 Sensitivity analysis 40

2.6 Summary 42

References 43

3 Comparative analysis between CO_2-EOR mechanisms in conventional reservoirs versus shale and tight reservoirs **45**

3.1 Introduction 46
3.2 Production data analysis (PDA) for CO_2-EOR in conventional reservoirs 49
3.2.1 CO_2-EOR huff-n-puff projects in conventional reservoirs 49
3.2.2 History matching for CO_2-EOR in conventional reservoirs 49
3.3 CO_2-EOR huff-n-puff operations in conventional reservoirs versus unconventional reservoirs 51
3.4 Production data analysis for CO_2-EOR in unconventional reservoirs 53
3.4.1 CO_2-EOR experiments in unconventional reservoirs 53
3.4.2 History matching with the experimental results of the shale-core samples 55
3.4.3 CO_2-EOR huff-n-puff pilot test in the Bakken formation 56
3.4.4 Reservoir modeling 58
3.4.5 History matching for the CO_2-EOR huff-n-puff pilot test in the Bakken formation 58
3.5 Summary 61
References 62

4 Natural gas-based EOR versus CO_2-EOR in shale and tight oil reservoirs **65**

4.1 Introduction 65
4.2 Experimental setups of miscible gas-based EOR 67
4.3 Simulation models of miscible gas-based EOR 70
4.4 Performance of natural gas EOR in pilot projects 72
4.5 Performance of CO_2-EOR in pilot projects 77
4.6 Lessons learned from miscible gas injection pilots 80
4.7 Summary 82
References 83

5 Air injection in shale and tight oil reservoirs **87**

5.1 Introduction 87
5.2 Types of air combustion 90
5.2.1 Dry forward combustion 90
5.2.2 Wet forward combustion 93
5.2.3 Reverse combustion 94
5.3 Modes of combustion reactions 95
5.3.1 Oxygen addition 95
5.3.2 Low-temperature oxidation or bond scission 95
5.3.3 Spontaneous combustion or high-temperature oxidation 96
5.4 Air injection EOR modeling 97
5.5 Performance of air injection in conventional reservoirs 98

5.6 Advantages and disadvantages of air injection 100
5.7 Performance of air injection in unconventional reservoirs 102
5.8 Air injection pilots in tight oil reservoirs 103
5.9 Pros and cons of applying air injection in shale-oil reservoirs 104
5.10 Numerical simulation for air injection in unconventional reservoirs 105
5.11 Summary 108
References 108

6 Water injection in unconventional reservoirs 113
6.1 Introduction 114
6.2 Oil recovery due to waterflooding 117
6.3 Factors impacting the performance of waterflooding 120
6.4 Waterflooding-derived techniques 121
6.4.1 Low-salinity water injection 121
6.4.2 Carbonated water injection 122
6.5 Water injection in unconventional reservoirs 124
6.6 EOR mechanisms for water injection in unconventional reservoirs 124
6.7 Smart water injection for shale-oil reservoirs 126
6.8 Comparison of water injection in huff-n-puff mode versus flooding mode in shale-oil reservoirs 129
6.9 Factors affecting the performance of water injection in unconventional reservoirs 130
6.10 Field applications for water injection in unconventional reservoirs 132
6.10.1 Case study I 132
6.10.2 Case study II 132
6.10.3 Case study III 134
6.10.4 Case study IV 134
6.10.5 Case study V 135
6.10.6 Case study VI 136
6.10.7 Case study VII 137
6.10.8 Case study VIII 137
6.11 Summary 137
References 138

7 Chemical enhanced oil recovery methods for unconventional reservoirs 141
7.1 Introduction 141
7.2 Chemical EOR methods 145
7.3 Surfactant types 146
7.4 Surfactant huff-n-puff EOR 150
7.5 Surfactant injection during hydraulic fracturing process 151
7.6 Surfactant selection 154

7.6.1 Introduction of surfactant selection criteria 154
7.6.2 Principles of surfactant selection criteria 154
7.6.3 Data compilation 155
7.6.4 Functionality of oil recovery factor with CA, IFT, and capillary pressure 158
7.7 Limitations of surfactant EOR for unconventional reservoirs 160
7.8 Summary 160
References 161

8 Selection criteria for miscible gases-based EOR in unconventional liquid-rich reservoirs (ULR) 165
8.1 Introduction 166
8.2 Significance of molecular diffusion 168
8.3 Numerical simulation setup 169
8.3.1 Model selection 169
8.3.2 Compositional model 171
8.3.3 Systematic methodology for diffusivity level 172
8.4 Results and discussion 174
8.4.1 Natural depletion 174
8.4.2 Flow-type determination 175
8.4.3 EOR stage for Bakken model 176
8.5 Closing remarks 182
8.6 Summary 182
References 183

9 Other enhanced oil recovery methods for unconventional reservoirs 185
9.1 Thermal enhanced oil recovery (EOR) techniques 185
9.2 Electromagnetic EOR techniques for shale oil reservoirs 187
9.3 Microbial EOR 188
9.3.1 Introduction of microbial EOR 188
9.3.2 Mechanisms of microbial EOR (MEOR) 189
9.3.3 Field applications of microbial EOR 191
9.3.4 Pros and cons of microbial EOR 191
9.4 Air-foam injection 192
9.5 Energized fluids for EOR 192
9.6 Other EOR methods 195
9.6.1 CO_2 huff-n-puff followed by surfactant assist spontaneous imbibition (SASI) technique 195
9.6.2 Chemical blend-based EOR for tight and shale oil reservoirs 196
9.6.3 Fracturing improved oil recovery (IOR) techniques 196
9.7 Summary 198
References 198

10 The effects of nanopore confinement on different enhanced oil recovery methods **201**
10.1 Introduction 201
10.2 Effect of pore confinement on primary production 204
10.3 Effect of pore confinement on enhanced oil recovery methods 206
10.3.1 Effect of nanopore confinement on air injection performance 207
10.3.2 Effect of nanopore confinement on CO_2-EOR performance 207
10.3.3 Effect of nanopore confinement on natural gas-based EOR 210
10.3.4 Effect of nanopore confinement on N_2-EOR 213
10.4 Summary 214
References 214

11 The impacts of geomechanics coupling on CO_2-EOR **217**
11.1 Theory 218
11.1.1 Flow type in unconventional liquid-rich reservoirs 218
11.1.2 Molecular diffusion 218
11.1.3 Dissolution 221
11.1.4 Multicomponent adsorption 221
11.1.5 Coupling between the geomechanics module and the reservoir flow model 222
11.2 Compositional models for the formation fluids 227
11.3 Reservoir modeling 228
11.4 Results and discussion 231
11.4.1 Molecular diffusion effect 231
11.4.2 Adsorption effect 233
11.4.3 Dissolution effect (water solubility effect) 236
11.4.4 Geomechanic coupling effects 237
11.5 Summary 241
References 242

12 Comparative and optimization of CO_2 and natural gas EOR methods **245**
12.1 Reservoir modeling 245
12.2 Sensitivity analysis 249
12.3 Results and discussion 251
12.3.1 Parameters which control CO_2-EOR 251
12.3.2 Parameters control NG-EOR 257
12.3.3 Comparison between CO_2-EOR and NGs-EOR 261
12.4 Summary 264
References 265

Index 267

Author's bio

Dr. Dheiaa Alfarge was born in 1989 in Iraq. He received his bachelor's degree in petroleum engineering from Baghdad University, Iraq; he was the valedictorian of Petroleum Engineering Department of class 2011. He received both his MSc and PhD degrees in petroleum engineering from Missouri University of Science and Technology, Rolla, Missouri, United States in 2016 and 2018 consequently with 4.0 GPA. His research interests are mainly in reservoir simulation and EOR methods for conventional and unconventional reservoirs. He published more than 23 papers in different competitive journals as well as conferences around the globe. He worked as a senior petroleum engineer in Iraqi Ministry of Oil in both of OPDC and MOC. Furthermore, he worked as an adjunct professor in University of Karbala, Iraq. Currently, he is serving as a research advisory board member in SRP-Center.iq, Karbala, Iraq. He has been recently appointed as an assistant professor in Izmir Katip Celebi University, Izmir, Turkey.

Dr. Mingzhen Wei, Associate Professor at the Department of Geosciences and Geological and Petroleum Engineering (GGPE), Missouri S&T. She was well trained in Petroleum Engineering (BS and MS) and Computer Science (PhD). Her PhD research was data management, data quality enhancement, applications of data mining, and advanced artificial intelligence computation in Petroleum Engineering. She had 2 years of experience in metadata modeling with Center of Excellence for Geospatial Information Science (CEGIS), US Geological Survey (USGS) as a post-doc. She has devoted her time on Enhanced Oil Recovery (EOR) and unconventional energy resources using reservoir simulation and data analytics methods. She has published more than 100 papers in peer-reviewed journals and SPE conferences.

Dr. Baojun Bai is a Professor and the holder of the Lester R. Birbeck Endowed Chair in Petroleum Engineering Program at Missouri University of Science and Technology (S&T). He holds PhD degrees in Petroleum Engineering from New Mexico Tech and in Petroleum Geology from China University of Geoscience-Beijing. He has 7 years of industry experience as a reservoir engineer and the head of conformance-control Team of RIPED, PetroChina. Dr. Bai was a post-doctoral scholar at the California Institute of Technology before he joined Missouri S&T as a faculty member in 2006. He has more than 20 years of experience in the research areas of enhanced oil recovery, unconventional reservoir development, conformance control, CO_2 sequestration, and Rock characterizations. Dr. Bai has published more than 200 papers in peer-reviewed journals and over 120 papers in SPE conferences by 2020. He served as the JPT Editorial Committee from 2007 to 2013. He serves as the associate editor for *Journal of Petroleum Sciences and Engineering, Geofluids,* and *Petroleum Sciences*. He is an SPE distinguished member.

Acknowledgments

The authors would like to express their sincere gratitude to the Society of Petroleum Engineers (SPE) for returning back the copyrights for some of their SPE publications to be used in some parts of this book. Also, we would like to thank our student, Ali Waqar, for his help in writing some tasks of this book.

Since this book contains some of our research results about the feasibility of different EOR methods in shale and tight oil reservoirs, we would like to thank our research collaborators; they are Dr. Todd Hoffman, Dr. Munqith Aldhaheri, Dr. Mortadha Al Saba, Dr. Abdullah Almansour, Dr. Ali Alhuraishawy, and Dr. Na Zhang. This book also cited some work, which has been published by other authors. Their contribution is sincerely acknowledged.

CHAPTER 1

Introduction to shale and tight oil reservoirs

Contents

1.1 Introduction 1
1.2 Concepts of unconventional resources 2
1.3 Geological characteristics of unconventional liquid-rich reservoirs 4
1.4 The necessity of IOR for unconventional oil reservoirs 6
1.5 Possible IOR methods in unconventional reservoirs 6
  1.5.1 EOR methods and their applicability 6
  1.5.2 Injectivity problems 10
  1.5.3 Conformance problems for IOR methods in unconventional reservoirs 11
1.6 Summary 12
References 12

Abstract

This chapter introduces the basic definitions for shale and tight oil reservoirs. It also describes the main geological characteristics of such unconventional resources and discusses the production techniques used to develop them. Moreover, it explains the main problems, which make the oil recovery in such reservoirs so low. This chapter presents the necessity of using improved oil recovery (IOR) methods, as well as some issues and options during the IOR process.

Keywords: shale oil reservoirs; tight oil reservoirs; unconventional liquid-rich reservoirs; unconventional reservoir development; hydraulic fracturing; unconventional EOR methods; IOR methods in shale oil reservoirs; huff-n-puff versus flooding mode

1.1 Introduction

Unconventional reservoirs have majorly influenced oil industry plans over the past decade (Fig. 1.1). Over 8 million STB/day oil production in the United States comes from these tight and shale oil reservoirs (EIA, 2019). In North America, unconventional liquid-rich reservoirs have cumulatively produced approximately 10 billion barrels of oil so far (Hoffman & Reichhardt, 2019). A dozen years ago, expecting a million barrels of oil production per day from such ultra-tight reservoirs was unheard of. However, due to advancements in drilling very long horizontal wells with multistage hydraulic fracturing, this success in unconventional reservoirs has been achieved and

Fundamentals of Enhanced Oil Recovery Methods for Unconventional Oil Reservoirs
http://dx.doi.org/10.1016/B978-0-12-818343-4.00001-2

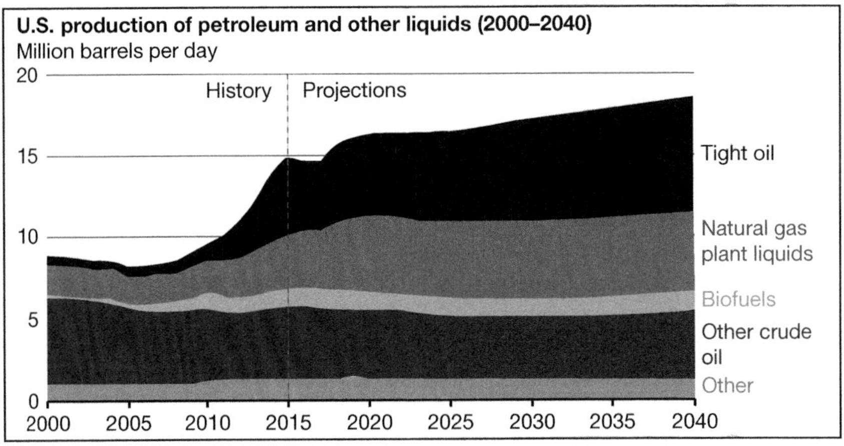

Figure 1.1 ***US oil production: history data and projections (EIA, 2015).***

is expected to increase in the future due to technology advances and production depletion in conventional reservoirs. Furthermore, the volume of oil-in-place in such unconventional reservoirs is enormous, with hundreds of billions of oil barrels estimated for each formation (Downey, Garvin, Lagomarsino, & Nicklin, 2011; Hoffman & Reichhardt, 2019).

Although this tremendous success has occurred for unconventional reservoirs over the last 2 decades, a major issue of the production sustainability associated with such reservoirs still exists, as shown in Fig. 1.2. The wells usually start producing at very high rates, which then drop off quickly, leading to low recovery factors, generally less than 10% (Clark, 2009; Hoffman & Reichhardt, 2019; King, 2014). Therefore, the need for enhanced oil recovery (EOR) methods is inevitable. In spite of the good understanding of EOR methods in conventional reservoirs, they are a new concept in unconventional reservoirs. Therefore, this book is devoted to highlighting and explaining the fundamentals of the EOR methods for unconventional reservoirs.

1.2 Concepts of unconventional resources

Unconventional resources are hydrocarbons (crude oil, natural gas, and condensates) found in very tight reservoirs. Tight reservoirs can be defined as the rocks with small or poorly connected pores in which the oil and natural gas cannot be produced using the traditional technologies applied in the conventional reservoirs. Tight reservoirs are petroleum resources produced

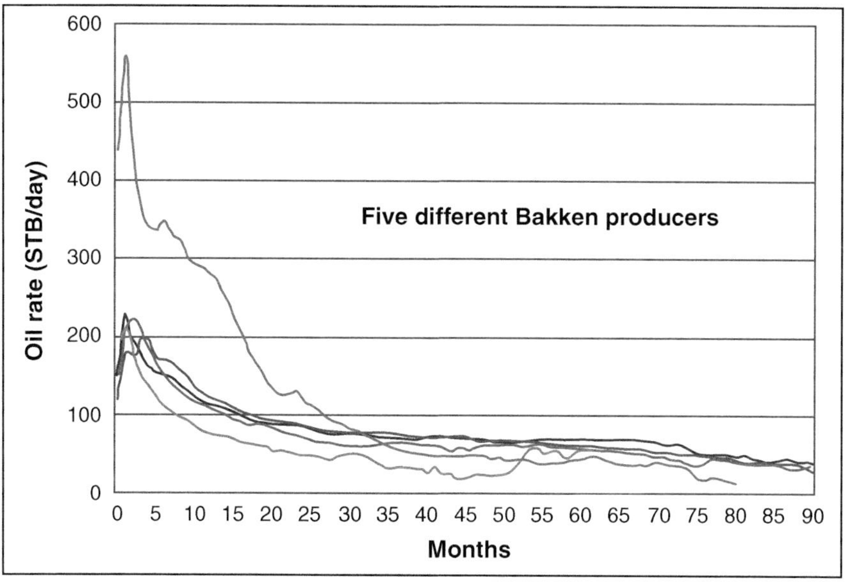

Figure 1.2 ***Production trends of different operators in the Bakken.*** *(Modified from Adekunle & Hoffman, 2016).*

from ultra-low permeability siltstone, sandstone, and carbonate, which are closely related to oil-source shales (Zhang et al., 2016). Hydrocarbons stored in such reservoirs are called "tight gas" or "tight oil." Shale reservoirs are common types of unconventional resources composed of extremely fine-grained sedimentary rocks, which contain oil or natural gas. The oil and gas contained in such rocks is called "shale oil" and "shale gas," respectively. Shale and tight resources have essentially the same type of oil and natural gas as their conventional counterparts. They are considered "unconventional" simply because of the methods used in their extraction, and the types of reservoirs from which they are produced (NRC website). According to the organic theory, unconventional hydrocarbons were formed over millions of years. When organic materials (plants and micro-organisms) were buried under high columns of rocks and subjected to extremely high temperatures and pressure, they slowly transformed into oil and natural gas. When the underground conditions allow the formed hydrocarbons to escape into adjacent rock layers, which are relatively easy to extract because of their high porosity and permeability, they are called conventional reservoirs (Fig. 1.3). However, when the majority of hydrocarbons remain locked in tighter (lower permeability) layers of their source rocks, they are called source rock reservoirs, unconventional liquid-rich reservoirs, or shale oil reservoirs.

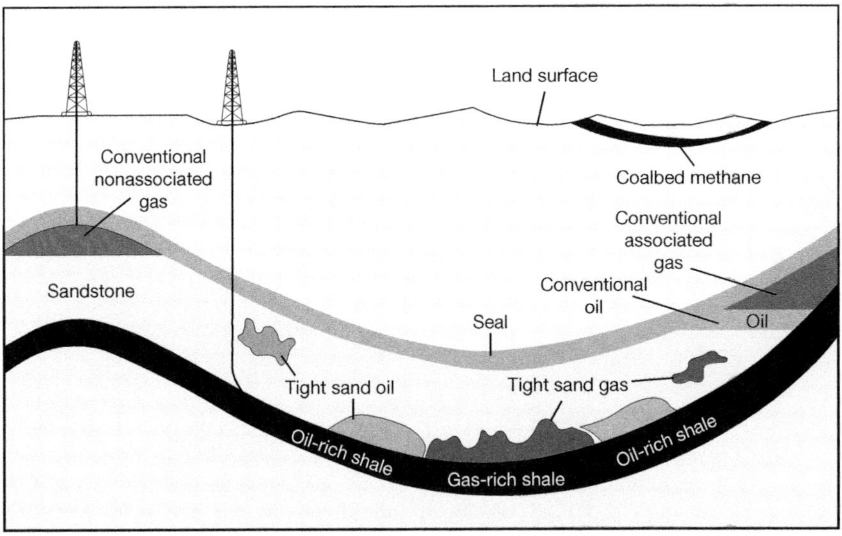

Figure 1.3 ***Diagram showing characteristic deposits of hydrocarbon resources below ground, including conventional oil and gas, shale and tight oil, and gas and coalbed methane.*** *(Modified from Natural Resources Canada, 2019).*

1.3 Geological characteristics of unconventional liquid-rich reservoirs

In conventional reservoirs, pore diameters are usually in the range of micrometers or even larger. The typical value for porosity in conventional reservoirs is generally more than 5% (Satter & Iqbal, 2016).

The permeability is usually more than 0.1 millidarcies, as shown in Fig. 1.4. However, in unconventional reservoirs, porosity is commonly

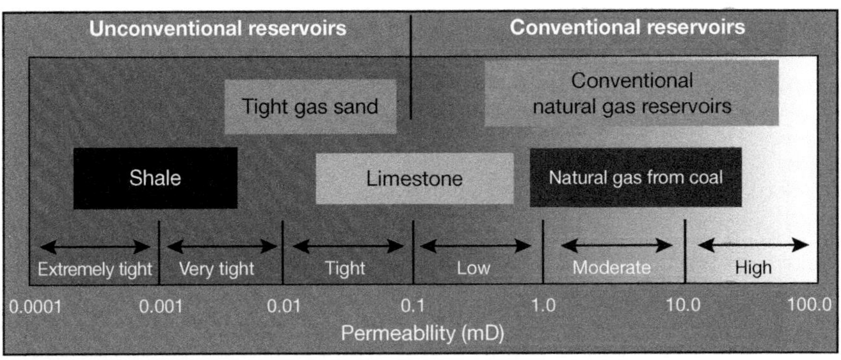

Figure 1.4 ***Types of oil and gas reservoirs according to the permeability cutoffs (CSUR, 2017).***

less than 10%. The typical cutoff for permeability is less than 0.1 millidarcies, whereas tight reservoirs usually have permeability in the range of 0.1–0.001 millidarcies; shale reservoirs are even less permeable, in the range of 0.001–0.0001 millidarcies (NRC, 2019). As a result, because the average permeability of tight and shale reservoirs is usually too small, commercial production using conventional methods, such as vertical wells, is impossible. Therefore, horizontal drilling and hydraulic fracturing are needed to allow commercial production from such ultra-tight reservoirs.

In conventional reservoirs, hydrocarbons migrate (usually upward) to the reservoir rocks until they become trapped against an impermeable rock called the cap rock. This process of migration leads to having conventional reservoirs in localized pools (discrete reservoirs). However, unconventional reservoirs are often found within the same source rocks and distributed over extensive continuous areas rather than concentrated in localized places. For example, the Bakken play extends from the northern area of the United States to the southern part of Canada, covering hundreds of thousands of square miles, as shown in Fig. 1.5. Chapter 2 of this book provides more

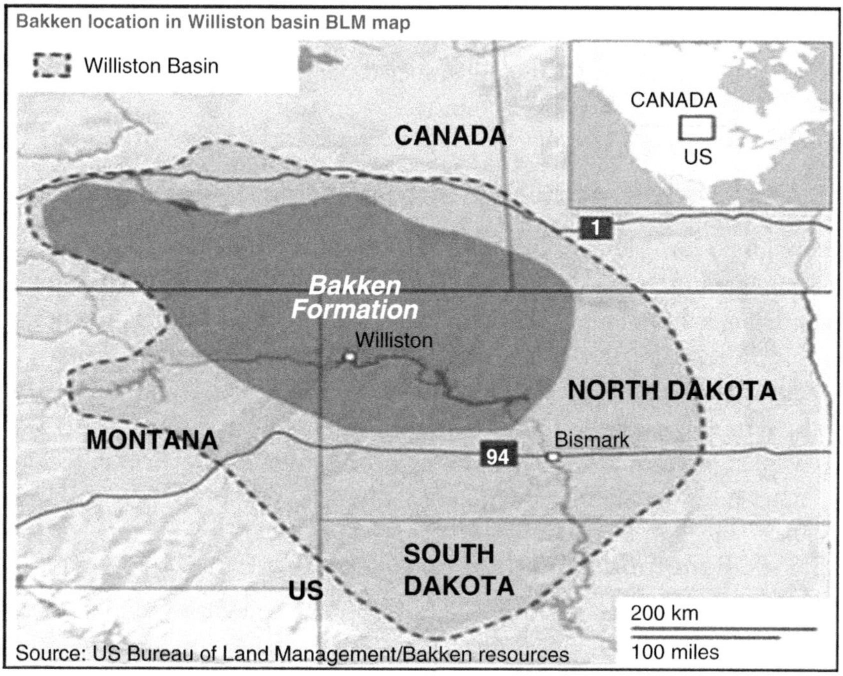

Figure 1.5 ***Diagram showing Bakken play extension.*** *(Modified from US Bureau of Land Management).*

details about the geological characteristics and fluid properties of unconventional reservoirs.

1.4 The necessity of IOR for unconventional oil reservoirs

After the production from the initial wells quickly declines, infill drilling is often used to develop these unconventional reservoirs and achieve a short-term increment in the oil production; however, this high oil rate from the new wells does not last long, just as seen in the previous wells. In addition, the cost of drilling new horizontal wells with long lateral lengths is very high. Therefore, infill drilling might not be the most economic practice in such reservoirs. Seeking different options is mandatory. The main drive mechanism in most such reservoirs is the depletion drive mechanism, which can recover up to 8%–12% of the original oil in place (OOIP). Such low oil recovery obtained in the primary stage is the main motivation to apply one of the improved oil recovery (IOR) methods in these reservoirs (Kurtoglu, Kazemi, Rosen, Mickelson, & Kosanke, 2014). Since these reservoirs have a huge OOIP, any improvement in the oil recovery factor would result in enormous produced oil volumes.

Therefore, IOR methods have great potential in these huge reserves. Although IOR methods are well understood in conventional reservoirs, they are new in unconventional ones.

1.5 Possible IOR methods in unconventional reservoirs

Although no formal definition exists, IOR typically refers to any process or practice that improves oil recovery. EOR refers to the oil recovery resulting from the injection of materials not normally present in the reservoir. Therefore, IOR includes EOR processes but can also include other practices, such as waterflooding, pressure maintenance, infill drilling, and horizontal wells. In this book, the water injection process in unconventional reservoirs has been considered as an EOR process because of the difference in recovery mechanisms as compared with conventional waterflooding.

1.5.1 EOR methods and their applicability

Over the past decade, extensive studies have been conducted to investigate the applicability of different EOR methods in unconventional reservoirs.

The reported studies used different approaches, such as experimental, simulation, and pilot tests. Also, they reported different mechanisms by which each method improves oil recovery. Some EOR methods are more applicable than others and major obstacles in applying the most reported EOR methods still exist, as shown in Table 1.1 (Alfarge, Wei, & Bai, 2017).

According to an extensive review by Alfarge et al. (2017), CO_2, natural gas, surfactant, and water are the most applicable IOR methods for unconventional reservoirs, as shown in Fig. 1.6. The most common mechanisms being reported for these IOR methods are listed in Table 1.2. CO_2 injection tops the list of these applicable EOR methods according to simulation studies and experimental work. However, natural gases performed better than CO_2 in pilot tests (Hoffman & Evans, 2016; Schmidt & Sekar, 2014), indicating that there is something missing which creates this gap between lab work and pilot tests for the CO_2 injection technique. Specifically, insufficient understanding of the physical and chemical mechanisms for CO_2 might be the main reason for such disappointing results from CO_2 pilot tests.

Although most of the investigators in the literature have believed that the right choice is to inject any EOR agent into these poor-quality plays using the huff-n-puff protocol, this is not always the proper practice. There are one or two primary reasons for this belief. The first reason is that the ultralow permeability in such tight reservoirs might prevent or impair any flooding process. The second is that due to the high diffusivity rate of miscible gases into shale cores observed in the lab scale conditions, most of the previous researchers likely recommended using the gas cyclic process over continuous flooding because the soaking period would better assist oil recovery. Therefore, the huff-n-puff protocol would be more economic for such reservoir conditions.

The choice between the cyclic protocol or continuous injection should technically depend on two main factors (Alfarge et al., 2017). The first and perhaps most important factor is the ratio of the reservoir permeability to the injector–producer spacing. When this ratio is higher than the critical economic value for the target reservoir, continuous flooding will perform better than the cyclic protocol. For example, when the reservoir permeability is high enough to conduct an economic sweeping process from the injector to the producer at their suggested spacing, the continuous mode might be highly recommended. However, when the reservoir permeability is too low to handle an economic sweeping process at the suggested spacing between the injector and producer, and the diffusion rate of the injected gas is significantly high through such reservoir conditions, the cyclic protocol should be selected. The second factor when choosing between the two methods is the diffusivity rate of the injected gases through unconventional reservoirs.

Table 1.1 Applicability of different EOR methods in unconventional reservoirs.

IOR category	IOR method	Applicability	Reservoir/fluid problems	Solution	Challenges
Thermal	All methods	Possible	Although the shale oils are light and low viscosity, the reservoir depletion and oil residuality will be the target for this method	Pressure maintenance and oil fluidity	Needs more investigation
Chemical	Surfactant	Yes	Oil wet, high IFT, negative PC	IFT reducer, and wettability alteration	Needs field tests
	Polymer	Unlikely	Reservoir heterogeneity	Improve sweep efficiency	Injectivity
	Alkaline	Not investigated	Wettability	Change wettability	Needs to be investigated
Microbial	Surfactant	Not investigated	Oil wet, high IFT, negative PC	Change wettability, IFT reducer	Not investigated
	Biopolymer	Likely	Reservoir heterogeneity	Would create problems, pore throat blocking	Not investigated
Air injection	Air injection: huff-n-puff	Likely	Depletion and poor sweep efficiency	Repressurization and oil extraction	Dangerous and not investigated in the pilot tests

	Air injection: flooding	Likely	Depletion and poor sweep efficiency	Pressure maintenance and improve oil fluidity	Dangerous and not investigated in the pilot tests
Gas injection	Natural gas huff-n-puff	Yes	Depletion and poor sweep efficiency	Repressurization and oil extraction	Produced back, not successful in field pilots
	Natural gas flooding	Yes	Depletion and poor sweep efficiency	Pressure maintenance and oil swelling	Conformance control
	CO_2 huff-n-puff	Yes	Depletion and poor sweep efficiency	Repressurization and oil extraction	Produced back, not successful in field pilots
	CO_2 flooding	Yes	Depletion and poor sweep efficiency	Pressure maintenance and oil swelling	Conformance control
Water	WAG	Yes	Depletion and poor sweep efficiency	Pressure maintenance	Needs field tests
	Water	Yes	Depletion and poor sweep efficiency	Pressure maintenance	Conformance control
	LSW	Likely	Negative PC, poor sweep efficiency	Enhancing water imbibition by osmosis	Needs field tests

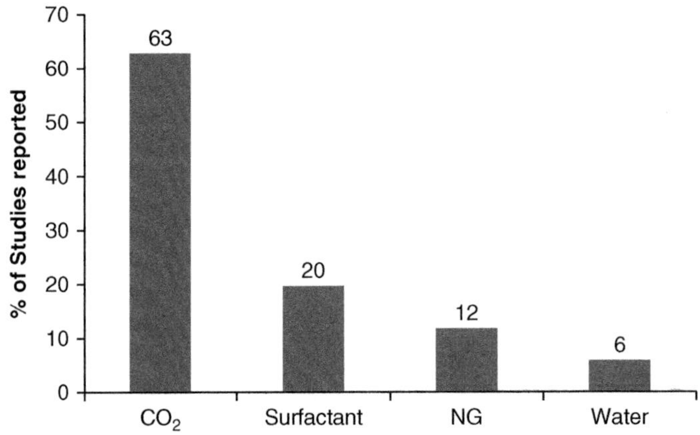

Figure 1.6 ***EOR methods with most potential for unconventional reservoirs.***

Table 1.2 EOR methods with most potential and their mechanisms in ULR.

IOR method	Mechanisms
CO_2	1. Diffusion mostly from lab work 2. Reduction in capillary forces 3. Repressurization 4. Extraction 5. Oil swelling and pressure maintenance 6. Oil viscosity reduction 7. Combination of some of the above
Surfactant	1. Wettability alteration 2. IFT reduction 3. Enhancement of water imbibition
NG	1. High compressibility to displace oil in matrix 2. Pressure maintenance 3. Oil swelling
LSW/water	1. Shale cracking 2. Osmotic effect 3. Wettability alteration by ionic layer 4. Changing PH

1.5.2 Injectivity problems

The main concern for all unconventional EOR methods is how to inject fluids into such ultra-low permeability reservoirs. Interestingly, some EOR pilot tests have reported that injectivity is not a significant issue for such complex plays. Having hydraulic fractures combined with many natural fractures might be the main reason behind the unexpected injectivity reported in such very tight reservoirs. However, some researchers speculate on whether this injectivity is natural or induced, resulting from the new fractures generated by the injection processes. Globally, some researchers argue that approximately 60% of the projects implementing a water flooding process create injection-induced fractures (IIFs) (Baker et al., 2016). The possibility of creating more IIFs in oil reservoirs increases when: (1) the reservoir permeability is very low, and (2) the viscosity of the injected fluids is very low, such as when injecting water and miscible gases (Baker et al., 2016). Therefore, the characteristics of unconventional reservoirs and the properties of the injectants tested in the reported IOR pilots indicate a significant chance of inducing new fractures. Moreover, it is not necessary to raise the injection pressure above the parting pressure to create fractures in shale plays; it is quite possible to induce new fractures through pressure depletion, thermal effects, or plugging effects (Baker et al., 2016).

1.5.3 Conformance problems for IOR methods in unconventional reservoirs

Conformance problems have been reported in several EOR pilots conducted in unconventional reservoirs (Hoffman & Evans, 2016). In reality, the reasons behind the conformance problems reported in these pilot tests have not been clearly diagnosed. Questions remain as to whether the conformance problems happen due to preexisting natural fractures, hydraulic fractures, and/or IIFs. Although unconventional reservoirs have a high incidence of natural fractures, some researchers do not think that preexisting natural fractures are the major contributors to such severe conformance problems as noticed in the US-Bakken and reported by Hoffman and Evans (2016). We suspect (Alfarge et al., 2017) that such conformance problems could happen due to IIFs. Although some pilots injected fluids at an injection pressure which was lower than the fracturing pressure, most injected at an injection pressure very close to the fracturing pressure (Alfarge et al., 2017). Furthermore, the reported EOR pilots were conducted on depleted reservoirs, so using the initial fracturing pressure, as a reference pressure might be a dangerous practice. In particular, these types of reservoirs have soft shales, which require geomechanic-coupling methods to calculate the new fracturing pressure at any new pore pressure (Alfarge, Wei, & Bai, 2018).

To determine whether or not these conformance problems resulted from IIFs, Baker et al. (2016) recommended a good strategy that can be applied for water flooding processes. This strategy includes injecting any fluid at different injection rates, then monitoring how the injection pressure changes. If the change in injection pressures versus the change in injection rates does not follow the diffusive Darcy law, the conformance problems could be due to IIFs. However, severe conformance problems could also happen due to completion interference among the hydraulically fractured horizontal wells (Schmidt & Sekar, 2014).

1.6 SUMMARY

- Unconventional reservoirs have significantly impacted the oil industry over the last decade and have a tremendous potential to be the main players in oil production in the future.
- The EOR methods for unconventional reservoirs are not yet mature, and the need to develop and understand them is critical.
- A review of unconventional reservoir properties suggested that wettability, heterogeneity, and depletion are the main targets for IOR methods.
- CO_2, natural gas, surfactant, and water are likely the most feasible IOR methods in shale reservoirs.
- Different mechanisms for each EOR method have been reported in the literature.
- The pilot tests reported that injectivity is not as big of an obstacle as conformance problems for IOR methods in unconventional reservoirs.

References

Adekunle, O., & Hoffman, B.T. (2016). Experimental and analytical methods to determine minimum miscibility pressure (MMP) for Bakken formation crude oil. *Journal of Petroleum Science and Engineering, 146*, 170–182.

Alfarge, D., Wei, M., & Bai, B. (2017). *IOR methods in unconventional reservoirs of North America: Comprehensive review*. Society of Petroleum Engineersdoi: 10.2118/185640-MS.

Alfarge, D., Wei, M., & Bai, B. (2018). *Mechanistic study for the applicability of CO_2-EOR in unconventional liquids rich reservoirs*. Society of Petroleum Engineersdoi: 10.2118/190277-MS.

Baker, R., et al. (2016). *The myths of waterfloods, EOR floods and how to optimize real injection schemes.* SPE-179536-MS paper was prepared for presentation at the SPE improved oil recovery conference held in Tulsa, Oklahoma, USA, 11–13 April 2016. Society of Petroleum Engineers.

Clark, A.J. (2009). *Determination of recovery factor in the Bakken formation, Mountrail County, ND.* Society of Petroleum Engineersdoi: 10.2118/133719-STU.

CSUR. (2017). *Canadian Society for Unconventional Resources*. https://www.csur.com/.

Downey, M.W., Garvin, J., Lagomarsino, R.C., & Nicklin, D.F. (2011). *Quick look determination of oil-in-place in oil shale resource plays, Eagle Ford Shale: American Association of Petroleum Geologists (AAPG) Search and Discovery Article No. 40764.*

EIA (2015). *US Energy Information Administration.* https://www.eia.gov/.

EIA (2019). *US Energy Information Administration.* https://www.eia.gov/.

Hoffman, B.T., & Reichhardt, D. (2019). *Quantitative evaluation of recovery mechanisms for huff-n-puff gas injection in unconventional reservoirs.* Unconventional Resources Technology Conference. https://doi.org/10.15530/urtec-2019-147.

Hoffman, B.T., & Evans, J. (2016). *Improved oil recovery IOR pilot projects in the Bakken formation.* SPE-180270-MS paper presented at the SPE low perm symposium held in Denver, Colorado, USA, 5–6 May 2016. Society of Petroleum Engineers.

King, G. E. (2014). *60 years of multi-fractured vertical, deviated and horizontal wells: What have we learned?* Society of Petroleum Engineersdoi: 10.2118/170952-MS.

Kurtoglu, B., Kazemi, H., Rosen, R., Mickelson, W., & Kosanke, T. (2014). *A rock and fluid study of middle Bakken formation: Key to enhanced oil recovery*. Society of Petroleum Engineersdoi: 10.2118/171668-MS.

Natural Resources Canada. (2019). https://www.nrcan.gc.ca/home. NRC Website. https://www.nrcan.gc.ca/energy/sources/shale-tight-resources/17675.

Satter, A., & Iqbal, G. (2016). *Reservoir engineering: The fundamentals, simulation, and management of conventional and unconventional recoveries*. Elsevier Book. ISBN: 978-0-12-800219-3.

Schmidt, M., & Sekar, B. K. (2014). *Innovative unconventional 2EOR-A Light EOR an unconventional tertiary recovery approach to an unconventional Bakken reservoir in Southeast Saskatchewan.* World Petroleum Congress.

Zhang, X. S., Wang, H. J., Ma, F., Sun, X. -C., Zhang, Y., & Song, Z. -H. (2016). Classification and characteristics of tight oil plays. *Petroleum Science, 13*, 18. doi: 10.1007/s12182-015-0075-0.

CHAPTER 2

CO_2-EOR in shale-oil reservoirs based on a laboratory database

Contents

2.1 CO_2-EOR in unconventional reservoirs 16
2.2 CO_2-EOR mechanisms 18
2.3 Rock and fluid properties of unconventional plays 19
2.4 Dataset for laboratory CO_2-EOR projects 27
2.5 Data analysis results and discussions 28
2.5.1 Effect of formation permeability on oil recovery obtained by CO_2 injection 28
2.5.2 Effect of total organic carbon content on oil recovery obtained by CO_2 injection 30
2.5.3 Effect of matrix porosity on oil recovery obtained by CO_2 injection 31
2.5.4 Effect of mean pore throat radius on oil recovery obtained by CO_2 injection 33
2.5.5 Effect of oil saturation on oil recovery obtained by CO_2 injection 34
2.5.6 Effect of water saturation on oil recovery obtained by CO_2 injection 36
2.5.7 Effect of exposure time on oil recovery obtained by CO_2 injection 37
2.5.8 Effect of core-sample bulk volume on oil recovery obtained by CO_2 injection 37
2.5.9 Effect of pressure and temperature on oil recovery obtained by CO_2 injection 39
2.5.10 Sensitivity analysis 40
2.6 Summary 42
References 43

Abstract

In this chapter, CO_2-EOR mechanisms in unconventional reservoirs are introduced, and the early lab and simulation studies, which investigated the fundamentals of CO_2-EOR in unconventional reservoirs, are presented. Data are analyzed to determine the feasibility of CO_2-EOR in 95 cases of natural preserved cores collected from different formations, including 44 cases from the Middle Bakken, 26 cases from the Lower Bakken, 17 cases from the Upper Bakken, 4 cases from Three Forks, and 4 cases from an unknown formation/formations. The relationship between the improved oil recovery achieved by the CO_2 injection and six rock properties, including porosity, permeability, mean pore throat radius, total organic carbon content (TOC), water saturation, and oil saturation, has been separately determined and is discussed here. Furthermore, the relationships between the improved oil recovery achieved by the injected CO_2 and four operating

Fundamentals of Enhanced Oil Recovery Methods for Unconventional Oil Reservoirs
http://dx.doi.org/10.1016/B978-0-12-818343-4.00002-4

parameters, including pressure, temperature, core sample bulk volume, and exposure time, have also been separately investigated. A proxy model to associate the functionality of the improved oil recovery by CO_2 injection and these 10 parameters has been constructed. Moreover, statistical methods for the design of experiments (DOEs) have been used to rank the most important parameters affecting CO_2-EOR performance in the microscale level (lab scale) of these unconventional reservoirs. This chapter provides an important set of lab-based data obtained from natural preserved cores to determine the applicability of CO_2-EOR in these unconventional reservoirs. Also, this research demonstrates some key points which could help in understanding CO_2-EOR mechanisms in shale plays because they are very complex and much different from conventional formations.

Keywords: CO_2-EOR lab database; CO_2-EOR mechanisms in shale and tight oil reservoirs; CO_2-EOR huff-n-puff; CO_2-EOR data analysis

Nomenclature

BPS Bakken petroleum system
DOE design of experiments
EIA Energy Information Administration
EOR enhanced oil recovery
IOR improved oil recovery
K permeability
LB lower Bakken formation
MB middle Bakken formation
OOIP original oil in place
RF oil recovery factor
SO oil saturation
SW water saturation
TF Three Forks formation
TOC total organic carbon content
UB upper Bakken
ULR unconventional liquid rich reservoirs

2.1 CO_2-EOR in unconventional reservoirs

CO_2-EOR is one of the most investigated methods for improving oil recovery in unconventional liquid-rich reservoirs. CO_2 injection has several advantages over other gas types. It is a supercritical fluid in all shale reservoir conditions with a reservoir temperature above 88°F and reservoir pressure above 1071 psi. As a supercritical gas, CO_2 has a density of liquids, which reduces the gravity segregation. CO_2 dissolves in shale oil easily, which swells the oil and lowers its viscosity, leading to improved oil mobility. CO_2 has a lower miscibility pressure with shale oil compared to other gases, such as N_2 and CH_4 (Zhang, 2016); therefore, CO_2 injection can take advantage of miscible drive. The low value of the acid number, which has been reported in unconventional reservoirs, lends hope to the successful application of CO_2-EOR without much concern about asphaltene precipitation (Kurtoglu et al., 2014).

Earlier published studies investigated CO_2-EOR in unconventional reservoirs using numerical modeling methods (Shoaib & Hoffman, 2009; Wang, Luo, Er, & Huang, 2010). Their results showed that 10%–20% of incremental oil could be recovered by continuous gas flooding, while 5%–10% could be recovered using the huff-n-puff protocol (Hoffman & Evans, 2016). Dong and Hoffman (2013) reported a numerical study evaluating CO_2-EOR performance for the Bakken interval in the Sanish field sector. They developed a model structure to increase CO_2 injectivity in that field by drilling more horizontal injection wells. They predicted the possibility of injecting 5000 Mscf/day at a maximum injection pressure of 8000 psi and found that using the CO_2 injection method might increase oil recovery from 5% to 24% in that field. Xu and Hoffman (2013) evaluated the reservoir performance of the Elm Coulee Field in Eastern Montana under CO_2 flooding with different hydraulic fracture orientations. They found that transverse fractures have a higher oil recovery factor, but they have a lower CO_2 utilization value than longitudinal fractures due to breakthrough problems. Zhu, Balhoff, and Mohanty (2015) constructed a model in which CO_2 could be injected into a hydraulic fracture along a horizontal well, and the production process could occur from an adjacent fracture intersecting with the same well. They found a substantial improvement in oil recovery when injecting CO_2 in reservoirs with a fluid flow from fracture to fracture. Pu and Li (2016) introduced a new model, which considered the capillarity and adsorption effect for the small pores of shale reservoirs and was able to properly simulate CO_2-EOR in unconventional reservoirs. The models, which considered capillarity in their simulation process for CO_2-EOR, predicted a higher oil recovery than the cases, which did not consider capillarity. Regarding experimental studies, Song and Yang (2013) conducted experimental investigations to compare results from injecting CO_2 and water in cores from the Canadian Bakken. They found that water flooding would enhance oil recovery better than immiscible CO_2 using the huff-n-puff protocol. However, the performance of miscible and near-miscible CO_2 huff-n-puff would exceed water performance in terms of enhancing oil recovery. Hawthorne et al. (2013) investigated the mechanisms causing incremental oil recovery when injecting CO_2 into the Bakken cores. They proved that diffusion is the main mechanism by which CO_2 increases oil recovery in these complex plays. However, extracting oil from the shale matrix using CO_2 requires long exposure times and large contact areas.

Gamadi et al. (2014) conducted experiments using shale cores from Mancos and Eagle Ford to investigate the EOR potential of CO_2-EOR in these reservoirs. Their laboratory results indicated that cyclic CO_2 injection could improve oil recovery in shale oil cores from 33% to 85%, depending on the shale core type and other operating parameters. Alharthy et al. (2015) compared EOR performance in the Bakken cores when injecting different types of gases, such as CO_2, C_1–C_4 mixtures, and N_2. They concluded that injecting gas composed of C_1, C_2, C_3, and C_4 could produce nearly as much oil as CO_2 injection did, which was 90% from several of the middle Bakken cores and nearly 40% from the lower Bakken cores. Also, they found that the counter-current mechanism was the main mechanism for these gases to recover more oil from shale cores. Finally, Yu and Sheng (2016) investigated the N_2 flooding process experimentally on Eagle Ford core plugs saturated with dead oil. They examined the effect of different flooding time ranges and different injection pressures on N_2 flooding performance, finding that more oil could be produced with a longer flooding time and higher injection pressure.

2.2 CO_2-EOR mechanisms

Different mechanisms have been proposed for the injected CO_2 to improve oil recovery in unconventional reservoirs, as shown in Table 2.1. However, because the matrix permeability in these unconventional reservoirs is in the range of 0.1–0.00001 mD, CO_2 would not be transported by convection

Table 2.1 Proposed CO_2-EOR mechanisms for improving oil recovery in unconventional reservoirs.

CO_2 mechanism	Approach tool
1. Molecular diffusion	Lab
2. Reduction in capillary forces	Lab and simulation
3. Repressurization	Lab
4. Extraction	Lab
5. Oil swelling and pressure maintenance	Lab and simulation
6. Oil viscosity reduction	Lab and simulation
7. Combination of more than one mechanism from above	–

flux from the fracture to the matrix (Yu et al., 2014). The main transportation method for CO_2 occurs through the difference in the CO_2 concentration gradient between the injected gases and the target oil. This process of transportation is subjected to Fick's law, which is called molecular diffusion transportation. Molecular diffusion is defined as the movement of molecules caused by Brownian motion or the composition gradient in a mixture of fluids (Mohebbinia & Wong, 2017). This type of flow would be the most dominant flow in fractured reservoirs with a low-permeability matrix when gravitational drainage is inefficient (Mohebbinia & Wong, 2017; Moortgat & Firoozabadi, 2013). Researchers have noticed that gas injection is the most common EOR process affected by the molecular-diffusion mechanism. Hawthorne et al. (2013) extensively investigated the CO_2 diffusion mechanism in the Bakken cores and proposed the following five conceptual steps to explain it: (1) CO_2 flows into and through the fractures, (2) an unfractured rock matrix is exposed to CO_2 at the fracture surfaces, (3) CO_2 permeates the rock driven by pressure, carrying some hydrocarbon inward; however, the oil is also swelling and extruding some oil out of the pores, (4) oil migrates to the bulk CO_2 in the fractures via swelling and reduced viscosity, and (5) as the CO_2 pressure gradient decreases, oil production is slowly driven by concentration-gradient diffusion from pores into the bulk CO_2 in the fractures.

2.3 Rock and fluid properties of unconventional plays

Unconventional plays have their own geological properties, and a general understanding of the target formation's rock and fluid properties is necessary to predict the applicability of any EOR method. According to US EIA (2013) reports, some of the major common properties of unconventional plays can be summarized as follows:

- Depositional Environment: One of the most important criteria for understanding shale reservoirs is the depositional environment of the shale, particularly whether it is marine or non-marine. Marine-deposited shales are usually classified as having lower clay content and tend to be high in brittle minerals, such as quartz, feldspar, and carbonates. In terms of hydraulic fracturing, brittle shales respond favorably to hydraulic stimulation, while shales deposited in non-marine settings (lacustrine, fluvial) tend to be higher in clay, more ductile, and less responsive to hydraulic stimulation.
- Depth: The depth criterion has a significant effect on the performance of the oil production in both the primary stage and the EOR stage. The

formation depth for unconventional plays is greater than 1,000 m but less than 5,000 m (3,300 ft. to 16,500 ft.). Reservoir areas, which are shallower than 1,000 m, have lower reservoir pressure and thus lower driving forces for oil and gas recovery. In addition, shallow shale formations usually carry the risk of being saturated with higher water content in their natural fracture systems. However, reservoir areas, which are deeper than 5,000 m usually, have nano-Darcy permeability and much higher drilling and development costs.

- Total Organic Content (TOC): In general, the average TOC of shale plays varies widely but seems usually to be greater than 2%. Organic materials, such as microorganism fossils and plant matter serving as the main resources for carbon, oxygen, and hydrogen atoms, are needed to produce natural gas and oil. Therefore, the TOC and carbon type (Types I and II) are very important factors, which control the potential of the oil in shale formations.
- Thermal Maturity: Thermal maturity is defined as the criteria measuring the degree to which a formation has been exposed to the high heat required to break down organic matter into hydrocarbons. The reflectance of certain types of minerals (Ro%) is generally used as the primary indicator of thermal maturity for hydrocarbon reservoirs. R_o can have a very wide range; for example, for some shale oil plays, R_o is greater than 0.7% but less than 1.0%. However, the wet gas and condensate areas in unconventional formations have an R_o between 1.0% and 1.3%, while dry gas areas typically have an R_o greater than 1.3%.

Shale oil basins and formations usually have very high clay content and/or very high geologic complexity (e.g., thrusted and high stress). They are usually abnormally pressurized with a high incidence of natural fractures. The major properties for shale plays can be found in Tables 2.2 and 2.3, which were gathered from a study conducted by "Tight Oil" Resources International and reported by the U.S. Energy Information Administration (EIA) in 2013.

The Bakken formation is the most productive oil-producing formation among unconventional plays in North America, with an estimated oil-in-place of 100–900 billion barrels and 7.4 billion barrels of technically recoverable oil (Gaswirth et al., 2013). Furthermore, approximately 90% of the data from the laboratory projects presented in this chapter were gathered from the Bakken Petroleum System (BPS). Therefore, the Bakken's geological structure is described here to help us understand its conditions before deciding the feasibility of CO_2-EOR. The most common rock and fluid

Table 2.2 Reservoir pressure, thermal maturity, and formation volume factor for major shale plays (US EIA, 2013).

Basin	Formation/Play	Age	Reservoir pressure	Thermal maturity (% R_o)	Formation volume factor (B_{oi})
Williston	Bakken ND Core	Mississippian-Devonian	Overpressured	0.80%	1.35
	Bakken ND Ext.	Mississippian-Devonian	Overpressured	0.80%	1.58
	Bakken MT	Mississippian-Devonian	Overpressured	0.75%	1.26
	Three Forks ND	Devonian	Overpressured	0.85%	1.47
	Three Forks MT	Devonian	Overpressured	0.85%	1.27
Maverick	Eagle Ford Play #3A	Late Cretaceous	Overpressured	0.90%	1.75
	Eagle Ford Play #3B	Late Cretaceous	Overpressured	0.85%	2.01
	Eagle Ford Play #4A	Late Cretaceous	Overpressured	0.75%	1.57
	Eagle Ford Play #4B	Late Cretaceous	Overpressured	0.70%	1.33
Ft. Worth	Barnett Combo - Core	Mississippian	Slightly Overpressured	0.90%	1.53
	Barnett Combo - Ext.	Mississippian	Slightly Overpressured	0.80%	1.41
Permian	Del. Avalon/BS (NM)	Permian	Slightly Overpressured	0.90%	1.70
	Del. Avalon/BS (TX)	Permian	Slightly Overpressured	0.90%	1.74
	Del. Wolfcamp (TX Core)	Permian-Pennsylvanian	Slightly Overpressured	0.92%	1.96
	Del. Wolfcamp (TX Ext.)	Permian-Pennsylvanian	Slightly Overpressured	0.92%	1.79
	Del. Wolfcamp (NM Ext.)	Permian-Pennsylvanian	Slightly Overpressured	0.92%	1.85
	Midl. Wolfcamp Core	Permian-Pennsylvanian	Overpressured	0.90%	1.67
	Midl. Wolfcamp Ext.	Permian-Pennsylvanian	Overpressured	0.90%	1.66
	Midl. Cline Shale	Pennsylvanian	Overpressured	0.90%	1.82
Anadarko	Cana Woodford - Oil	Upper Devonian	Overpressured	0.80%	1.76
	Miss. Lime - Central OK Core	Mississippian	Normal	0.90%	1.29
	Miss. Lime - Eastern OK Ext.	Mississippian	Normal	0.90%	1.20
	Miss. Lime - KS Ext.	Mississippian	Normal	0.90%	1.29

(Continued)

Table 2.2 Reservoir pressure, thermal maturity, and formation volume factor for major shale plays (US EIA, 2013). (*Cont.*)

Basin	Formation/Play	Age	Reservoir pressure	Thermal maturity (% R_o)	Formation volume factor (B_{oi})
Appalachian	Utica Shale - Oil	Ordovician	Slightly Overpressured	0.80%	1.46
D-J	D-J Niobrara Core	Late Cretaceous	Normal	1.00%	1.57
	D-J Niobrara East Ext.	Late Cretaceous	Normal	0.70%	1.26
	D-J Niobrara North Ext. #1	Late Cretaceous	Normal	0.70%	1.37
	D-J Niobrara North Ext. #2	Late Cretaceous	Normal	0.65%	1.28

Table 2.3 Age, oil-in-place, and oil recovery factor for major shale plays (US EIA, 2013).

Basin	Formation/Play	Age	Oil in-place (MBbls/Mi^2)	Oil recovery (MBbls/Mi^2)	Oil recovery efficiency (%)
Williston	Bakken ND Core	Mississippian-Devonian	12,245	1,025	8.4%
	Bakken ND Ext.	Mississippian-Devonian	9,599	736	7.7%
	Bakken MT	Mississippian-Devonian	10,958	422	3.9%
	Three Forks ND	Devonian	9,859	810	8.2%
	Three Forks MT	Devonian	10,415	376	3.6%
Maverick	Eagle Ford Play #3A	Late Cretaceous	22,455	1,827	8.1%
	Eagle Ford Play #3B	Late Cretaceous	25,738	2,328	9.0%
	Eagle Ford Play #4A	Late Cretaceous	45,350	1,895	4.2%
	Eagle Ford Play #4B	Late Cretaceous	34,505	2,007	5.8%
Ft. Worth	Barnett Combo - Core	Mississippian	25,262	377	1.5%
	Barnett Combo - Ext.	Mississippian	13,750	251	1.8%
Permian	Del. Avalon/BS (NM)	Permian	34,976	648	1.9%
	Del. Avalon/BS (TX)	Permian	27,354	580	2.1%
	Del. Wolfcamp (TX Core)	Permian-Pennsylvanian	35,390	1,193	3.4%
	Del. Wolfcamp (TX Ext.)	Permian-Pennsylvanian	27,683	372	1.3%
	Del. Wolfcamp (NM Ext.)	Permian-Pennsylvanian	21,485	506	2.4%
	Midl. Wolfcamp Core	Permian-Pennsylvanian	53,304	1,012	1.9%
	Midl. Wolfcamp Ext.	Permian-Pennsylvanian	46,767	756	1.6%
	Midl. Cline Shale	Pennsylvanian	32,148	892	2.8%
Anadarko	Cana Woodford - Oil	Upper Devonian	11,413	964	8.4%
	Miss. Lime - Central OK Core	Mississippian	28,364	885	3.1%
	Miss. Lime - Eastern OK Ext.	Mississippian	30,441	189	0.6%
	Miss. Lime - KS Ext.	Mississippian	21,881	294	1.3%
Appalachian	Utica Shale - Oil	Ordovician	42,408	906	2.1%
D-J	D-J Niobrara Core	Late Cretaceous	33,061	703	2.1%
	D-J Niobrara East Ext.	Late Cretaceous	30,676	363	1.2%
	D-J Niobrara North Ext. #1	Late Cretaceous	28,722	1,326	4.6%
	D-J Niobrara North Ext. #2	Late Cretaceous	16,469	143	0.9%

properties, which have been reported previously in different resources for the Bakken formation, have been gathered in Tables 2.4 and 2.5, respectively. The Bakken formation consists of two portions located in the Williston Basin, as shown in Fig. 2.1A. The first portion, which has a vertical

Table 2.4 Common rock properties of BPS.

Reservoir characteristics	Common quality	Common quantity	References
Porosity (%)	Low	5–10	SPE-178659-MS; SPE-178489-PA
Permeability (mD)	Ultralow	0.0001–0.1	SPE-168915-MS
Temperature (°F)	High	240	SPE-184486-STU
Wettability	Poor	Oil-wet to intermediate	SPE-179688-MS; URTeC: 2461651; SPE-153853-PA
Natural fracture intensity (#/ft.).	High	0–32	SPE-168915-MS
Grain density (g/cm3)	Usual	2.55–2.75	URTeC 2461651
Drive mechanism	Poor	Depletion	SPE-171668-MS
Oil saturation (%)	Good	50–75	SPE-179533-MS
Median pore radius (μm)	Poor	0.034–0.010	SPE-179688-MS; URTeC: 2461651
Dominant grain size (μm)	Tiny	<62.5	SPE-179533-MS
Total organic content (wt.%)	Rich	0.1–5	URTeC 2461651
Reservoir depths (ft.)	Deep	5045–12150	URTeC: 2433692; 21-1921 WPC
Pressure (psi)	Abnormal	0.78	SPE-169575-MS
Bulk density (g/cm3)	Usual	2.3–2.5	URTeC 2461651
Net thickness (ft.)	Intermediate	10–40	URTeC: 2433692
Formation type	Complex	Silt, limestone, sand, shale	URTeC 1619698
Clay content (%)	High	7%–30%	URTeC 2461651; SPE-180378-MS

Table 2.5 Common fluid properties of BPS.

Fluid properties	Common quality	Common quantity	Reference
Oil density (API)	Excellent	38–42	21-1921 WPC
Brine-specific gravity	Heavy	1.9	SPE-171668-MS
Saturation pressure (psi)	High	2500–3403	SPE-175034-MS
Contact angle	High	81–142	URTeC 2461651
Brine TDS (mg/L)	High salinity	228,500–285,000	SPE-171668-MS; SPE-178489-PA
Oil viscosity (cP)	Very low	<4.2	URTeC: 2433692; SPE-178489-PA
Total acid number (KOH/g)	Low	0.02–0.36	SPE-171668-MS; URTeC 2461651
Crude oil polarity	Favorable	More paraffinic	SPE-171668-MS
Total base number (KOH/g)	Low	0.12–1.16	SPE-171668-MS; URTeC 2461651
PH	More acidic	5.7	SPE-171668-MS
GOR, SCF/STB	High	507–1712	URTeC: 2433692; SPE-171668-MS
MMP for CO_2 (psi)	Achievable	2450–2650	SPE-175034-MS
IFT W/O (mN/m)	High	17.2–34	URTeC 2461651

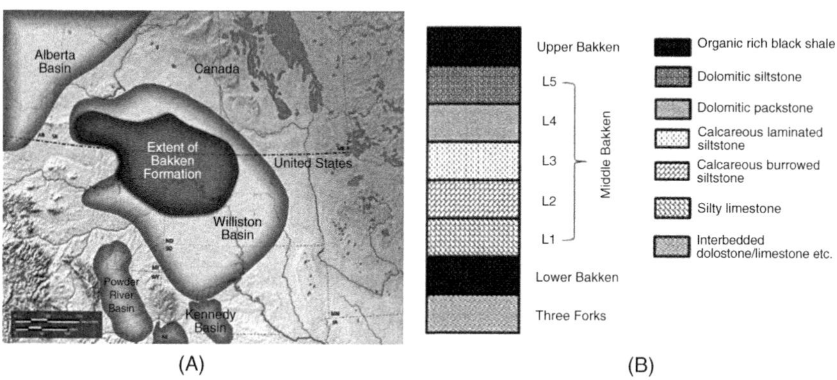

Figure 2.1 (A) Map of Bakken formation; (B) Schematic of BPS stratigraphy. *(Part A: Modified from Pu and Li (2016); Part B: Modified from Jin et al. (2016)).*

depth beyond 8000 ft., is located in the United States, mainly in North Dakota, South Dakota, and Montana. The second part is in Canada, mainly in Saskatchewan and Manitoba, and has a vertical depth of approximately 5050 ft. The two parts of the Bakken have different characteristics; therefore, we present them separately.

In the United States, the Bakken formations include the Upper Bakken, Middle Bakken, Lower Bakken, and Three Forks, as shown in Fig. 2.1B. The Middle Bakken is composed of facies with a highly diverse mineral composition, including mainly limestone, siltstone, dolomite, and sandstone, while the upper member and lower member are mainly shale. The Middle Bakken is the most productive formation among the BPS formations. The average porosity of the Middle Bakken ranges from 5% to 10% (Sorensen et al., 2015); the average matrix permeability has a large range of 0.01–0.001 mD (Pu & Li, 2016); the formation temperature is approximately 230–240°F; and the net pay is approximately 40 ft. The upper and lower members of the Bakken are organic rich, with a TOC in the range of 12–36 wt.% (Tran, Sinurat, & Wattenbarger, 2011), while the middle members have a TOC content of 0.1–0.3 wt.% (Kurtoglu, Sorensen, Braunberger, Smith, & Kazemi, 2013). These two members of shale have a high concentration of Type II kerogen and are considered the source rocks for the petroleum in the Bakken system (Zhang, 2016). The natural fracture intensity ranges from 1–32/ft. These unconventional reservoirs are usually abnormally pressurized with an initial pressure of 7500 psi. Water saturation is between 25% and 50% in the Middle Bakken (Pu & Li, 2016).

The Bakken formations in Canada are shallower than those in the United States, with a higher porosity of 9%–12%, higher matrix permeability of 0.01–0.1 mD, and net pay of 23–26 ft. (Schmidt & Sekar, 2014). The formation has a total of three members and is typically composed of siltstone and black shale. The upper and lower members are dominated by shale, while the middle one is dominated by siltstone and sandstone. The pore pressure in the Canadian Bakken is normally pressurized with an average pressure of approximately 2320 psi.

The common fluid properties for these shale plays are presented in Table 2.5. Unlike the rock properties, not much difference exists in the fluid properties between the Canadian Bakken and the US Bakken. In those reservoirs, the TDS of the formation brine is approximately 285,000 mg/L, with a specific gravity of 1.9 (Kurtoglu et al., 2014). The total acid number, total base number, and water content in these oils are 0.09 mg KOH/g, 1.16 mg KOH/g, and 0.02 wt.%, respectively (Kurtoglu et al., 2014). The most common API is approximately 40 degrees and with a viscosity of less than 3 cP at a pressure of 2600 psi and temperature of 170°F (Schmidt & Sekar, 2014; Zhang, 2016). Furthermore, the oil type is more paraffinic than aromatic (Kurtoglu et al., 2014). The shale oils usually contain a high gas oil ratio (GOR) varying from 507 to 1712 scf/ bbl, and a bubble point pressure in the range of 1617–3403 psi (Pu & Li, 2016). The minimum miscibility pressure of CO_2 in these types of oils ranges widely, between 2500 psi and 3300 psi (Kurtoglu et al., 2014). In summary, the shale oils are usually characterized to be light oil, low viscosity, more paraffinic, and high GOR.

2.4 Dataset for laboratory CO_2-EOR projects

In order to survey and investigate the CO_2-EOR mechanisms in unconventional resources, a dataset has been constructed from 95 cases of cores collected from different formations, with more than 90% of these cases coming from the Bakken Petroleum System (BPS). In this dataset, all cores are naturally preserved; we include six rock parameters, namely, porosity, permeability, pore throat radius, total organic carbon content (TOC), water saturation, and oil saturation, and four operating parameters, namely, CO_2 bath pressure, CO_2 bath temperature, core sample bulk size, and exposure time. We present our data survey and analysis results based on data collected, revealing some interesting relationships among the improved oil recovery by CO_2 injection and the above-mentioned parameters. Furthermore, the parameters, which play a significant role in the success of CO_2-EOR in shale formations, have been investigated at the lab scale level.

The characteristics of the experimental dataset used in the data analysis system are presented as follows:

1. 95 cases of natural preserved core samples were collected from the literature, technical papers, and Department of Energy (DOE) reports from 2013 to 2017. Alharthy et al. (2015) and Jin et al. (2016) are examples of the sources from which this database was constructed.
2. All core samples were naturally saturated with a live crude oil.
3. More than 90% of the experiments were conducted during the huff-n-puff process (CO_2 bathing process). However, the other 10% of the experiments were conducted during the flooding process (concurrent flow).
4. There were some parameters missing in the sources. We handled the missing data in two ways. The first way consisted of running our analysis using the data the way it was. The second way involved using some

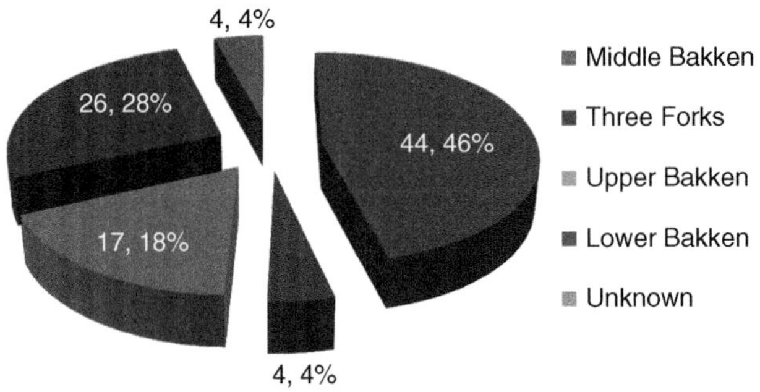

Figure 2.2 ***Laboratory data distribution based on different formations.***

techniques to impute the missing data before proceeding to the analysis stage. Both methods resulted in the same conclusions for this study.

5. The core samples gathered in this work were from different formations, including the Upper Bakken (UB), Middle Bakken (MB), Lower Bakken (LB), Three Forks (TF), and unspecified formation(s).
6. Boxplot, histogram, and scatter plot were used to visualize the data distribution and some important relationships among the data. Fig. 2.2 illustrates the number of samples and percentage of data availability for each formation.

2.5 Data analysis results and discussions

2.5.1 Effect of formation permeability on oil recovery obtained by CO_2 injection

First, a boxplot has been used to determine the five statistical values for the permeability of the shale cores, including: (1) the lowest value (minimum), (2) the highest value (maximum), (3) the first quartile (25th percentile), (4) the second quartile (50th percentile), and (5) the third quartile (75th percentile), as shown in Fig. 2.3A. Second, the scatter plot shown in Fig. 2.3B has been used to explore the potential association of oil recovery obtained by CO_2 injection (RF) and the core sample permeability. The results indicate that CO_2-EOR performance increased with the formation permeability, possibly because formations with a high permeability would have a larger contact area between the injected CO_2 and the formation oil. As a result, when the contact area between the injected CO_2 and the formation oil increased, more CO_2 diffused into the formation oil and therefore was able to extract more oil by means of the counter-current mechanism. However, the relationship is not significantly strong ($R^2 < 20\%$). Third, histograms

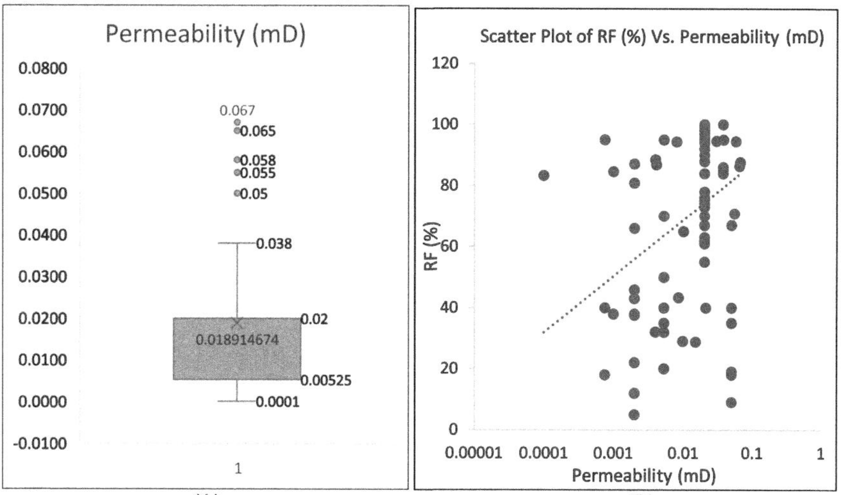

(A) (B)

Figure 2.3 (A) Boxplot for core-sample permeability (mD), (B) Scatter plot for the oil recovery factor extracted by CO_2 injection versus formation permeability (mD).

have been used to plot the average permeability for each formation, as well as the oil recovery factor obtained by CO_2 injection, as shown in Fig. 2.4. There is not good agreement between the two parameters, which provides more evidence that the relationship between the formation permeability and the oil recovery factor obtained by CO_2 is not that strong.

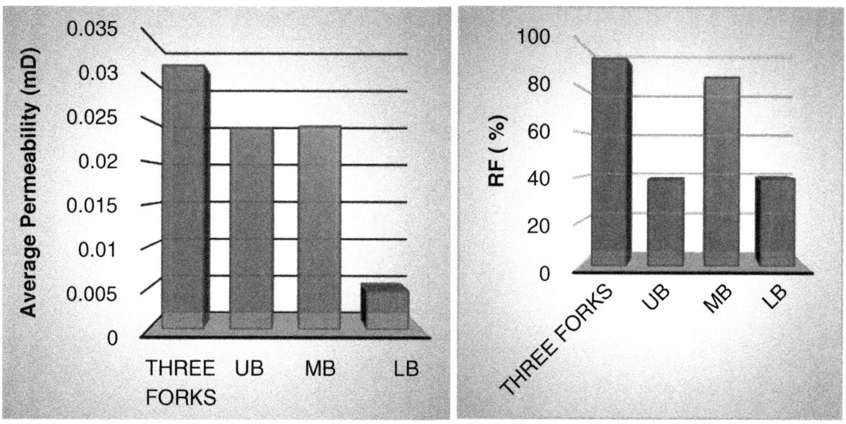

(A) (B)

Figure 2.4 (A) Average permeability (mD) for each formation, (B) Average RF obtained by CO_2 injection for each formation.

2.5.2 Effect of total organic carbon content on oil recovery obtained by CO_2 injection

Fig. 2.5A shows the five statistical values for the TOC of the shale cores. The scatter plot in Fig. 2.5B reports the relationship between the oil recovery obtained by CO_2 injection and the TOC of the core samples. The results suggest that CO_2-EOR performance decreases as the TOC increases.

The physical processes causing this relationship are unclear. However, Jin et al. (2016) reported the same finding when using experimental data from 21 core samples, explaining that it might be due to the strong affinity kerogen has for oil. Kerogen tends to be more oil-wet, so CO_2 would have difficulty displacing the hydrocarbon molecules in formations with a high TOC content.

A histogram is used to visualize the distribution of the average TOC in different Bakken formations, as shown in Fig. 2.6A, which agrees with the investigation conducted by Jin et al. (2016), as shown in Fig. 2.6B. Even if only one average value of TOC were taken for each formation, good agreement would still exist for the relationship between the average TOC content of each formation and the oil recovery factor obtained by CO_2 injection at that formation, as shown in Fig. 2.6.

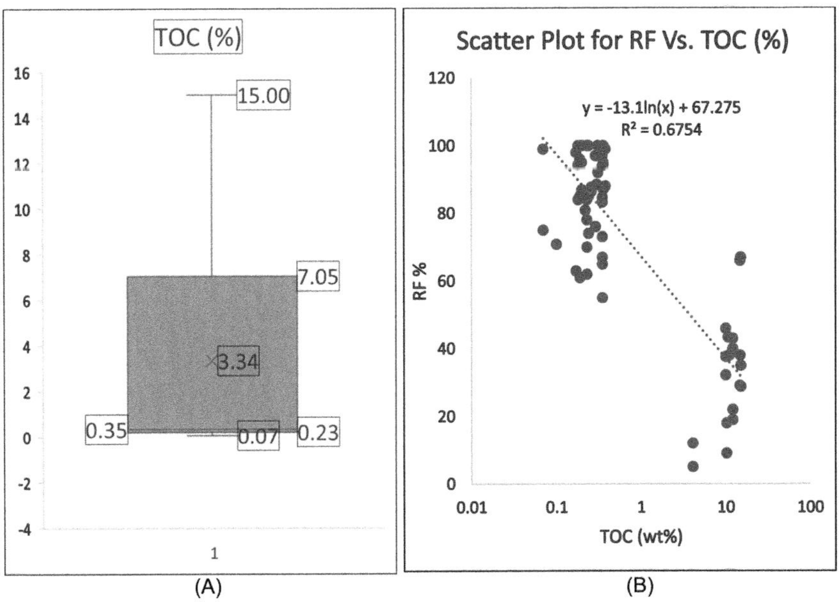

Figure 2.5 (A) Box plot for TOC (%), (B) Scatter plot for the oil recovery factor extracted by CO_2 injection versus TOC (%).

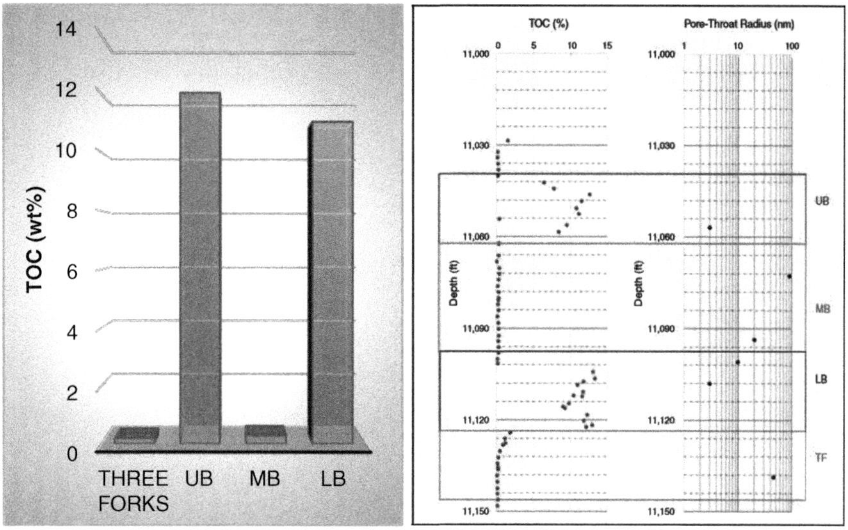

(A) (B)

Figure 2.6 (A) Average TOC (%) for each formation in this study, (B) TOC (%) distribution for each formation (Jin et al., 2016).

2.5.3 Effect of matrix porosity on oil recovery obtained by CO_2 injection

The five statistical values for the porosity of shale cores are shown in Fig. 2.7A. The scatter plot in Fig. 2.7B suggests a weak relationship between the oil recovery obtained by CO_2 injection and the core sample porosity. The results confirm the diffusivity prediction, which is that CO_2-EOR performance increases as the formation porosity increases, as shown. This happens perhaps because the contact area between CO_2 and the formation oil increases when the formation porosity is large enough, and vice versa. When the contact area between the injected CO_2 and the formation oil increases, the effective diffusivity of the CO_2 into the formation oil increases, so CO_2 can extract more oil by means of the counter-current mechanism.

This relationship shares similarities with the relationship between the formation permeability and the RF obtained by CO_2 injection. The histogram shows the average porosity for each formation, as shown in Fig. 2.8A. The scatter plot shown in Fig. 2.8B presents the relationship between the porosity and the permeability of the core samples. This relationship has a curved shape, but it is not very strong due to the high incidence of natural fractures in these types of formations.

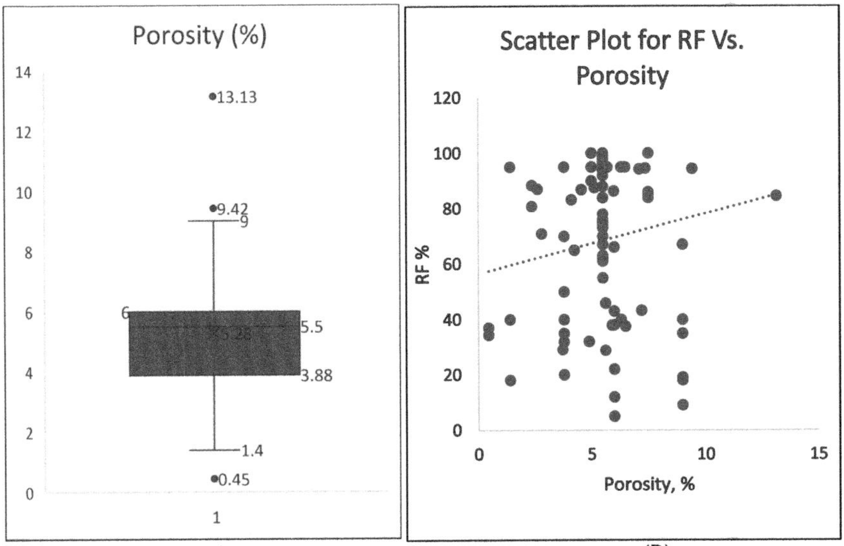

(A) (B)

Figure 2.7 (A) Box plot for core sample porosity (%), (B) Scatter plot for the oil recovery factor extracted by CO_2 injection versus formation porosity (%).

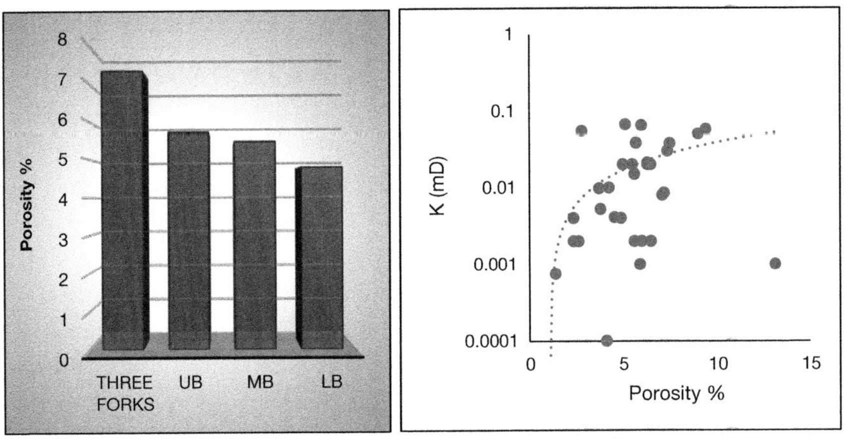

(A) (B)

Figure 2.8 (A) Average porosity (%) for each formation, (B) Porosity and permeability cross plot for different formations.

2.5.4 Effect of mean pore throat radius on oil recovery obtained by CO_2 injection

The box plot in Fig. 2.9A shows the five statistical values for the mean pore throat radius of the shale core samples. The scatter plot in Fig. 2.9B reports the relationship between the oil recovery and the mean pore throat radius of the core samples. The results indicate that CO_2-EOR performance increases as the mean pore throat radius of the core samples increases, which confirms the molecular diffusivity prediction according to Fick's law. However, the enhancement in the RF with the increase in the pore throat radius would be limited at a specific value of the pore throat radius. For the 95 cases in this study, the cut-off for the increase in the RF with the increase in the pore throat radius occurs at approximately 26 nm, perhaps due to the relative ratio of the CO_2 molecule size to the pore throat radius. Third, a good relationship has been found between the mean pore throat radius and the TOC for all different formations, as shown in Fig. 2.10A. This relationship indicates that the TOC decreases as the pore throat radius increases, even though they are from different formations. This result is consistent with the results reported by Jin et al. (2016), as shown in Fig. 2.10B.

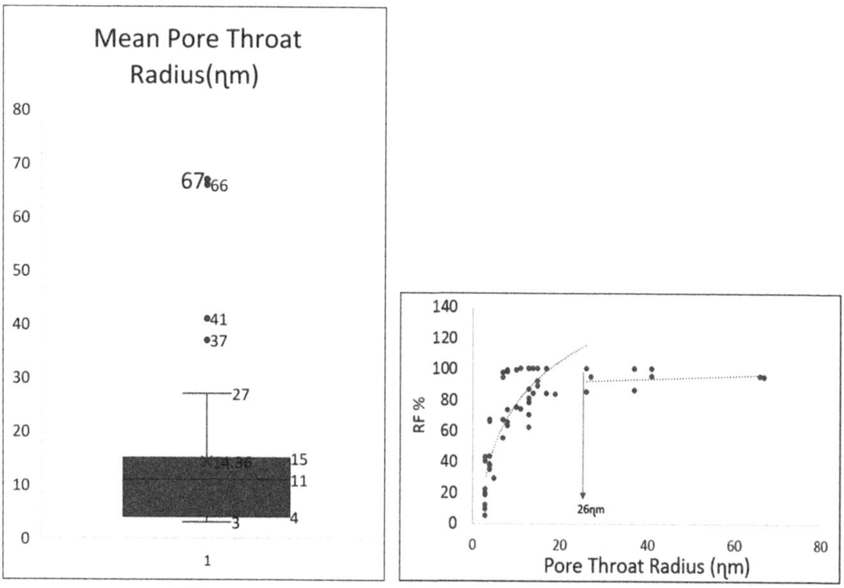

(A) (B)

Figure 2.9 (A) Box plot for mean pore throat radius (¼ m), (B) Scatter plot for the oil recovery factor extracted by CO_2 injection versus mean pore throat radius (¼ m).

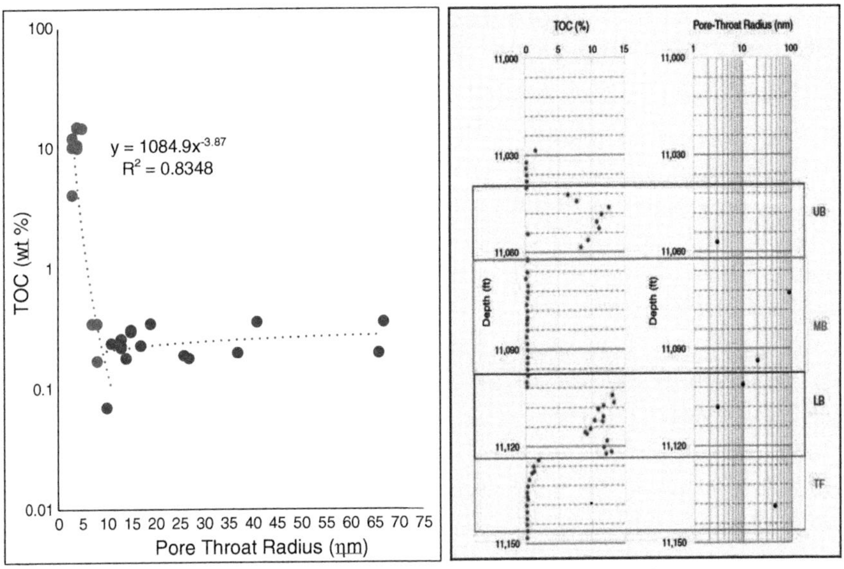

(A) (B)

Figure 2.10 (A) Cross plot of TOC versus the mean pore throat radius for different formations from this study, (B) TOC versus the mean pore throat radius correlated for different formations of the Bakken Petroleum System (BPS) (Jin et al., 2016).

2.5.5 Effect of oil saturation on oil recovery obtained by CO_2 injection

Fig. 2.11A shows the five statistical values for the oil saturation of the shale core samples. The scatter plot in Fig. 2.8B presents the relationship between the oil recovery obtained by CO_2 injection (RF) and the oil saturation of the core samples.

Surprisingly, the results indicate that CO_2-EOR performance decreases with the increase in the oil saturation in the core samples. However, the relationship is not significantly strong. The histograms in Fig. 2.12A show the average oil saturation for each formation. The average oil saturation distribution was compared with the results of the oil saturation distribution for the different formations reported by Jin et al. (2016) and shown in Fig. 2.12B. The trends revealed in the results of this study have good agreement with those reported by Jin et al. (2016), although our study has 95 cases while their study only has 21 cases.

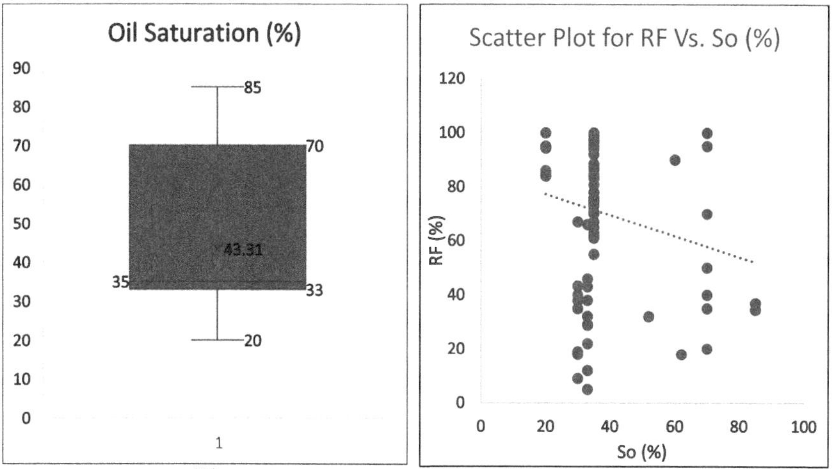

(A) (B)

Figure 2.11 (A) Box plot for the oil saturation (%) in the core samples, (B) Scatter plot for the oil recovery factor extracted by CO_2 injection versus oil saturation (%).

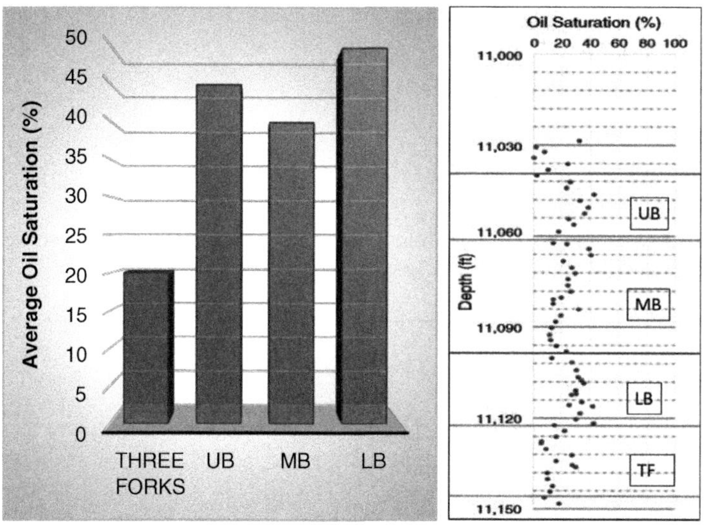

(A) (B)

Figure 2.12 (A) Average oil saturation (%) for each formation, (B) Oil saturation (%) distribution for each formation (Jin et al., 2016).

2.5.6 Effect of water saturation on oil recovery obtained by CO_2 injection

Fig. 2.13A gives the five statistical values for the water saturation of the shale core samples. The scatter plot in Fig. 2.13B suggests a relationship between the oil recovery and the water saturation of the core samples. Surprisingly, the results indicate that CO_2-EOR performance increases with the increase in water saturation in the core samples. However, the enhancement in the RF with the increase in the water saturation would be limited at a specific value of water saturation. For the 95 cases in this study, the cut-off of the water saturation controlling the increase in the RF occurs at approximately 30%, as shown in Fig. 2.14B. The histograms in Fig. 2.14A plot the average

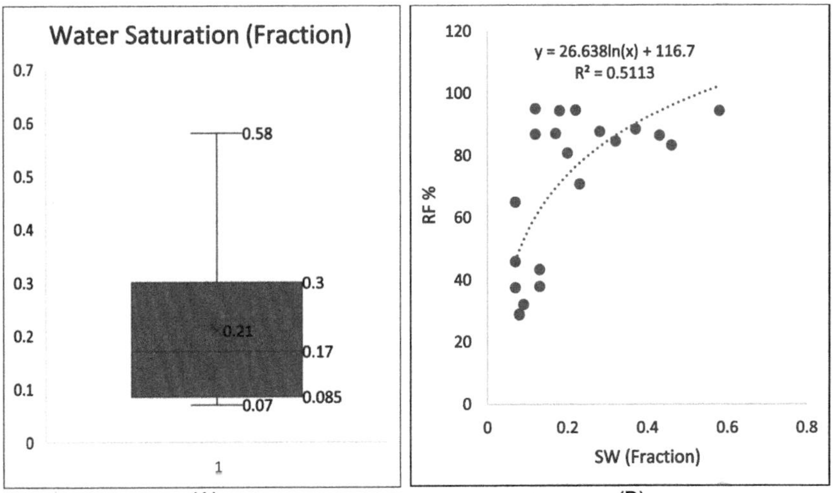

(A) (B)

Figure 2.13 (A) Box plot for water saturation (%), (B) Scatter plot for the oil recovery factor extracted by CO_2 injection (*y*-axis) versus water saturation (*x*-axis).

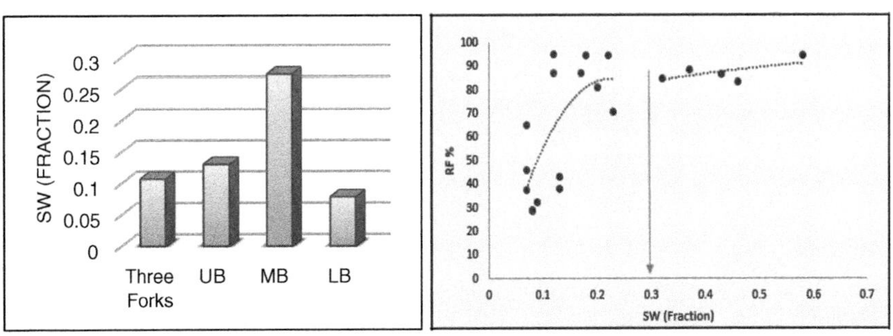

(A) (B)

Figure 2.14 (A) Average water saturation (fraction) for each formation of this study, (B) Changes in RF by CO_2 as water saturation increases for different core samples.

water saturation for each formation. Even if only one average water saturation value is taken for each formation, the conclusion could still be drawn that the oil recovery factor obtained by CO_2 injection increases with the increase in the water saturation, as shown in Fig. 2.14B.

2.5.7 Effect of exposure time on oil recovery obtained by CO_2 injection

Fig. 2.15A shows the five statistical values for the exposure time used for the 95 experiments. The scatter plot in Fig. 2.15B shows the relationship between the oil recovery and the exposure time. The results show that CO_2-EOR performance increases with the exposure time, which is also consistent with the molecular diffusivity prediction. However, the relationship is not strong enough because most of the core samples were very small. Therefore, the maximum RF from the core samples would be obtained very quickly under CO_2 bathing, as shown in Fig. 2.16.

2.5.8 Effect of core-sample bulk volume on oil recovery obtained by CO_2 injection

Fig. 2.17A presents the five statistical values for the bulk volume of the core samples. The scatter plot in Fig. 2.17B suggests the relationship between the oil recovery and the core-sample bulk volume. The results indicate better CO_2-EOR performance when small bulk volumes of core samples are used. However, the relationship between the RF and the core-sample bulk

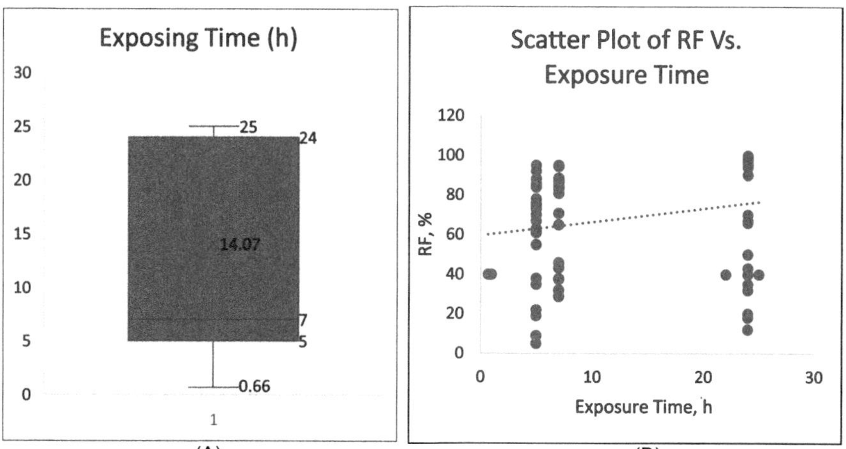

(A) (B)

Figure 2.15 (A) Box plot for the exposure time (h) between core samples and the injected CO_2, (B) Scatter plot for the oil recovery factor extracted by CO_2 injection with different exposure times.

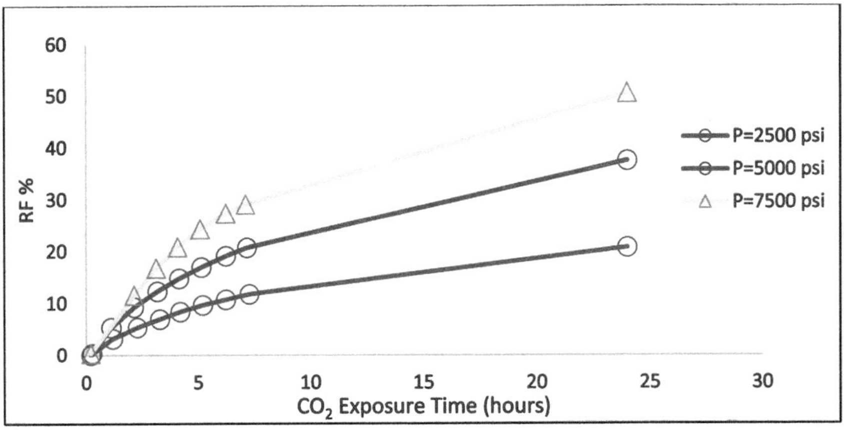

Figure 2.16 *Changes in oil recovery by CO_2 injection with different exposure times (Hawthorne et al., 2017).*

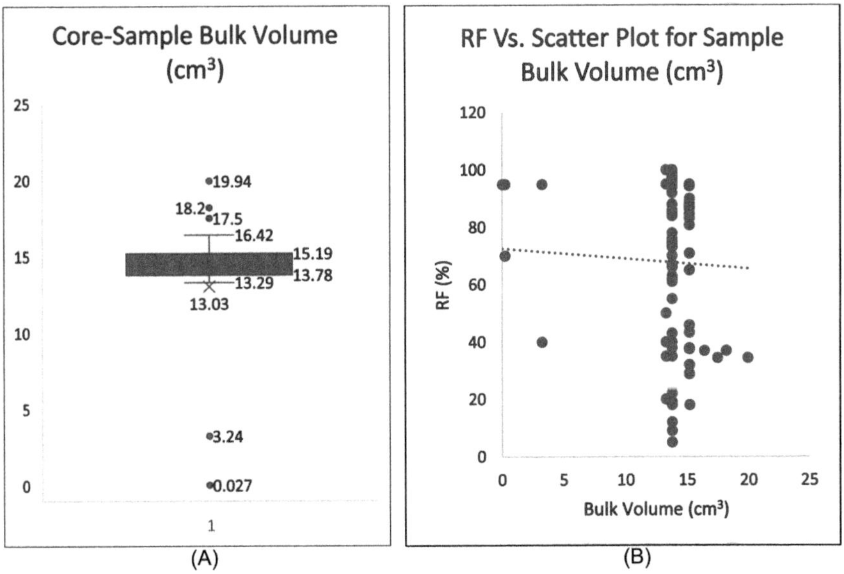

Figure 2.17 (A) Box plot for the core-sample bulk volume used in the 95 cases, (B) Scatter plot for the oil recovery factor extracted by CO_2 injection with different core-sample bulk volumes.

volume is not strong either (i.e., the R^2 is very low). This weak relationship is caused by the very small size of most of the core samples used in the experiment. Very small chips of cores were used in the experimental data, making the contact area between the injected CO_2 and the formation oil very large, thereby resulting in much more volume of the injected CO_2

invading the core matrix and interacting with the formation oil. The same conclusion could hold true for cores cut in a design that gives them larger surface areas exposed to the injected CO_2.

2.5.9 Effect of pressure and temperature on oil recovery obtained by CO_2 injection

Fig. 2.18 presents the five statistical values for the CO_2 bath pressure and temperature. The scatter plot in Fig. 2.19 reports the relationship between the oil recovery and both the injection pressure and the temperature, separately. The results suggest that CO_2-EOR performance increases slightly with an increase in both the injection temperature and pressure. However, the relationship between the RF and temperature and between the RF and pressure is not very significant (i.e., the R^2 is too low). This happens perhaps due to the following two main reasons. First, the lab database for this study was gathered from different resources with differing experimental designs. Second, the reported experiments were conducted under the lab conditions of 230°F and 5000 psi, Bakken formation conditions.

With such a limited range of pressure and temperature data, the relationship with oil recovery cannot be fully explored. For example, many more experimental cases have been conducted under a pressure of 5000 psi, with

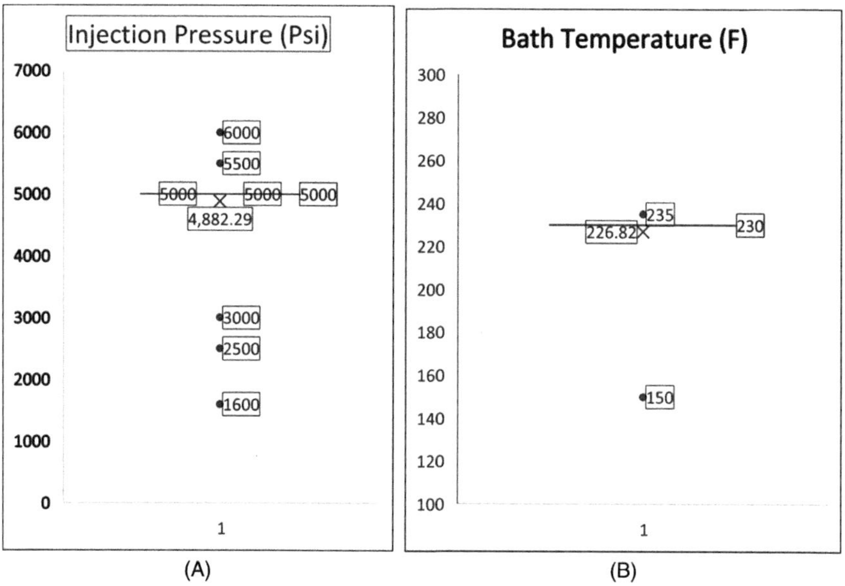

Figure 2.18 (A) Box plot for the injection pressure used in the CO_2 bath for the 95 cases, (B) Box plot for the lab temperature used in the CO_2 bath for the 95 cases.

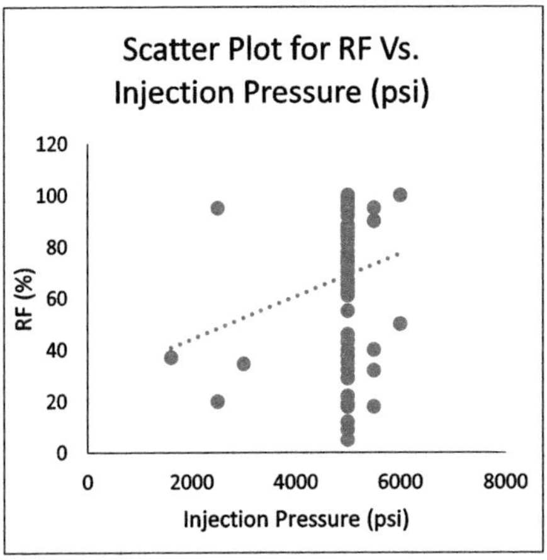

(A)

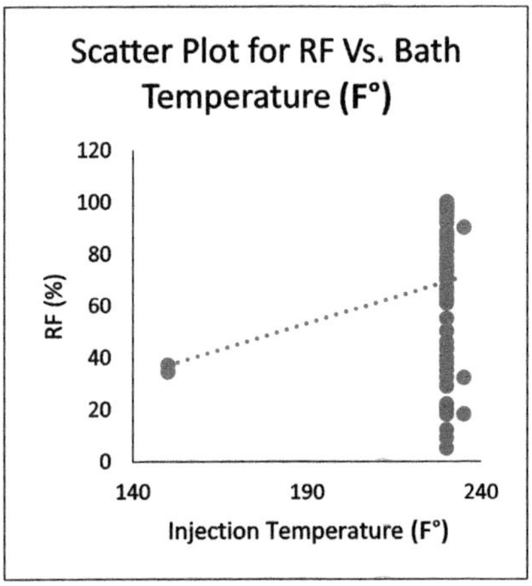

(B)

Figure 2.19 (A) Scatter plot for the oil recovery factor extracted by CO_2 injection with different CO_2 bath pressures, (B) Scatter plot for the oil recovery factor extracted by CO_2 injection with different CO_2 bath temperatures.

fewer cases reported for 2000 psi. These kinds of data lead to difficulty drawing a comprehensive relationship between the objective functions and the independent parameters.

2.5.10 Sensitivity analysis

The purpose of sensitivity analysis (SA) is to determine how sensitive an objective function is to different parameters qualitatively and quantitatively. Identifying the parameters that have a high impact on CO_2-EOR huff-n-puff performance at the lab scale would give a good prediction for CO_2-EOR success or failure prior to the field application, depending on reservoir properties. Also, it would help in optimizing the operating parameters at the field scale. In this study, the design of experiments (DoEs) for the factors affecting the performance of the CO_2-EOR huff-n-puff process at the lab scale has been conducted using JMP software (SAS Institute Inc., 2015). JMP is a statistical tool, which can be used to determine the sensitivity of an objective function for different parameters. JMP also produces variable-importance outputs, which reflect the importance of factors in the prediction model regardless of the fitting method. Several estimation methods for the correlation options are available in the JMP software. JMP estimates the parameters of the model numerically through an iterative fitting process. The dispersion parameter is also estimated by dividing the Pearson goodness-of-fit statistic by its degrees of freedom. Covariances, standard errors, and confidence limits are computed for the estimated parameters based on the asymptotic normality of maximum likelihood estimators. However, REML (restricted maximum likelihood), the method used in this study, and pairwise are the methods used most frequently. REML estimates are less biased than the ML (maximum likelihood) estimation method. The REML method also maximizes marginal likelihoods based upon error contrasts. More info about the basics of this technique and algorithms can be found in the reference SAS Institute Inc. (2015).

In this chapter, the objective function used is the oil recovery factor obtained by CO_2 injection. The parameters investigated are porosity, permeability, TOC, mean pore throat radius, oil saturation, water saturation, exposure time, injection pressure, CO_2-bath temperature, and core-sample bulk volume. The functionality of the oil recovery obtained by injecting CO_2 into the 10 parameters listed previously has been fitted in a linear regression model shown in Fig. 2.20A. The sorted qualitative and quantitative effects for all 10 parameters on RF obtained by CO_2 injection are

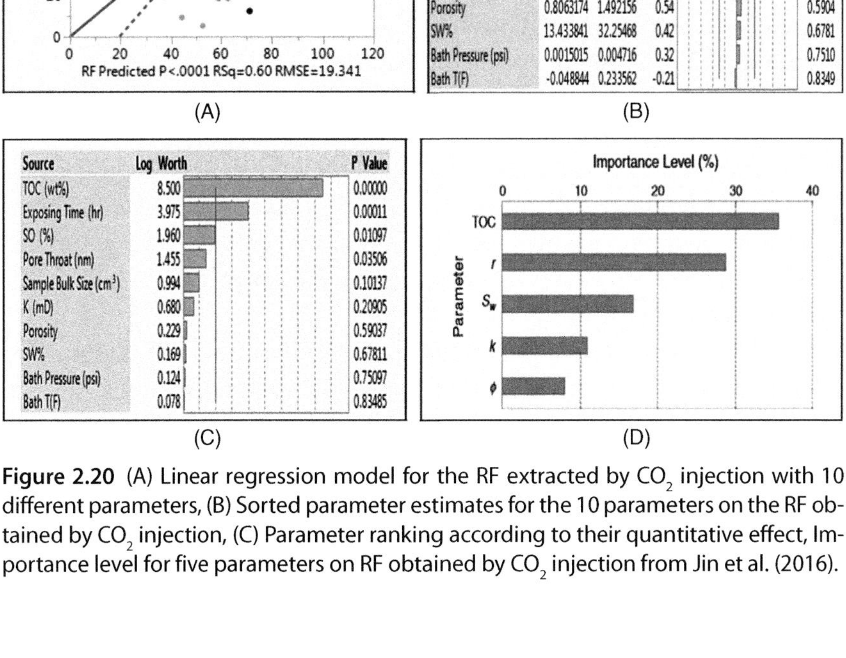

Figure 2.20 (A) Linear regression model for the RF extracted by CO_2 injection with 10 different parameters, (B) Sorted parameter estimates for the 10 parameters on the RF obtained by CO_2 injection, (C) Parameter ranking according to their quantitative effect, Importance level for five parameters on RF obtained by CO_2 injection from Jin et al. (2016).

shown in Fig. 2.20B. The results clearly indicate that TOC and exposure time are the most effective parameters controlling the RF obtained by CO_2 injection, as shown in Fig. 2.20C. The results of this study have been compared with those of the study conducted by Jin et al. (2016), as shown in Fig. 2.20D, and good agreement exists between the results. However, they did not consider the effect of the operating parameters. Furthermore, they used only water saturation instead of both oil saturation and water saturation in their model. In their paper, they explained that they did so because the light components in their natural cores could have evaporated over time. Therefore, using water saturation would include the effect of oil saturation as well because the water saturation has good stability in terms of evaporation effects.

2.6 SUMMARY

- In this chapter, data have been analyzed for the reported experimental results obtained from 95 cases of naturally preserved core samples to investigate the feasibility of CO_2-EOR in shale-oil reservoirs.
- Although the CO_2-EOR performance at the lab scale is encouraging, this study investigated the effect of 10 parameters, which could enhance or downgrade the CO_2-EOR performance at the microscale level.
- The experimental design indicated that the TOC and exposure time are the two main parameters controlling the success of CO_2-EOR in shale reservoirs. The sensitivity of the oil recovery factor obtained by CO_2 to the TOC might suggest that the CO_2-EOR mechanisms in shale-oil reservoirs are different from those in conventional reservoirs.
- The diffusion flow is the most dominant flow in these nano-pore formations, so the operating parameters for CO_2-EOR would be different from its operating parameters in conventional reservoirs. For example, according to the data results presented in this chapter, CO_2 requires a long exposure time in shale formations.
- The molecular diffusion mechanism has a significant impact on the success of CO_2-EOR in shale-oil reservoirs. Therefore, the results of this study show good sensitivity of the RF obtained by CO_2 injection to the bulk size of the core samples. When the core sample size is small, CO_2-EOR performance is better, as shown in Fig. 2.19B. This means that CO_2 requires a high incidence of natural fractures to perform well at the field scale.

References

Alharthy, N., Teklu, T., Kazemi, H., et al. (2015). *Enhanced oil recovery in liquid-rich shale reservoirs: Laboratory to field*. Society of Petroleum Engineers. https://doi.org/10.2118/175034MS.

Dong, C., & Hoffman, B. T. (2013). Modeling gas injection into shale oil reservoirs in the Sanish Field, North Dakota. Society of Petroleum Engineers. https://doi.org/10.1190/URTEC2013-185.

Gamadi, T. D., Sheng, J. J., Soliman, M.Y., Menouar, H., Watson, M. C., & Emadibaladehi, H. (2014). *An experimental study of cyclic CO_2 injection to improve shale oil recovery*. Society of Petroleum Engineers. doi: 10.2118/169142-MS.

Gaswirth, S.B., Marra, K.R., Cook, T.A., Charpentier, R.R., Gautier, D.L., Higley, D. K., Klett, T.R., Lewan, M.D., Lillis, P.G., Schenk, C.J., Tennyson, M.E., & Whidden, K.J. (2013). Assessment of undiscovered oil resources in the Bakken and Three Forks formations, Williston Basin Province, Montana, North Dakota, and South Dakota, 2013. U.S. Geological Survey Fact Sheet 2013–3013. U.S. Geological Survey: Denver, CO.

Hawthorne, S.B., et al. (2017). Measured crude oil MMPs with pure and mixed CO_2, methane, and ethane, and their relevance to enhanced oil recovery from middle Bakken and Bakken shales. SPE-185072-MS paper was prepared for presentation at the SPE Unconventional Resources Conference held in Calgary, Alberta, Canada, 15–16 February 2017.

Hawthorne, S. B., Gorecki, C. D., Sorensen, J. A., Steadman, E. N., Harju, J. A., & Melzer, S. (2013). *Hydrocarbon mobilization mechanisms from upper, middle, and lower Bakken reservoir rocks exposed to CO*. Society of Petroleum Engineers.

Hoffman, B. T., & Evans, J. (2016). Improved oil recovery IOR pilot projects in the Bakken formation. SPE-180270-MS paper presented at the SPE Low Perm Symposium held in Denver, Colorado, USA, 5–6 May 2016.

Jin, L., Sorensen, J. A., Hawthorne, S. B., Smith, S. A., Bosshart, N. W., Burton Kelly, M. E., Miller, D. J., Grabanski, C. B., & Harju, J. A. (2016). *Improving oil recovery by use of carbon dioxide in the Bakken unconventional system: A laboratory investigation*. Society of Petroleum Engineers. doi: 10.2118/178948-PA.

Kurtoglu, B., Kazemi, H., Rosen, R., Mickelson, W., & Kosanke, T. (2014). *A rock and fluid study of middle Bakken formation: Key to enhanced oil recovery*. Society of Petroleum Engineers. doi: 10.2118/171668-MS.

Kurtoglu, B., Sorensen, J. A., Braunberger, J., Smith, S., & Kazemi, H. (2013). *Geologic characterization of a Bakken reservoir for potential CO_2 EOR*. Society of Petroleum Engineers. doi: 10.1190/URTEC2013-186.

Mohebbinia, S., & Wong, T. (2017). Molecular diffusion calculations in simulation of gasfloods in fractured reservoirs. SPE-182594-MS Paper presented at the SPE Reservoir Simulation Conference held in Montgomery, TX, USA, 20–22 February 2017.

Moortgat, J., & Firoozabadi, A. (2013). Fickian diffusion in discrete-fractured media from chemical potential gradients and comparison to experiment. *Energy & Fuels*, *27*(10), 5793–5805. doi: 10.1021/ef401141q.

Pu, H., & Li, Y. (2016). Novel capillarity quantification method in IOR process in Bakken shale oil reservoirs. SPE-179533-MS presented at the SPE Improved Oil Recovery Conference held in Tulsa, Oklahoma, USA, 11–13 April 2016.

SAS Institute Inc. (2015). JMP® 12 basic analysis. SAS Institute Inc., Cary, NC.

Schmidt, M., & Sekar, B. K. (2014). Innovative unconventional 2EOR-A light EOR an unconventional tertiary recovery approach to an unconventional Bakken reservoir in Southeast Saskatchewan. World Petroleum Congress.

Shoaib, S., & Hoffman, B. T. (2009). *CO_2 flooding the Elm Coulee Field*. Society of Petroleum Engineers. doi: 10.2118/123176-MS.

Song, C., & Yang, D. (2013). Performance evaluation of CO_2 huff-n-puff processes in tight oil formations. Society of Petroleum Engineers. https://doi.org/10.2118/167217-MS.

Sorensen, J. A., Braunberger, J. R., Liu, G., Smith, S. A., Hawthorne, S. A., Steadman, E. N., & Harju, J. A. (2015). *Characterization and evaluation of the Bakken petroleum system for CO enhanced oil recovery*. Society of Petroleum Engineers. doi: 10.2118/178659MS.

Tran, T., Sinurat, P., & Wattenbarger, R. A., 2011. Production characteristics of the Bakken shale oil. Presented at the SPE annual technical conference and exhibition, Denver, CO, USA, 30 October–2 November. SPE-145684.

US EIA (2013). US Energy Information Administration. Available from: https://www.eia.gov/analysis/studies/worldshalegas/pdf/methodology_2013.pdf.

Wang, X., Luo, P., Er, V., & Huang, S. -S. S. (2010). *Assessment of CO_2 flooding potential for Bakken formation, Saskatchewan*. Society of Petroleum Engineers. doi: 10.2118/137728-MS.

Xu, T., & Hoffman, T. (2013). *Hydraulic fracture orientation for miscible gas injection EOR in unconventional oil reservoirs*. Society of Petroleum Engineers. doi: 10.1190/URTEC2013-189.

Yu, Y., & Sheng, J.J. (2016). Experimental evaluation of shale oil recovery from eagle ford core samples by nitrogen gas flooding. SPE-179547-MS Paper presented at the SPE Improved Oil Recovery Conference held in Tulsa, Oklahoma, USA, 11–13 April 2016.

Yu, W., Lashgari, H., & Sepehrnoori, K. (2014). *Simulation study of CO_2 huff-n-puff process in Bakken tight oil reservoirs*. Society of Petroleum Engineers. doi: 10.2118/169575-MS.

Zhang, K. (2016). Experimental and numerical investigation of oil recovery from Bakken formation by miscible CO_2 injection. Paper SPE 184486 presented at the SPE international Student Paper Contest at the SPE Annual Technical Conference and Exhibition held in Dubai, UAE, 26–28 September 2016.

Zhu, P., Balhoff, M. T., & Mohanty, K. K. (2015). *Simulation of fracture-to-fracture gas injection in an oil-rich shale*. Society of Petroleum Engineers. doi: 10.2118/175131-MSSPE-180270-MS.

CHAPTER 3

Comparative analysis between CO_2-EOR mechanisms in conventional reservoirs versus shale and tight reservoirs

Contents

3.1 Introduction 46
3.2 Production data analysis (PDA) for CO_2-EOR in conventional reservoirs 49
  3.2.1 CO_2-EOR huff-n-puff projects in conventional reservoirs 49
  3.2.2 History matching for CO_2-EOR in conventional reservoirs 49
3.3 CO_2-EOR huff-n-puff operations in conventional reservoirs versus unconventional reservoirs 51
3.4 Production data analysis for CO_2-EOR in unconventional reservoirs 53
  3.4.1 CO_2-EOR experiments in unconventional reservoirs 53
  3.4.2 History matching with the experimental results of the shale-core samples 55
  3.4.3 CO_2-EOR huff-n-puff pilot test in the Bakken formation 56
  3.4.4 Reservoir modeling 58
  3.4.5 History matching for the CO_2-EOR huff-n-puff pilot test in the Bakken formation 58
3.5 Summary 61
References 62

Abstract

This chapter discusses how the recovery mechanisms of CO_2 injection differ between unconventional and conventional reservoirs. A methodology which combines real production data with the outputs of reservoir numerical simulation is introduced to diagnose the mechanisms of CO_2-EOR huff-n-puff in conventional versus shale and tight oil reservoirs. Two reverse engineering approaches have been integrated to produce a unique type of curve for the performance of CO_2-EOR huff-n-puff in shale-oil reservoirs. First, numerical simulation methods were used to upscale the reported experimental outcomes to the field conditions. As a result, forward diagnostic plots have been generated by combining different physical and chemical mechanisms of CO_2 with different shale-reservoir conditions. Second, different backward diagnostic plots have been produced from the history match with the CO_2 performances of pilot tests conducted in some portions of the Bakken formation in North Dakota and Montana. Finally, the backward and forward diagnostic plots were fitted to produce another unique type of curve, which agrees with the typical general simulation solutions curve, to represent

Fundamentals of Enhanced Oil Recovery Methods for Unconventional Oil Reservoirs
http://dx.doi.org/10.1016/B978-0-12-818343-4.00003-6

CO_2-EOR performance in shale-oil reservoirs. This chapter clearly reports that the CO_2-EOR mechanisms in unconventional reservoirs are different from those in conventional reservoirs. Also, the performance behavior of CO_2-EOR in the puff stage is totally different in conventional reservoirs. This chapter provides a thorough explanation of how CO_2-EOR performance differs at the field scale of conventional reservoirs versus shale formations.

Keywords: CO_2-EOR mechanisms; shale and tight oil reservoirs; huff-n-puff; conventional reservoirs; history matching; field production data

3.1 Introduction

Unconventional plays have very complex nano pore throats at the microscopic level and very heterogeneous formations at the megascopic scale. The heterogeneity and uncertainty of unconventional reservoir characteristics lead to significant difficulty in understating the mechanisms of fluid flow in ULRs. It is known that the tight porous reservoir structure is dense, and the pore throat is narrow (Bai et al., 2019). At the nanometer scale, the fluid flow is also affected by the boundary layer effect (Bai et al., 2019). Flooding is the most common application process for most EOR methods in conventional reservoirs, as shown in Fig. 3.1. However, given the aforementioned

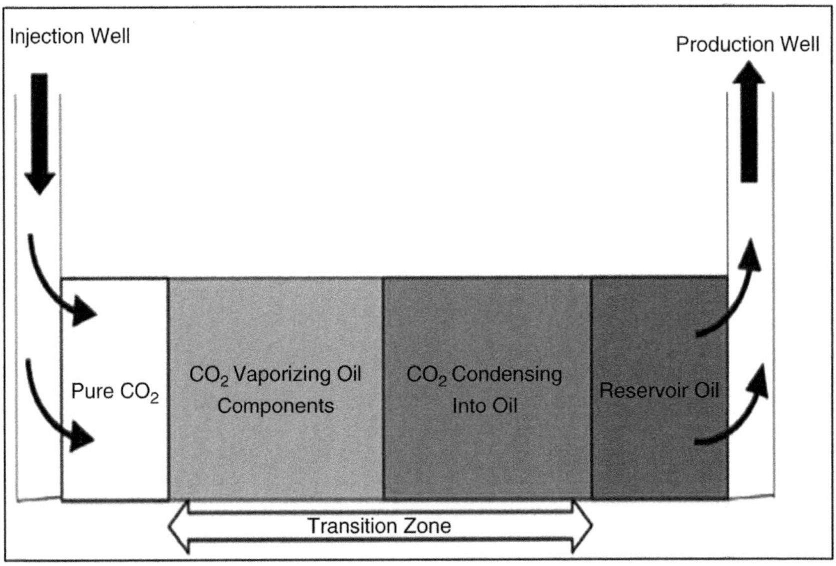

Figure 3.1 *Schematic for flooding mode during CO_2-EOR process (Mohammadian et al., 2019).*

characteristics of unconventional reservoirs, the traditional flooding application is very difficult to perform due to higher-pressure gradients through the tight matrix during the flooding process (Sohrabi et al., 2001). In addition, with the existence of fractures, gas/water breakthrough usually occurs during the tight reservoir flooding process, so the huff-n-puff method is considered a promising EOR mode (Bai et al., 2019). In this method, three actions are performed through the same well, as shown in Fig. 3.2. The fluid is firstly injected into a producer well, then the well is shut off for soaking, and finally, the well is reopened for fluids to be produced. Interestingly, the EOR mechanisms for any EOR technique might be totally different between the huff-n-puff and flooding modes. Therefore, to achieve the fullest understanding, the EOR mechanisms for most of the EOR methods in huff-n-puff need to be re-investigated and thoroughly re-evaluated.

CO_2 is one of the most effective injection gases for enhanced oil recovery in conventional reservoirs. In addition, this method serves the dual purpose of greenhouse gas storage, so it has been an attractive EOR technique for experimentation in unconventional reservoirs. This chapter focuses on the comparison of CO_2-EOR mechanisms in huff-n-puff operations for conventional reservoirs versus unconventional formations.

The mechanisms of CO_2-EOR in tight-oil reservoirs are considered to be significantly different from those in conventional reservoirs due to the special petrophysical properties, reservoir fluid thermodynamics, and mass transport mechanisms (Bai et al., 2019). The conventional CO_2-EOR mechanisms include: (1) viscosity reduction, (2) oil swelling, (3) solution

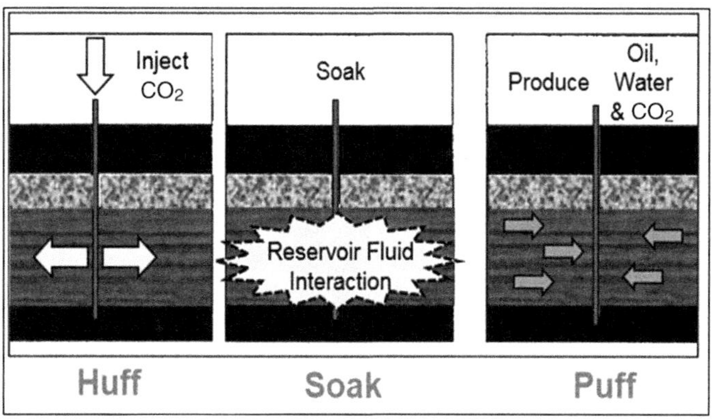

Figure 3.2 ***Schematic for three stages of actions performed during CO_2-EOR huff-n-puff process.*** *(Modified from Kawahara, 2016).*

gas drive, and (4) hydrocarbon extraction by CO_2. The most important operating parameters in CO_2 huff-n-puff include the injection rate and time, number of cycles, soaking time, and pressure (Bai et al., 2019).

Hawthorne et al. (2013) explained a detailed mechanism for CO_2-EOR in shale and tight reservoirs, as shown in Fig. 3.3. Their mechanism is the most acceptable among mechanisms reported by different investigators to explain CO_2-EOR at the lab scale of unconventional reservoirs. They reported their mechanism, explaining the CO_2-EOR in shale and tight oil reservoirs as follows: (1) CO_2 is preferentially injected into the fractures, (2) the CO_2 contacts the matrix rock through the fractures, (3) the CO_2 seeps further into the matrix, dominated by pressure gradients; when the CO_2 dissolves in the crude oil, the oil swells and its viscosity decreases, (4) the oil drains into the fractures under the effect of the CO_2, and (5) when the pressure gradient drops, oil transport is dominated by the concentration gradient across the matrix and the fractures.

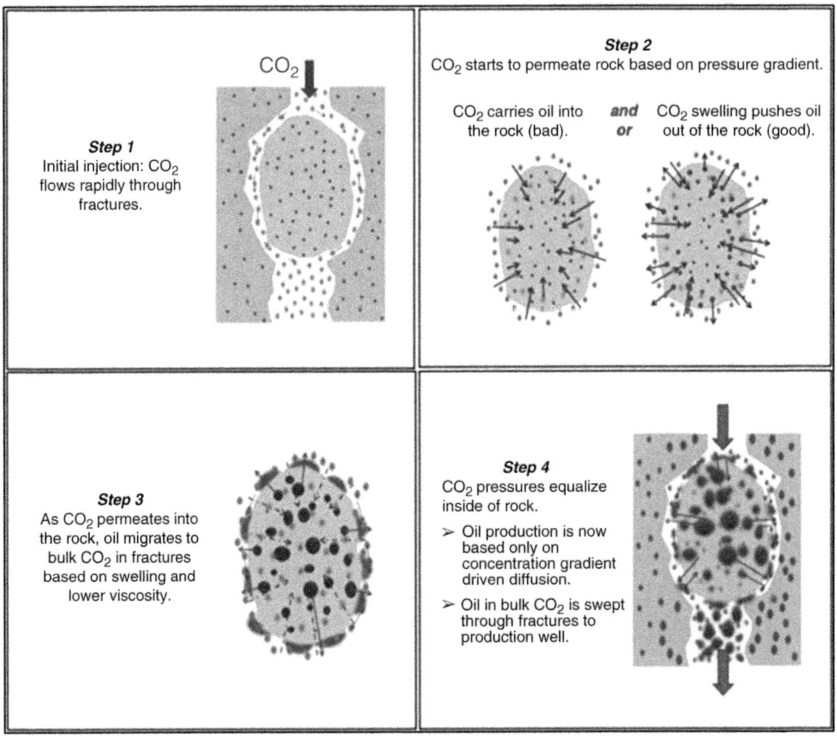

Figure 3.3 ***Schematic for the molecular diffusion mechanism of CO_2-EOR huff-n-puff at the lab scale.*** *(Modified from Hawthorne et al., 2013).*

Other researchers have reported that the main mechanism by which CO_2 extracts oils from shale core samples at the lab scale is the molecular diffusion mechanism. For example, Bai et al. (2019) reported that diffusion is an important mechanism affecting gas transport under tight reservoir conditions, and its effect on the enhanced oil recovery of tight oil cannot be neglected based on the core scale experiments. However, other researchers have reported that oil swelling, viscosity reduction, and repressurization are the main mechanisms by which CO_2 extracts oils from unconventional core samples. To identify the major CO_2-EOR mechanisms under different reservoir conditions, a combination of numerical simulation methods and production data analysis (PDA) was used to investigate the CO_2-EOR performance in pilot tests in unconventional formations and field projects in conventional reservoirs. This chapter also investigates how CO_2-EOR mechanisms differ between the lab scale and the field scale.

3.2 Production data analysis (PDA) for CO_2-EOR in conventional reservoirs

3.2.1 CO_2-EOR huff-n-puff projects in conventional reservoirs

The CO_2-based EOR technique has been effectively applied to improve oil recovery in the tight formations of conventional reservoirs. Since the first CO_2-EOR huff-n-puff project was conducted in conventional reservoirs in Trinidad and Tobago in 1984, more than 130 additional projects have been put in operation around the world, mainly located in the United States, Turkey, Trinidad, and Tobago (Alfarge et al., 2018). In this chapter, we combined PDA for the production data of these projects with numerical simulation methods to produce a generalized decline curve graph accounting for the main mechanisms controlling CO_2-EOR performance in conventional reservoirs.

3.2.2 History matching for CO_2-EOR in conventional reservoirs

Thomas and Monger-McClure (1991) conducted history matches of the field performance for CO_2-EOR huff-n-puff projects in conventional reservoirs using numerical simulation, suggesting that the principal enhanced oil recovery mechanisms are oil swelling, oil viscosity reduction, and gas relative permeability hysteresis. Interestingly, most CO_2 huff-n-puff field projects conducted in conventional reservoirs have the same production decline curve during the puff stage, as shown in Fig. 3.4. Furthermore, the authors conducted a numerical simulation study combining different possible mechanisms for the injected CO_2 with a specific range of uncertainty

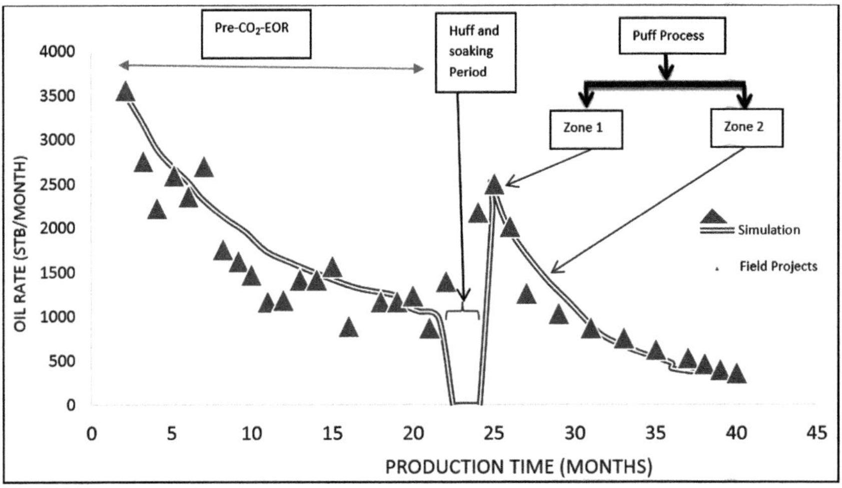

Figure 3.4 ***Typical plot for CO_2-EOR huff-n-puff in conventional reservoirs.***

in the reservoir properties for some of the CO_2-EOR huff-n-puff projects reported by Thomas and Monger-McClure (1991). The fluid compositions were produced using compositional models generated via WinProp software, with input and output parameters described in detail in Alfarge, Wei, and Bai (2017b). The numerical simulation investigation was able to match the unique decline curve in the puff process of CO_2-EOR projects conducted in conventional reservoirs, as shown in Fig. 3.4. The simulation scenarios, which matched the CO_2-EOR projects in conventional reservoirs, suggest that oil swelling, oil viscosity reduction, and repressurization are the main mechanisms responsible for enhancing oil recovery in these conventional formations.

In the puff process of CO_2-EOR projects conducted in conventional reservoirs, there are two main zones in both the simulated and field data, as shown in Fig. 3.4. These zones are as follows:

- Zone 1 (Real Response Zone). In zone 1, the oil production rate responds quickly after the well is put back into production. This is clearly shown in both the field data and simulation results. This quick response in the oil production rate, resulting from the CO_2 injection in conventional reservoirs, occurs for two main reasons. The first is the build-up effect resulting from the CO_2 injection in the reservoir areas situated close to the production well. This build-up stimulates fluid flow and enhances the reservoir pressure in small areas surrounding the wellbore. The second reason is the good communication between the wellbore and the formation matrix, which happens for the following two main reasons:

- Conventional formations usually have high porosity and permeability compared with shale formations. The large pore throats of conventional reservoirs allow for fast and good contact between the injected CO_2 and the oils in the formation matrix. As long as the permeability of the target formation is higher, the peak production zone happens very quickly when the well is put back into production during the puff process. The same logic holds true for the porosity, where the time span of the real response zone is very short for formations with a higher porosity.
- Injecting CO_2 in conventional formations usually removes some damage in the formation areas situated close to the wellbore, especially in heavy oil reservoirs. Therefore, the injected CO_2 works like a stimulation technique. This function causes the oil production to peak very quickly if there is too much damage in high permeability formations.

- Zone 2 (Decline Zone): The next highest oil production rate is achieved in the post CO_2 injection period, where there is a clear decline trend in the production rate (Decline Zone). The Decline Zone usually starts after the peak oil production period in the Real Response Zone has been reached. Zone 2 may be delayed for a long time depending on the permeability of the reservoir and the volume of CO_2 injected into the target formation. When the formation's permeability is lower but all other conditions are the same, it will take longer for the Decline Zone to begin. The radius of investigation for the production wave during the depletion process in oil and gas reservoirs is analog for this reason. Pseudo steady-state conditions take too long to be observed in ultra-tight reservoirs. However, such conditions can be achieved much more quickly in high permeability reservoirs. Regarding the volume of CO_2 injected into the target formation, when the injection volume is higher, the Decline Zone appears later during the puff period. This happens mainly because the high volume of injectant results in a significant stimulation effect throughout the target formation.

3.3 CO_2-EOR huff-n-puff operations in conventional reservoirs versus unconventional reservoirs

Extending the CO_2-EOR approach to unconventional formations has been extensively investigated over the last decade because CO_2 has unique properties, which make it the first EOR option to be tried. However, the applications and mechanisms for CO_2-EOR in unconventional reservoirs are not necessarily the same as in conventional reservoirs due to the complex and poor-quality properties of these plays. Some researchers have reported that the oil recovery obtained by injecting CO_2 into lab core samples of conventional reservoirs has an exponential relationship with the exposure time (Hawthorne et al., 2013). Under the same conditions, they report that the oil recovery obtained by injecting CO_2 into lab core samples of shale-oil reservoirs has a polynomial relationship with the exposure time, as shown in Fig. 3.5. They attribute this difference to the difference in the principal mechanisms by which CO_2 extracts oils from these cores. Researchers have speculated that molecular diffusion serves as the main mechanism by which CO_2 extracts oil from shale cores (Alfarge et al., 2017cAlfarge, Wei, & Bai, 2017c). Therefore, these shale core samples require a longer exposure time and larger contact areas with the injected CO_2 so that good oil recovery can be produced in the same time frame. However, the permeability and pore throats in conventional core samples are much bigger than those in shale core samples, so the convective flow in conventional reservoirs dominates the diffusion flow. Therefore, oil swelling, oil viscosity reduction, and

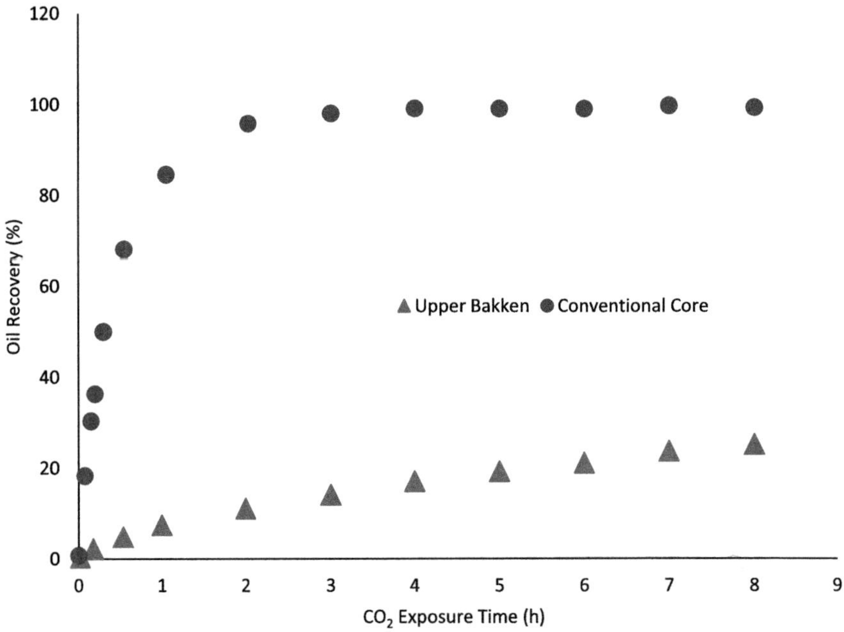

Figure 3.5 ***CO_2 mobilization of hydrocarbons from Upper Bakken and a conventional reservoir rock with 8 hours of exposure time at 5000 psi and 110 Co in round rods of core samples (cylinders with a diameter of c. 10 mm × c. 40 mm long).*** *(Modified from Hawthorne et al., 2013).*

repressurization are the main mechanisms by which CO_2 extracts oil from conventional core samples.

Some CO_2-EOR huff-n-puff pilots were conducted in the Bakken formation between 2008 and 2016. In this chapter, two reverse engineering approaches are integrated to produce a unique decline curve for the performance of CO_2-EOR during the huff-n-puff process in shale-oil reservoirs. First, a numerical simulation study is performed to upscale the reported experimental outcomes to the field conditions. As a result, different forward diagnostic plots can be generated by combining different CO_2 physical mechanisms with different shale reservoir conditions. Second, different backward diagnostic plots are produced from the history match with CO_2 performances in pilots performed in some portions of the Bakken formation located in North Dakota and Montana. Finally, the backward and forward diagnostic plots are fitted to produce a unique type of curve to represent CO_2-EOR performance in shale-oil reservoirs. This combination of field data and simulation methods reveals that the delayed response in the incremental oil production results from the CO_2 injection in shale reservoirs being primarily a function of CO_2 molecular diffusion. On the other hand, the CO_2 diffusion mechanism has approximately no effect on CO_2-EOR performance in conventional reservoirs, which results in a quick response to CO_2 injection. This conclusion is consistent with the reported experiments regarding the role of diffusion in conventional cores versus shale cores. In addition, this chapter reports that the kinetics of the oil recovery process in productive areas and the CO_2 diffusivity level are both key in performing a successful CO_2-EOR project in shale formations.

3.4 Production data analysis for CO_2-EOR in unconventional reservoirs

3.4.1 CO_2-EOR experiments in unconventional reservoirs

A core model of the Bakken shale was constructed by the Energy and Environmental Research Center in North Dakota to perform CO_2-EOR experiments (Alharthy et al., 2015). In their experiments, a cylindrical, natural, preserved core of the Bakken formation was placed in an extraction vessel. The apparatus design for their experiment is shown in Fig. 3.6. The space between the core and the extraction vessel was considered a fracture surrounding the matrix in the real reservoir conditions. An ISCO pump was used to inject CO_2 at a constant pressure of 5000 psi to the inlet valve, and oil samples were collected throughout the entire

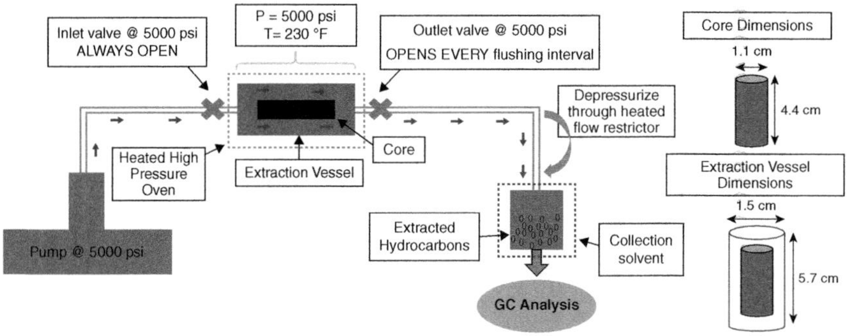

Figure 3.6 ***Enhanced oil recovery experiments on Bakken cores performed at EERC.*** *(Modified from Alharthy et al., 2015).*

experiment. The outlet valve was closed for 50 minutes to soak the core after CO_2 was injected. Then, the outlet valve was opened for 10 minutes to extract the crude oil. This cycle was repeated for 500 minutes. The procedure and design of these experiments have been adopted and simulated in a numerical core model. Fig. 3.7 shows the schematic views of the Bakken shale-core model. The external areas are the space between the extraction vessel and the core matrix, while the internal areas are the core matrix itself. Table 3.1 shows the basic core sample dimensions and properties.

Most of the experimental studies reported that CO_2 diffusion served as the mechanism behind the increment in oil recovery obtained under lab conditions. Both the observed increment in oil recovery from lab

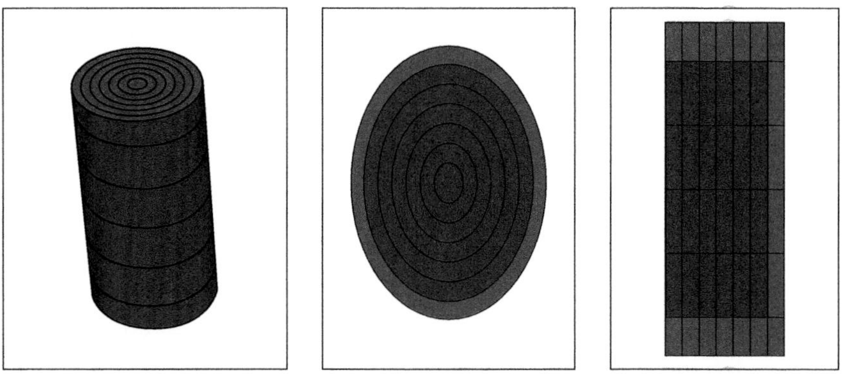

(A) (B) (C)

Figure 3.7 ***Schematic views of Bakken shale core simulated in the core model.*** (A) 3D core model. (B) Top view of the core model. (C) Cross section in the core model.

Table 3.1 Properties of Bakken shale core model.

Parameter	Value
Core length	4.4 cm^3
Core diameter	1.3 cm
Vessel diameter	1.5 cm
Vessel length	5.7 cm

experiments and the CO_2 diffusion rate under lab conditions were upscaled directly to the field scale using numerical simulation methods. This direct upscaling methodology may be optimistic given that the lab cores have a higher contact area and longer exposure time to CO_2 than what would likely be the case under real unconventional reservoir conditions. As a result, both the simulation studies and the experimental work might be overly optimistic in predicting a quick improvement in oil recovery from injecting CO_2 into these tight formations (Alfarge, Wei, & Bai, 2017b). These optimistic oil recovery improvement results reported from the lab work and simulation studies led the operators to conduct CO_2 pilot tests. However, the results from the pilot tests using CO_2 as an injectant were disappointing (Hoffman & Evans, 2016).

A careful examination of the conditions and procedures of the reported experiments reveals that very small core chips of the Bakken or other formations were used in the lab experiments, and those core chips were exposed to CO_2 for a long time, up to 96 hours, to achieve an increment in the oil recovery (Hawthorne et al., 2013). Therefore, acquiring a good understanding of the CO_2 diffusion rate in ultralow permeability reservoirs requires a different approach to providing or estimating the CO_2 diffusivity level from pilot test performance. To achieve this aim, we combine three different tools, namely, experimental reports, simulation methods, and pilot tests, to determine which mechanism primarily controls CO_2-EOR performance in shale formations, and this is discussed in this chapter. Also, this study would determine the CO_2 molecular diffusivity level at the field scale of unconventional reservoirs.

3.4.2 History matching with the experimental results of the shale-core samples

Different scenarios have been run to obtain the best match. Firstly, any scenarios not considering the molecular diffusion have been simulated. Then, two different correlations for molecular diffusion have been simulated, the

Sigmund correlations (Sigmund, 1976a,b) and the Wilke-Chang correlation (Wilke and Chang, 1955). Furthermore, co and counter-diffusive transport were simulated because the experimental procedure did not clearly represent the flooding process or the huff-n-puff protocol. A good history match has been obtained between the experimental data and the models in which the molar diffusion mechanism is enabled. Both the Sigmund and Wilke–Chang correlations gave the same results for the core model, as shown in Fig. 3.8. However, the counter-diffusive transport model matches the lab results better than the co-diffusive transport model, as shown in Fig. 3.8.

3.4.3 CO_2-EOR huff-n-puff pilot test in the Bakken formation

In 2009, three companies (Continental Resources, Enerplus, and XTO) jointly conducted a CO_2-EOR pilot test in the Burning Tree–State 36-2H well in the north central part of the Elm Coulee Field, the location of which is shown in Fig. 3.9. This pilot test was conducted in a horizontal well with a lateral leg of 1592 ft, drilled in 2000, and stimulated with a single-stage, hydraulic fracturing operation with 20–40 sand proppant. The top of the Bakken formation at the well location is at 9740 ft. The horizontal well penetrated through the middle Bakken formation, which is dominated by sandy to silty dolostone in this area. The target formation contains light oil with 40.1 API and has a thickness of 40 ft. The average porosity in this area

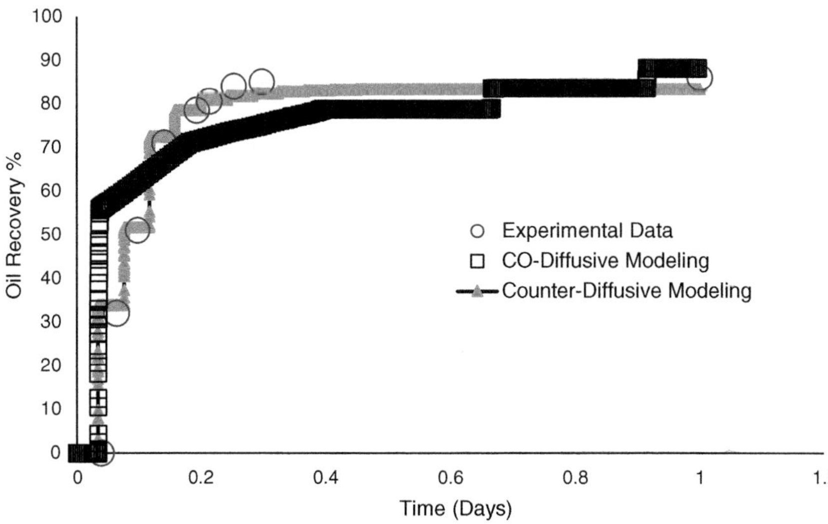

Figure 3.8 ***Comparison of CO_2 flooding and huff-n-puff methods in the Bakken core model.***

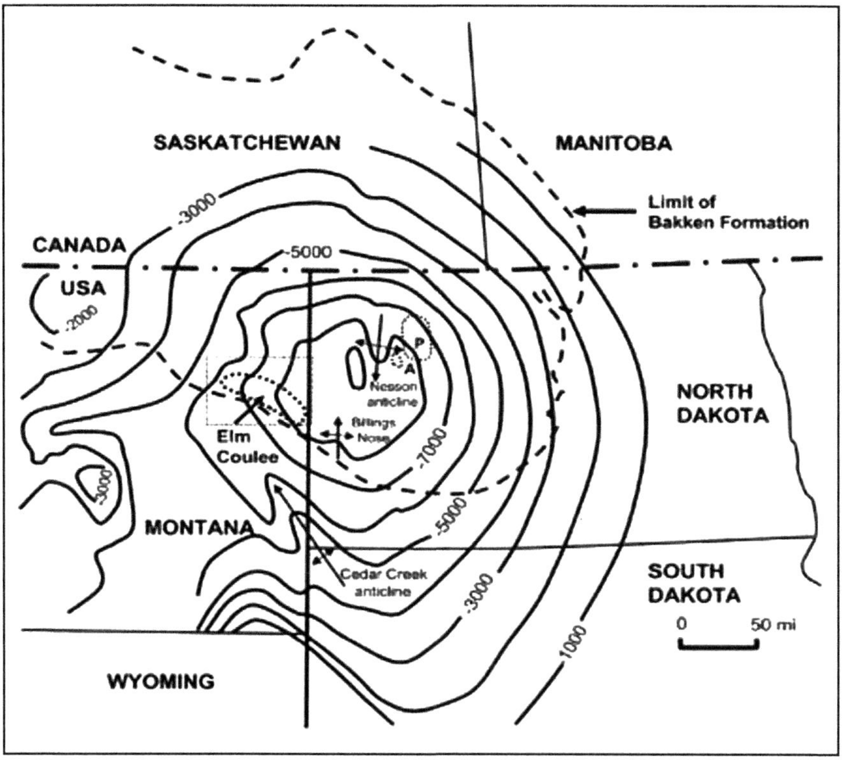

Figure 3.9 ***Location of Elm Coulee Field.*** *(Modified from Todd et al., 2017).*

ranges from 1% to 9%. The average permeability in this part of the Bakken formation ranges from 0.0009 to 0.5 mD. The average oil saturation in this area ranges from 20% to 70%. This well started producing at 195 bbl/day; however, it dropped to 30–40 bbl/day in August 2008.

From January 16, 2009 to February 28, 2009, 45,000 mcf (2570 tons) of CO_2 were injected. The average daily injection rate was 1000 mcf/day, with rates ranging from 0 to 3000 mcf/day. The maximum injection pressure was 1848 psi (BHP), clearly way below the fracture pressure in that formation. After that, the well was shut in (soaking period) for 64 days, from March 1 to May 3, 2009. Then, the well was opened for production. It produced 160 bbl/day for the first 8 days, then 20 bbl/day for 30 days; after that, the well was no longer producing naturally and was put on a pump. In March 2010, the well reached a peak in the post-injection period, approaching 44 bbl/day. After that, the production rate declined constantly. Half of the CO_2 volume injected in the well was recovered in the production process from May 2009 to August 2010. Oil production (in the post-injection period) peaked approximately 1 year after the start of CO_2 injection. Some of the previous researchers were unsure whether the incremental oil production was due to CO_2 injection or not. Therefore, to help make this determination, numerical simulation methods were used to upscale the outcomes of the lab model presented in the previous section to the field scale of this pilot test.

3.4.4 Reservoir modeling

The core model, as described previously in the "CO_2-EOR Experiments in Unconventional Reservoirs" section, was upscaled to the field scale through the construction of a LS-LR-DK (logarithmically spaced, locally refined, and dual permeability) model. The LS-LR-DK method can accurately simulate the fluid flow in fractured shale-oil reservoirs (Wan & Sheng, 2015). In this study, an advanced general equation-of-state compositional simulator was used to build the formation's fluid model.

Then, the LS-LR-DK model and the fluid model were combined to simulate compositional interactions between the reservoir fluid and the injected CO_2 during EOR processes. Furthermore, the diffusion model was implemented in both models. In this study, we attempted to build a numerical model having the same typical fluid and rock properties of the Bakken formation areas used in the pilot test. In the models used in our study, we injected CO_2 as a huff-n-puff process under the same operating parameters as in the pilot test conducted through a hydraulically fractured well. All of the mechanisms proposed in Table 2.1 have been considered in this model. The completed model description, model input parameters, and 3D graph of the reservoir model used in this study have been explained in detail by Alfarge et al. (2017b). Also, the compositional models used in this study were described previously in Alfarge et al. (2017b).

3.4.5 History matching for the CO_2-EOR huff-n-puff pilot test in the Bakken formation

Different scenarios were simulated and different mechanisms were combined into different models until the best match was obtained with the real production data from the CO_2-EOR pilot test conducted in the Elm Coulee Area, Bakken Formation, Montana. Fig. 3.10 depicts how closely the simulation results match the field pilot performance. In both the simulated and field data, the puff process stage of CO_2-EOR contains the following four zones:

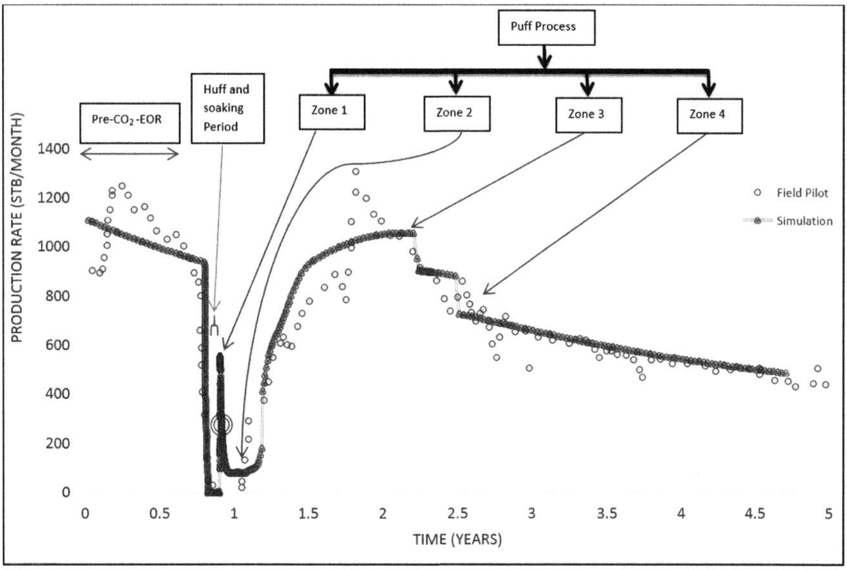

Figure 3.10 ***Typical plot for CO_2-EOR huff-n-puff in shale-oil reservoirs.***

- Zone 1 (Pseudo Response Zone). In Zone 1, the oil production rate responds quickly after the well is put back into production. This behavior, which is clearly shown in both the field data and the simulation results, is primarily due to the build-up effect resulting from CO_2 injection into the fractures of the horizontal well. This build-up occurs because of the poor communication between the hydraulic fractures and the tight matrix of these formations, which is caused by the permeability gel.
- Zone 2 (Real Response Zone). In Zone 2, the oil production rate responds slowly to the injected CO_2. This is clearly shown in both the field data and the simulation results. This increment in the oil production rate grows very slowly with time, but it is much slower than the response of conventional reservoirs to CO_2-EOR.

This slow response suggests that molecular diffusion is the main mechanism by which CO_2 enhances oil recovery in unconventional formations.

- Zone 3 (Peak Zone): In Zone 3, oil production in the post CO_2 injection period peaks approximately 1 year after the start of CO_2 injection. It may seem difficult to believe that this peak occurs due to CO_2 injection given this long delay. However, both the simulation and field data have confirmed that this peak in the oil production rate is due to the CO_2-EOR. During the puff process in conventional reservoirs, this peak would occur much more quickly. This slow response to the injected CO_2 suggests that CO_2-EOR mechanisms in unconventional formations differ greatly from CO_2-EOR mechanisms in conventional reservoirs.

- Zone 4 (Decline Zone): After the highest oil production rate is achieved in the post CO_2 injection period, there is a clear decline trend in the production rate (Decline Zone). The Decline Zone usually starts after the oil production peak period in the Real Response Zone has been reached. The timing for Zone 4 might be delayed for a long time depending on two main factors. The first factor is the permeability of the reservoir. When the formation's permeability is lower but all other conditions are the same, the Decline Zone takes a long time to begin. The second parameter is the CO_2 volume injected into the formation. When the injection volume is higher, the Decline Zone appears at a later time during the puff period.

According to these results, the most important question is under which conditions the CO_2-EOR curve in unconventional reservoirs would look like the CO_2-EOR curve in conventional reservoirs. Therefore, we enabled some of the CO_2 mechanisms in the simulator and disabled others. We also conducted a parametric study by changing some reservoir properties. We found that Zones 1, 2, and 3 in the puff process of CO_2-EOR in unconventional reservoirs can be merged together into one zone, as shown in Fig. 3.11. This would allow the CO_2-EOR performance in unconventional

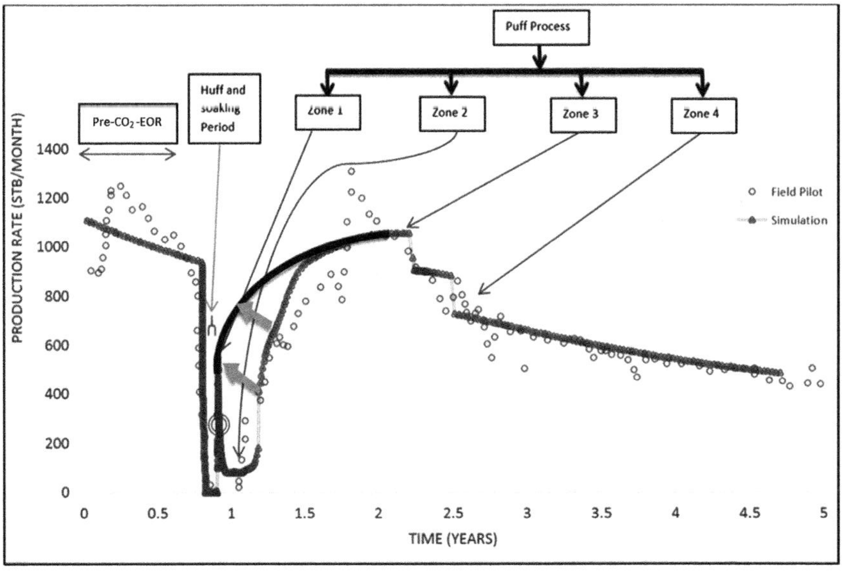

Figure 3.11 ***Typical plot for CO_2-EOR huff-n-puff in shale-oil reservoirs using a high permeability reservoir model with the molecular diffusion mechanism disabled.***

reservoirs to have the same type of curve as in conventional reservoirs. Two scenarios exist, each with two conditions, by which the three zones can be merged into one. In the first scenario:

1. The CO_2 molecular diffusion mechanism is disabled.
2. A reservoir model with a permeability greater than 10 mD is used.

On the other hand, the two conditions in the second scenario would cause the transition from Zone 2 to Zone 3 to take a very long time, as shown in Fig. 3.12. In this scenario:

1. The CO_2 molecular diffusion mechanism is enabled, but the CO_2 has a low molecular diffusivity rate.
2. Reservoirs are used that have very tight permeability, less than 0.1 mD.

3.5 SUMMARY

- This chapter has combined three different investigative tools, namely, lab tools, simulation methods, and field data, to investigate the principal mechanisms controlling CO_2-EOR performance at the field scale of conventional reservoirs versus unconventional formations. The molecular diffusion mechanism is clearly the main controller of CO_2-EOR performance in shale formations, while oil swelling, oil viscosity reduction, and repressurization are the main mechanisms responsible for enhancing oil recovery in conventional reservoirs.

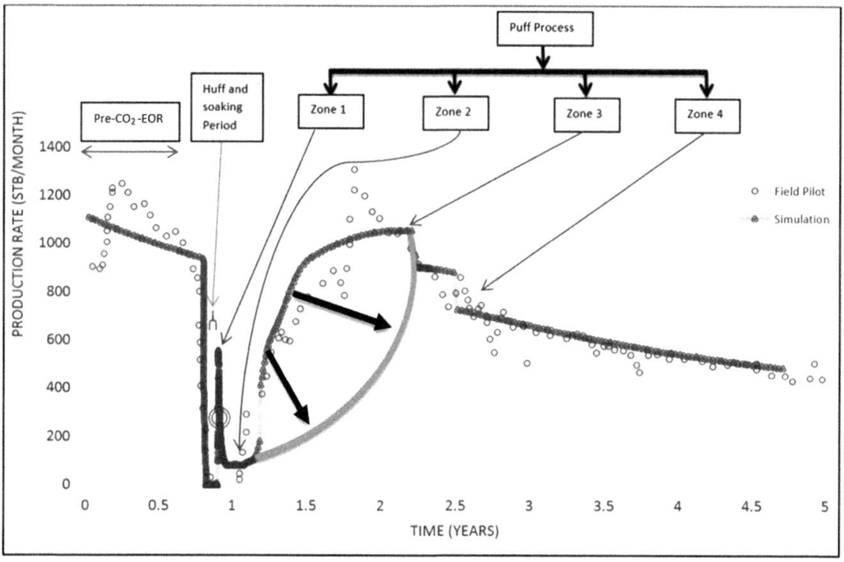

Figure 3.12 *Typical plot for CO_2-EOR huff-n-puff in shale-oil reservoirs using a very tight permeability reservoir model with a low CO_2 molecular diffusivity rate.*

- This chapter combined numerical simulation methods with the decline curve analysis of field data (CO_2-EOR pilot tests in shale formations and CO_2-EOR projects in conventional reservoirs) to understand the mechanisms most responsible for CO_2-EOR performance in conventional reservoirs versus unconventional formations at the field scale level.
- One unique decline curve has been generated to reflect the CO_2-EOR performance in conventional reservoirs containing two distinct zones in the puff process. Another unique type of curve has been generated to reflect the CO_2-EOR performance in shale-oil reservoirs containing four distinct zones in the puff process.
- This study found that molecular diffusion is the main mechanism controlling CO_2-EOR performance in shale-oil formations. This conclusion has been confirmed by integrating three different tools, including experimental reports, field pilot data, and a numerical simulation study. However, oil swelling, oil viscosity reduction, and repressurization are the main mechanisms controlling CO_2-EOR performance in conventional reservoirs.
- In the shale reservoirs, the delayed response of the oil production rate to the injected CO_2, as indicated by the pilot test conducted in the Bakken, matched the general simulation solutions, which have a low molecular diffusivity for CO_2 molecules. This suggests that the exposure time and the contact areas between the injected CO_2 and the target formations need to be significantly increased for more successful results.
- Decreasing the kinetics of the oil recovery process in the productive areas of shale formations during the CO_2 post-injection period would lead to a successful CO_2-EOR project.

References

Alfarge, D., Wei, M., & Bai, B. (2018). CO_2-EOR mechanisms in huff-n-puff operations in shale oil reservoirs based on history matching results. *Fuel, 226*, 112–120. 2018, ISSN 0016 2361, https://doi.org/10.1016/j.fuel.2018.04.012.

Alfarge, D., Wei, M., & Bai, B. (2017b). Factors affecting CO_2-EOR in shale-oil reservoirs: Numerical simulation study and pilot tests. *Energy & Fuel, 31*(8), 8462–8480. https://doi.org/10.1021/acs.energyfuels.7b01623.

Alfarge, D., Wei, M., & Bai, B. (2017c). Data analysis for CO_2-EOR in shale-oil reservoirs based on a laboratory database. *Journal of Petroleum Science and Engineering, 162*, 697–711. https://doi.org/10.1016/j.petrol.2017.10.087.

Alharthy, N., Teklu, T., Kazemi, H., et al. (2015). *Enhanced oil recovery in liquid rich shale reservoirs: Laboratory to field*. Society of Petroleum Engineers. https://doi.org/10.2118/175034MS.

Bai, et al. (2019). CO_2, water and N_2 injection for enhanced oil recovery with spatial arrangement of fractures in tight-oil reservoirs using huff-n-puff. *Energies, 12*, 823. doi: 10.3390/en12050823. 2019.

Hawthorne, S. B., Gorecki, C. D., Sorensen, J. A., Steadman, E. N., Harju, J. A., & Melzer, S. (2013). *Hydrocarbon mobilization mechanisms from upper, middle, and lower Bakken reservoir rocks exposed to CO*. Society of Petroleum Engineers.

Hoffman, B.T., & Evans J. (2016). *Improved oil recovery IOR pilot projects in the Bakken formation*. SPE-180270-MS paper presented at the SPE Low Perm Symposium held in Denver, Colorado, USA, 5–6 May 2016.

Kawahara (2016). *Presentation given about* CO_2*-EOR Pilot in Vietnam*. Available from: https://www.cslforum.org/cslf/sites/default/files/documents/tokyo2016/Kawahara-Vietnam-EORProject-Workshop-Session1-Tokyo1016.pdf.

Mohammadian et al. (2019). CO_2-EOR/Sequestration: Current Trends and Future Horizons, Enhanced Oil Recovery Processes—New Technologies, Ariffin Samsuri, IntechOpen, DOI: 10.5772/intechopen.89540. Available from: https://www.intechopen.com/books/enhanced-oil-recovery-processes-new-technologies/co-sub-2-sub-eor-sequestration-current-trends-and-future-horizons.

Sheng, J. J. (2015). Enhanced oil recovery in shale reservoirs by gas injection. *Journal of Natural Gas Science and Engineering*, *22*, 252–259. https://doi.org/10.1016/j.jngse.2014.12.002.

Sigmund, P. M. (1976a). Prediction of molecular diffusion at reservoir conditions. Part 1—Measurement and prediction of binary dense gas diffusion coefficients. *Journal of Canadian Petroleum Technology*, *15*(2), . https://doi.org/10.2118/76-02-05.

Sigmund, P. M. (1976b). Prediction of molecular diffusion at the reservoir conditions. Part II—Estimating the effects of molecular diffusion and convective mixing in multicomponent systems. *Journal of Canadian Petroleum Technology*, *15*(3), 53–62. https://doi.org/10.2118/76-03-07, PETSOC-76-03-07.

Sohrabi, M., Tehrani, D. H, Danesh, A, Henderson, G.D. (2001). Visualisation of Oil Recovery by Water Alternating Gas (WAG) Injection Using High Pressure Micromodels - Oil-Wet & Mixed-Wet Systems. In: *Proceedings of the SPE Annual Technical Conference and Exhibition*, New Orleans, LA, USA, 30 September–3 October 2001.

Thomas, G. A., & Monger-McClure, T. G. (1991). *Feasibility of cyclic* CO_2 *injection for light-oil recovery*. Society of Petroleum Engineers. https://doi.org/10.2118/20208-PA.

Todd, B. J., Reichhardt, D. K., & Heath, L. A. (2017). *An evaluation of EOR potential in the Elm Coulee Bakken Formation, Richland County, Montana*. Society of Petroleum Engineers. https://doi.org/10.2118/185028-MS.

Wan, T., & Sheng, J. (2015). *Compositional modelling of the diffusion effect on EOR process in fractured shale-oil reservoirs by gas flooding*. Society of Petroleum Engineers. https://doi.org/10.2118/2014-1891403-PA.

Wilke, C. R., & Chang, P. (1955). Correlation of diffusion coefficients in dilute solutions. *AIChE Journal*, *1*(2), 264–270. https://doi.org/10.1002/aic.690010222.

CHAPTER 4

Natural gas-based EOR versus CO_2-EOR in shale and tight oil reservoirs

Contents

4.1 Introduction 65
4.2 Experimental setups of miscible gas-based EOR 67
4.3 Simulation models of miscible gas-based EOR 70
4.4 Performance of natural gas EOR in pilot projects 72
4.5 Performance of CO_2-EOR in pilot projects 77
4.6 Lessons learned from miscible gas injection pilots 80
4.7 Summary 82
References 83

Abstract

In this chapter, a comprehensive and critical review has been conducted to evaluate the feasibility of the miscible gas-based enhanced oil recovery (EOR) technique in unconventional liquid-rich reservoirs (ULRs). Reports and studies from three different approaches (lab, simulation, and pilot tests) were summarized and combined to provide in-depth insights and lessons learned from the application of miscible gas-based EOR in ULRs. First, the main problems in the lab and simulation approaches used to investigate the viability of different EOR methods have been diagnosed. Second, the EOR performance resulting from the injection of different miscible gases has been discussed extensively. Third, the physical and chemical reasoning behind the performance gap for the injected gases at the lab scale versus the field scale of ULRs has been diagnosed.

Keywords: CO_2-EOR mechanisms in shale and tight oil reservoirs; natural gas-based EOR; CO_2-EOR mechanisms in conventional reservoirs; EOR history matching; combination of reservoir simulation with field production data

4.1 Introduction

Unconventional reservoirs have a very tight matrix, so increasing the EOR displacement efficiency is one major target. This is the main reason that many researchers initially tried to investigate the feasibility of miscible gas-based EOR in ULRs. Surprisingly, in lab studies, the oil recovery increment does

Fundamentals of Enhanced Oil Recovery Methods for Unconventional Oil Reservoirs
http://dx.doi.org/10.1016/B978-0-12-818343-4.00004-8

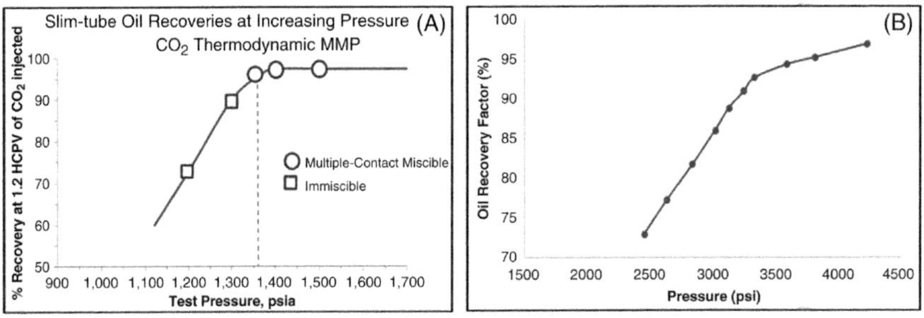

Figure 4.1 ***Effect of MMP on oil recovery performance in conventional reservoirs versus unconventional reservoirs.*** (A) Conventional reservoirs, (B) Shale reservoirs. *(Part A: Modified from Verma, 2015; Part B: Modified from Mohammad et al., 2018.)*

not level off after the minimum miscibility pressure (MMP) is achieved, as usually happens in conventional reservoirs (Mohammad et al. 2018). The recovery factor continuously increases as the pressure increases for shale cores, although the injection pressure is much higher than the MMP, as shown in Fig. 4.1. This observation has been reported by different investigators, with no clear explanation. Injecting a fluid to sweep the oil through the shale matrix to improve the sweep efficiency is a very difficult task in ultralow permeability reservoirs. Furthermore, due to the fracture complexity caused by the mix of natural fractures, hydraulic fractures, and injection-induced fractures because of the extremely low injectivity in shale-oil reservoirs, the injected gas easily breaks through producers in gas flooding mode. For the aforementioned reasons, huff-n-puff injection has become the proper injection mode in unconventional reservoirs. Different types of gases have been proposed to be injected in huff-n-puff mode to improve oil recovery in shale-oil reservoirs.

In the United States, miscible gas-based EOR commonly uses CO_2, N_2, or natural gas to enhance oil recovery in conventional reservoirs (Wang et al., 2017). Each gas type has some advantages and disadvantages in terms of the operations involved, as shown in Table 4.1 (Wang et al., 2017). Injecting CO_2 into shale reservoirs not only enhances oil recovery through pressure

Table 4.1 Comparison of gases used for miscible gas injection (Wang et al., 2017).

Gases	MMP	Sweep efficiency	Cost	In situ oil
CO_2	Lowest Medium	Highest	Medium	>25 API
Natural gas	Highest	Medium	Highest	>30 API
N_2		Lowest	Lowest	>40 API

maintenance, multi-contact miscible displacement, molecular diffusion, and/or desorption mechanisms, but also mitigates climate change due to its permanent sequestration within the small pores in an adsorbed state (Du and Nojabaei, 2019). Although N_2 is an economic and eco-friendly alternative, it may displace oil mostly through an immiscible displacement process due to its high MMP (Du and Nojabaei, 2019). Furthermore, due to the low viscosity of N_2, viscous fingering may be significant during the displacement process if it is used in the flooding mode (Du and Nojabaei, 2019). Flue gas is abundant in shale reservoirs and has been successfully injected in other unconventional reservoirs (Ratner and Tiemann, 2018). Moreover, enormous quantities of produced gas associated with oil production in shale-oil reservoirs is flared or vented into the air (Hoffman and Evans, 2016). Such a practice not only wastes energy, but also poses a danger to the environment (Du and Nojabaei, 2019). A win-win strategy would be to recycle the produced natural gases to enhance oil recovery and reduce gas-flaring problems.

Based on research presented in the literature, CO_2 injection is the most investigated EOR method in ULRs (Alfarge, Wei, Bai, & Alsaba, 2017i). However, natural gas-based EOR has shown better performance than CO_2-EOR in pilots in some situations (Hoffman & Evans, 2016; Schmidt & Sekar, 2014). The difference in the performance of the injected CO_2 at the lab scale versus the field scale has indicated that there is something missing (Alfarge, Wei, & Bai, 2018d). Understanding the physical and chemical mechanisms for CO_2 at the field scale versus the lab scale is crucial for diagnosing this gap (Alfarge, Wei, & Bai, 2017c). This chapter is dedicated to describing the integrated methodology for determining the feasibility of miscible gas-based EOR in ULRs and providing comprehensive applicability guidelines for unconventional miscible gas-based EOR selection based on experimental data investigation results, numerical simulation methods, and pilot test data.

4.2 Experimental setups of miscible gas-based EOR

In the literature, several studies have been conducted through lab experiments on huff-n-puff gas injection for unconventional reservoirs (Yu et al., 2017; Jin et al., 2016; Wan et al., 2015). Fig. 4.2 illustrates an experimental setup for gas flooding and huff-n-puff gas injection in the conducted studies. The common procedures for the reported experimental studies conducted on outcrop cores are as follows (Du and Nojabaei, 2019):

1. Saturate core plugs with oil samples at a given pressure for a reasonably long time;

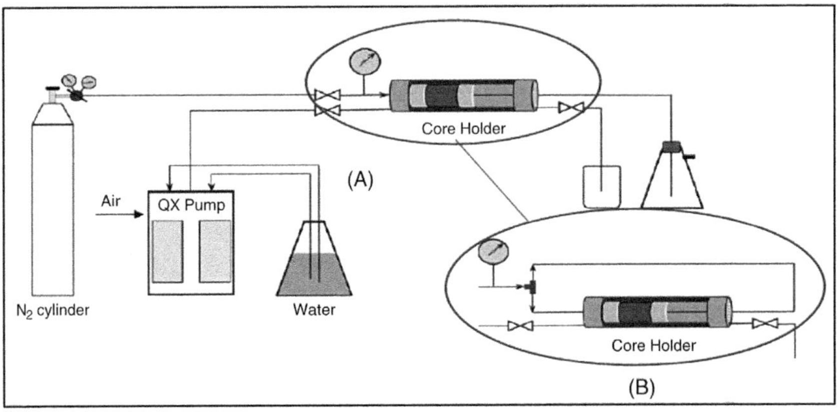

Figure 4.2 Schematic diagram of experimental setup for (A) gas flooding and (B) huff-n-puff gas injection. *(Modified from Yu et al., 2017.)*

2. Place the pre-saturated cores in a core holder and then inject high-pressure gas;
3. Shut in for the soaking process, during which the gas is expected to penetrate into the matrix and make sufficient contact with the oil;
4. Resume production after the soaking period, decreasing the pressure in the system over several cycles so that the oil seeps out from the matrix; and
5. Calculate the oil recovery by re-weighting the core sample or collecting the recovered oil using organic solvents.

When conducting experiments on natural preserved cores, the first step can be ignored. The commonly reported injection gases or solvents in such experiments are N_2, CO_2, CH_4, C_2H_6, and a CH_4/C_2H_6 mixture (Du and Nojabaei, 2019).

Alfarge et al. (2017c) presented a compressive review of laboratory studies, which investigated the performance of injecting different miscible gases into shale cores to enhance oil recovery. The literature review reveals that miscible gas-based EOR in shale oil reservoirs is a new concept in the oil industry. Furthermore, although there have been extensive lab studies investigating the applicability of miscible gas-based EOR in ULRs over the last decade, the reported experimental procedures still suffer from significant problems. What follows are the two main aspects of the faulty procedures used in previous experiments that might complicate the understanding of EOR mechanisms for miscible gases:

- Most of the investigators used outcrop cores and/or synthetic oils to simulate the conditions of shale reservoirs, which likely would not be representative of the oil properties in shale formations. This methodology

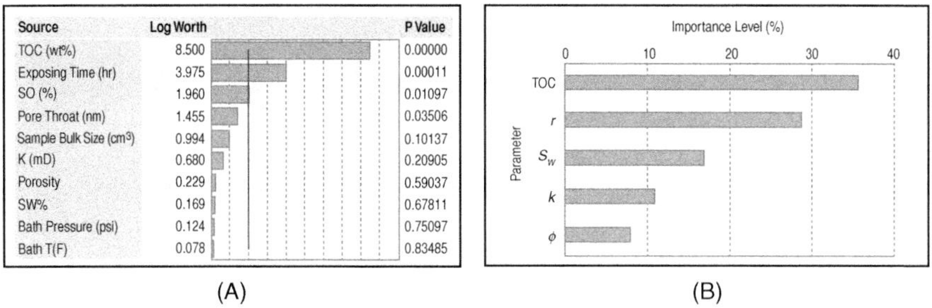

Figure 4.3 ***Design of experiments (DOEs) results for ranking the factors affecting CO_2-EOR in ULRs.*** *(Part A: Study results of Alfarge et al., 2018f, Part B: Study results of Jin et al., 2016.)*

has resulted in significant ambiguity around fluid flow in the nanopores of shale media, and therefore the EOR mechanisms of the injected gases in these unconventional reservoirs. For example, Jin et al. (2016) used 27 natural preserved cores from the Bakken formation in their experiments to investigate the parameters affecting CO_2-EOR in these natural cores. They found that the total organic carbon content (TOC) is the main factor impacting oil recovery during CO_2 injection, as shown in Fig. 4.3A. Alfarge, Wei, & Bai, 2018f arrived at the same conclusion when surveying the experimental results of 95 natural preserved core samples, as shown in Fig. 4.3B. They posited that any outcomes or conclusions from experimental studies that used outcrop cores would have clear problems because they do not represent what would happen in the real conditions of shale reservoirs. This means that the findings and conclusions produced from lab studies that used synthetic oils and outcrop cores lack credibility. The majority of the lab studies conducted in this area of research used synthetic oils and outcrop cores.

- For those experimental studies that used natural shale cores, the cores were in the shape of very small chips or rods, in a bulk volume of 0.027–19.94 cm^3 (Alfarge et al., 2018f), as shown in Fig. 4.4. In the lab experiments, these small chips were immersed in a bath of the injected gas under high pressure (average of 5000 psi) and a high temperature (average of 230°F), for a soaking period of much longer than 40 minutes for each cycle, which resulted in a single experiment taking much longer than 400 minutes. Therefore, most of the previous researchers reported that 60%–95% oil recovery could be obtained under lab conditions by injecting one of the miscible gases into these small core chips due to the large contact area between the injected gas and the formation cores, and the relatively large soaking period applied.

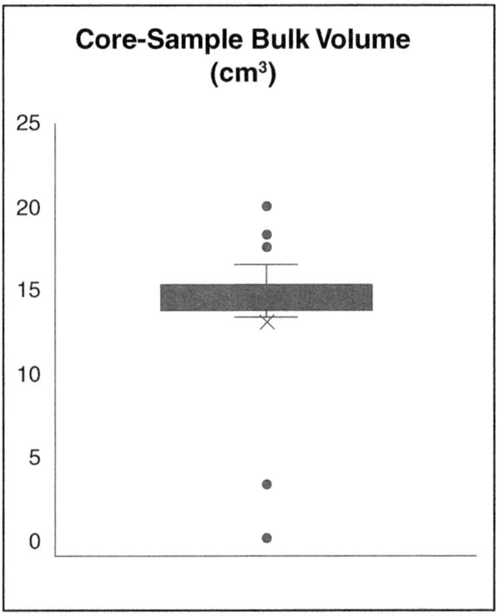

Figure 4.4 ***Box plot for the core-sample bulk volume, which was used in the 95 cases reported in the study conducted by Alfarge et al. (2018f).***

4.3 Simulation models of miscible gas-based EOR

In addition to experimental studies, some numerical simulation studies have been conducted to investigate the feasibility of miscible gas-based EOR at the field scale of shale-oil reservoirs using in-house simulation approaches or commercial software tools. Most of the reported simulation studies simulated these naturally fractured shale reservoirs using a combination of discrete fractures with a tight formation matrix, as shown in Table 4.2. They used grid refinement for the grids around the discrete fractures to cause the numerical calculations to eventually converge. We posit that combining discrete fractures with a tight formation matrix would not capture the real physics of fluid flow in these fractured shale reservoirs. Furthermore, some of the simulation studies used the same molecular diffusion coefficients in their models that were reported from the lab. However, due to the large contact area and long exposure time between the injected CO_2 and the matrix oil, the diffusion coefficients at the lab scale should be considered bulk diffusion coefficients, not effective coefficients. At the field scale, the effective diffusion coefficients should dominate because they account for the properties of the porous media, such as the tortuosity. Therefore, adopting diffusion coefficients directly from the lab scale would over-predict the

Table 4.2 Reported simulation studies investigating miscible gas-based EOR in ULRs.

Authors	Model	Simulator
Shoaib and Hoffman (2009)	Single porosity	ECLIPSE
Vega, O'Brien, and Kovscek (2010)	Single porosity	GEM
Hoffman (2012)	Single porosity	ECLIPSE
Dong and Hoffman (2013)	Single porosity	ECLIPSE
Chen, Balhoff, and Mohanty (2014)	Single porosity	IMEX
Wan, Sheng, and Soliman (2013)	Single porosity	IMEX
Kurtoglu (2013)	Double porosity	In-house
Pu and Hoffman (2014)	Single porosity	IMEX
Fai-Yengo, Rahnema, and Alfi (2014)	Single porosity	NA
Chen et al. (2014)	Single porosity	NA
Sanchez (2014)	Single porosity	GEM
Yu, Lashgari, and Sepehrnoori (2014)	Single porosity	GEM
Xiong (2015)	Hybrid of double and single porosity	In-house
Alharthy, Teklu, and Kazemi (2015)	Double porosity	GEM
Jia, Tsau, and Barati (2017)	Dual permeability	GEM
Alfarge, Wei, Bai, and Alsaba (2017j)	Different models (single porosity; dual permeability)	GEM

diffusion impact at the field scale. The rest of the simulation studies used the direct upscaling philosophy for the obtained oil recovery increment, which resulted from injecting one of the miscible gases into the lab cores. As a result, the molecular diffusion coefficients were eventually reflected in that oil increment. Therefore, both previous experimental reports and simulation studies overpredicted the oil recovery obtained by injecting one of the miscible gases into shale cores because the molecular diffusion mechanism has been overestimated.

To our knowledge, no simulation or experimental study has combined or validated its results and/or conclusions with the outcomes of pilot tests conducted in unconventional reservoirs. This has led to a clear gap in understanding the performance differences of the miscible gases at the lab scale and the field scale, and the mechanisms behind those differences. Certain factors prevented the previous investigators from combining their conclusions with the field data. One is that most of the investigators in this area of research are working in academia, while the pilot tests are reported to government agencies (Hoffman & Evans, 2016). Second, understanding miscible gas-based EOR at the field scale of shale-oil reservoirs requires extensive effort and time due to the extreme uncertainties regarding the properties of these unconventional plays.

Although extensive simulation and lab studies have been conducted to investigate the feasibility of miscible gas-based EOR in ULRs, the approaches used in the previous studies clearly suffered from many weak points and/or were not well integrated. The experimental studies used either nonrepresentative core samples or improper methodology to simulate the performance of the injected miscible gases in these plays. Meanwhile, the available simulation studies used speculative procedures to upscale the experimental outcomes.

4.4 Performance of natural gas EOR in pilot projects

Although a limited number of pilots have been conducted to investigate the applicability of NG-based EOR in shale-oil reservoirs, this section provides some of the published results for these pilots, which were mainly conducted in the United States and Canada.

Beginning with the pilots conducted in the Canadian Bakken, they interestingly have approximately the same well pattern as the Toe-Heel pattern (Alfarge et al., 2017c). Also, the spacing of the horizontal injection and production wells is very short, 200 ft with a lateral length of about one mile. Although the injection process in these pilots was sporadic, no injectivity problems were ever reported. Schmidt and Sekar (2014) reported a successful pilot project covering 1280 acres, which were developed using a combination of 80-acre and 160-acre spacing. They designed their project using a one-mile horizontal injector and nine perpendicular horizontal producers. The wells followed the Toe-Heel pattern. Table 4.3 shows this project's fluid and rock properties. Natural gas (primary methane) was used as an injectant in these reservoirs due to its availability, high compressibility, and low viscosity. They injected a lean gas (with C2–C7 content in the range of 138 bbl/MMCF to 145 bbl/MMcf) at an injection rate of 350–1000 Mscf/day. The reported results of their pilot were very encouraging in all nine offset producers, where the oil production rate increased from 135 bbl/day to 295 bbl/day, as shown in Fig. 4.5. However, there were some problems related to conformance control in which some injected gases experienced early breakthrough in some of the producers. The gas utilization value improved from 10 MCF/bbl to 6.5 MCF/bbl, which is very consistent with the model prediction provided by Alfarge, Wei, Bai, & Almansour (2017g).

Table 4.3 Summary of fluid and rock properties of the project area (Schmidt & Sekar, 2014).

Parameter	Value	Unit
Pilot area	1280	acres
Net pay	23–26	Ft
Porosity	9–10	%
Permeability	0.01–0.1	Md
Water saturation	55–59	%
Original formation volume factor	1.328	Rb/STB
Bubble point pressure	990	psi
Oil viscosity	2–3	cP
Oil gravity (stock tank)	42	API
OOIP (pilot area)	8000	MSTB

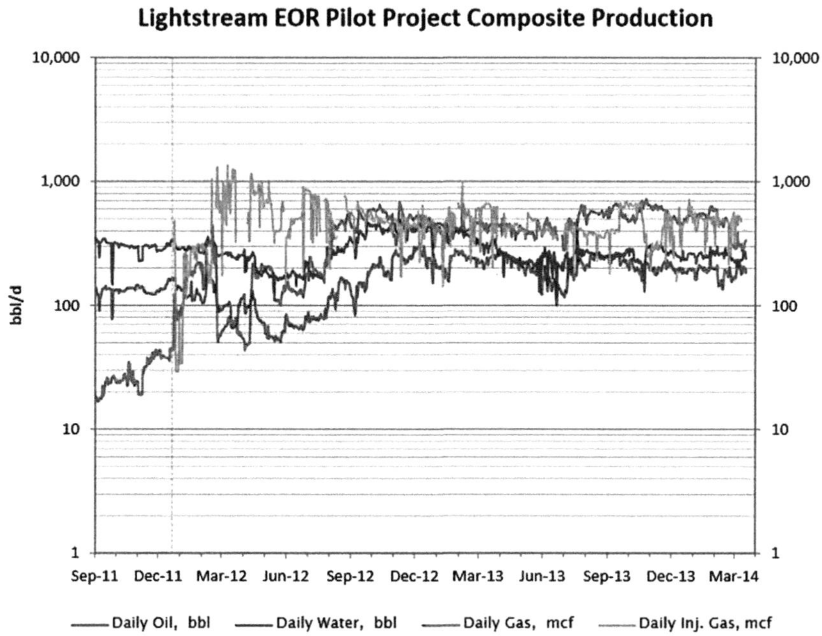

Figure 4.5 ***Performance of natural gas EOR in Canadian Bakken (Schmidt & Sekar, 2014).***

Hoffman and Evans (2016) reported a pilot test conducted in the Bakken formation in North Dakota in 2014. In that pilot, enriched natural gas was injected in a continuous mode using an injector in the center surrounded by four offset wells. Two of the producers were located 2300 ft away from the injector to the east and west, while the other two, which were to the north and south, were located 900 ft and 1200 ft, respectively, away from the injector. They injected enriched natural gas with approximately 55% methane, 10% nitrogen, and 35% $C2_+$ fractions. The injection rate was 1600 Mscf/day for 55 days at a target surface injection pressure of 3500 psi. As a result, all four offset wells had an increment in the oil production rate. However, whether that oil increment stemmed from the injection process or from the frac hits going on in the neighboring wells remained a matter of argument. Once again, the natural gas EOR, like in the Canadian Bakken, proved to be a promising technique in these reservoirs (Alfarge, Wei, & Bai, 2018c).

In the spring of 2016, EOG resources reported great success with a huff-n-puff natural gas injection project in the Eagle Ford (Thomas et al., 2016). The company issued a statement indicating that the operating wells could produce 30%–70% more oil after huff-n-puff gas injection, as shown in Fig. 4.6 (Hoffman, 2018). However, the company did not provide detailed information about those pilots due to the competitive investment among operators in the Eagle Ford. Hoffman (2018) collected data about the gas-injection pilots from the Texas Railroad Commission (TRC) and reported seven pilot projects conducted in the Eagle Ford, occurring in three different counties in Texas. In the EOG Earning Calls (2016), the company mentioned reporting the production data on a lease basis, as required under

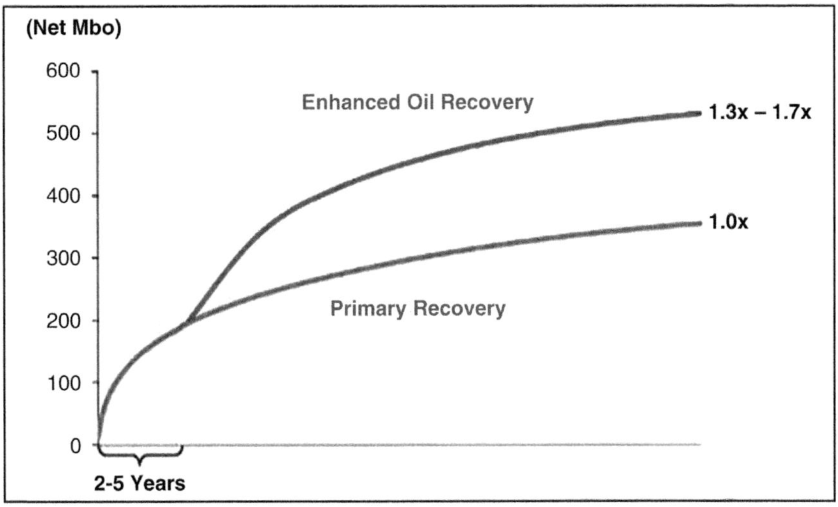

Figure 4.6 ***Reported expected recovery from huff-n-puff gas injection into the Eagle Ford (Thomas et al., 2016).***

the Railroad Commission rules. Therefore, the pilot data reported by Hoffman (2018) had been skewed by the production data from other wells not involved in the pilot test area. The 7 pilots, which contained 49 wells, had all been performed over the last 5 years. In all the pilots, natural gases were injected in huff-n-puff mode.

The first of the seven pilots was performed in late 2012 with three cycles of the lean gas huff-n-puff process. The injection and soaking period was approximately 4–6 weeks (Hoffman, 2018). Hoffman (2018) speculated that the injection rate was about 2–3 million scf/day, and the surface injection pressure was around 6000 psi based on the compressor size. The results of this pilot were encouraging because each cycle increased the production rate by approximately 50% of the initial rate, as shown in Fig. 4.7.

The second pilot contained four injection (cyclic) wells and another four production wells. As with the first pilot, the results of this pilot were encouraging. Although natural gas was only injected for 1½ years, the second pilot showed a 17% increment in the cumulative oil produced over more than a 2½-year span. The third pilot, designed as a Toe-Heel pattern, contained 14 wells, 6 cyclic injectors, and 8 production wells. This pilot also clearly showed successful results, increasing the cumulative production by 20% through natural gas injection over only 2½ years. The fourth project was located in La Salle county, over 100 miles to the southwest of the other three previous pilots. This project contained four wells, all of which were injected at the same time. The injection period was 6 months, with an injection rate of 2–4 million scf/day. After that, the wells were soaked for a period of time (no information is available about the duration of the soaking

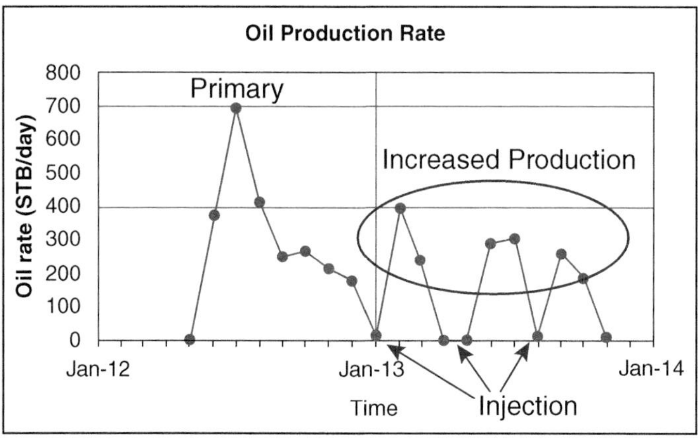

Figure 4.7 ***Oil production rate for single well in Eagle Ford Pilot A (Hoffman, 2018).***

period). Then, the wells were opened for production for 2–3 months. Later, a new huff-n-puff cycle was performed. This process was repeated for 4 cycles. After that, the wells were subjected to a shorter injection/soak time of 4–6 weeks, with a production period of 2 months. The reduced injection/soak period positively impacted the production profile of the wells, as shown in Fig. 4.8, where the slope of the cumulative production changes. This pilot lasted for 3 years and doubled the incremental production due to the injection process.

The fifth and sixth pilots were also conducted as a huff-n-puff protocol in Atascosa County, ND, USA. Both pilots used richer gas with approximately 70% methane and 30% $C2_+$ (Hoffman, 2018). The injection time was one month, with an injection rate of 2–2½ million scf/day, followed by a short soaking period (no information is available about the exact duration of the soaking period), and then around 1 month of production. The performance results of these pilots were unclear due to the overwhelming ratio of production wells to injection wells in the reported lease data. The seventh and last project, which contained 32 wells, is considered an extension of the third pilot. This injection project began recently, and evaluating it in these early stages would be difficult; however, it is showing clear incremental oil production.

Although most of the pilots which injected natural gas to enhance oil recovery in the Eagle Ford were successful, some people think the successful application of this process is unique to the Eagle Ford formation because of

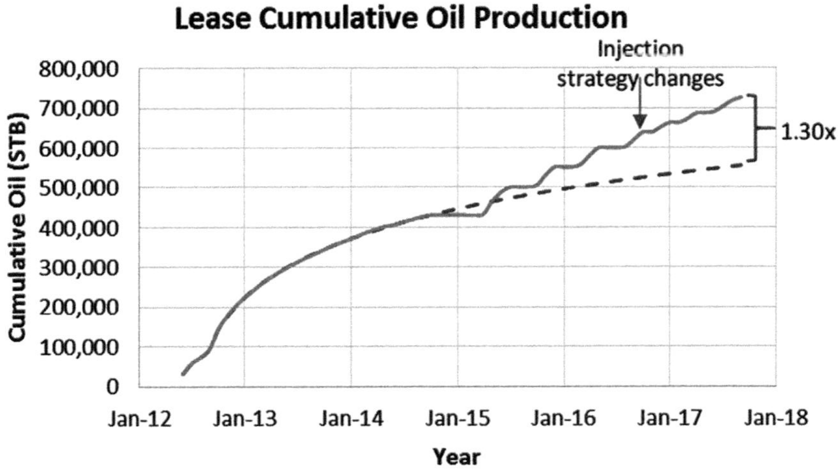

Figure 4.8 *Cumulative lease production for NG-EOR performed in Eagle Ford, Pilot D (Hoffman, 2018).*

its significant vertical containment. Vertical confinement clearly is required to achieve positive results when injecting natural gas for EOR. However, the performance of NG-based IOR pilots conducted in the US Bakken also exceeds the performance of the pilots injecting CO_2. Furthermore, the NG-EOR project conducted in the Canadian Bakken showed excellent performance. Therefore, while vertical confinement helps to improve the performance of NG-EOR, it is not mandatory.

4.5 Performance of CO_2-EOR in pilot projects

Hoffman and Evans (2016) reported seven EOR pilot tests conducted in the US Bakken, in North Dakota and Montana. Three of those seven pilots injected CO_2. Table 4.4 shows the pilot distribution and the fluid type injected. The three pilots that injected CO_2 will be discussed in detail in this section.

In late 2008, EOG conducted a CO_2 injection test via huff-n-puff in the Bakken formation in the Parshall Field in Mountrail County. The well is named Austin #1-02H. The well was horizontally drilled with a lateral leg of 4951 ft. The drainage area of the well is 634 acres. The thickness of the Bakken formation at this pilot area is 33 ft, with a porosity of 7.5% and a water saturation of 25%. The estimated OOIP is 352 Mbbl/acre ft. The solution gas drive is the primary drive mechanism in this area of the reservoir, with a predicted recovery factor of 10%. This well produced 208,322 STB of oil, 4514 bbl of water, and 78,839 Mscf of gas through June 2008. The operator proposed injecting CO_2 through the horizontal well; the affected reservoir area spanned 14.7 acres and reached as far as 74 ft from the lateral. The operator injected 30 MMscf over 30 days at an injection rate of 1 MMscf/day. No increment in oil production was observed, as shown in

Table 4.4 Summary of pilot tests in the Bakken-North America (Hoffman & Evans, 2016).

Name	State	Year	Fluid	Type
Pilot test #1	ND	2008	CO_2	Huff-n-puff
Pilot test #2	MT	2009	CO_2	Huff-n-puff
Pilot test #3	ND	2012	Water	Huff-n-puff
Pilot test #4	ND	2012–13	Water	Flood
Pilot test #5	ND	2014	CO_2	Vertical inj.
Pilot test #6	MT	2014	Water	Flood
Pilot test #7	ND	2014	Nat. gas	Flood

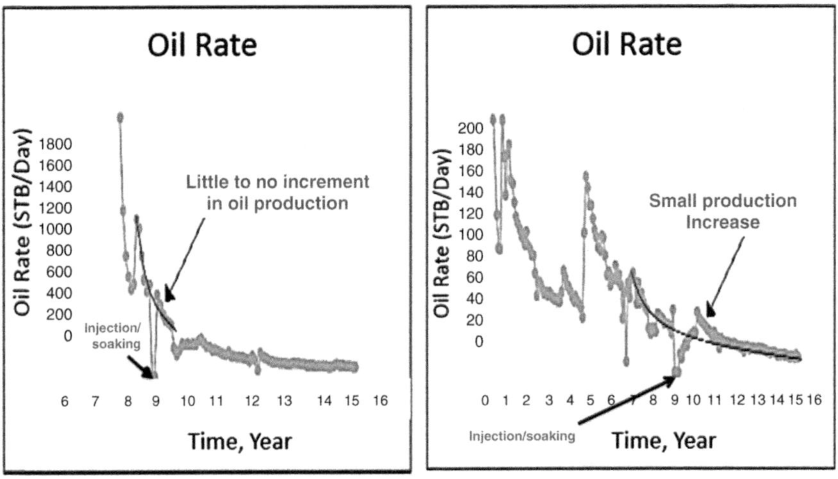

Pilot test 1 Pilot test 2

Figure 4.9 ***Oil production from two Bakken wells undergoing CO_2-EOR (Hoffman & Evans, 2016).***

Fig. 4.9. After 11 days of injection, CO_2 breakthrough was observed in an offset well-located one mile west of the injector. The breakthrough rate was about 4 Mscf/day in an offset well-located 5000 ft away from the injector.

In 2009, three companies jointly conducted a CO_2-EOR huff-n-puff pilot test (pilot test #2 in Table 4.4) in the north central part of the Elm Coulee Field. This pilot test was conducted in the Bakken formation, but not in the same area of pilot test #1. Also, pilot test #2 was conducted by a different operator. This pilot test was conducted in a horizontal well with a lateral leg of 1592 ft drilled and stimulated with a single stage of hydraulic fractures. This pilot was conducted in the Bakken formation, at a point where the top of the formation reaches 9740 ft. The initial production rate from this well was 195 bbl/day; however, it dropped to 30–40 bbl/day in 2008. About 45,000 mcf (2570 tons) of CO_2 was injected from January 16, 2009 to February 28, 2009. The average daily injection rate was 1000 mcf/day, with rates ranging from 0 to 3000 mcf/day at a maximum injection pressure of 1848 psi (BHP). This injection pressure was below the fracturing pressure of the formation. Then, the well was shut in for 64 days (soaking period) from March 1 to May 3, 2009. After that, the well was opened for production.

The well started producing at a rate of 160 bbl/day over the first 8 days, then 20 bbl/day for 30 days; after that, the well stopped naturally producing and therefore was put on a pump. In March 2010, about one year after the injection, the well reached a production peak in the post-injection period, 44 bbl/day, and then the production rate declined continuously. One theory is that this peak in production was due to the frac hits. Dit actually occurred due to the CO_2 injection performed approximately 1 year prior. This may seem difficult to believe. However, history matching between the field data and the numerical simulation has diagnosed this scenario (Alfarge, Wei, & Bai, 2018e). The takeaway from this pilot test is that CO_2-EOR huff-n-puff operations take a very long time to enhance oil recovery in shale-oil reservoirs because the molecular diffusion mechanism is the most dominant, as shown in Fig. 4.8. Also, this pilot indicates that the CO_2 diffusivity is very slow in shale-oil reservoirs as compared with conventional reservoirs, as shown in Fig. 4.10.

The third pilot was conducted in 2014 through a vertical well with 60 ft of the middle Bakken pay-thickness to perform a CO_2 cyclic process. The operators injected 300–500 Mscf/day of CO_2 over 20–30 days before shutting in the well for 20 days; then, the production process was resumed. The injected CO_2 was produced in an offset well-located 900 ft away from the injection well. The operators had clearly fractured the vertical well at a high flow rate, so they ceased their operations.

To sum up, CO_2-EOR did not prove successful in the huff-n-puff process, which indicates that the proposed CO_2 diffusion mechanism under lab

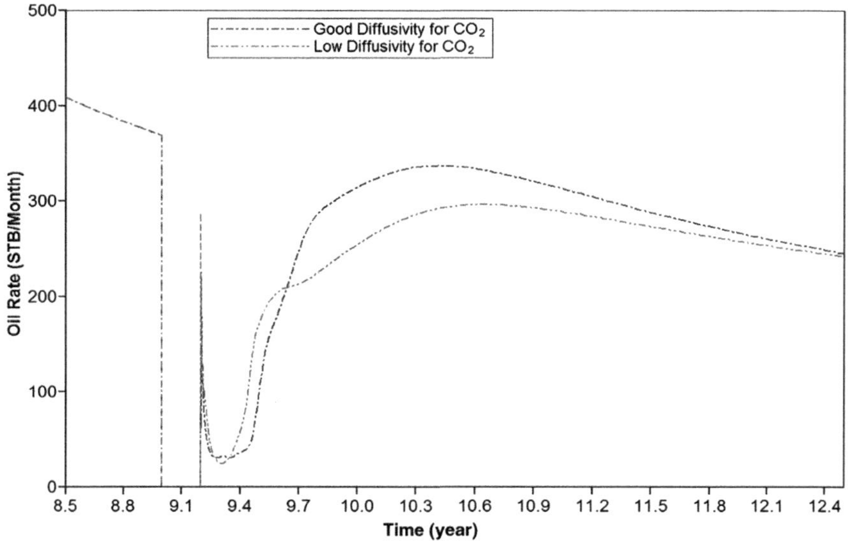

Figure 4.10 ***Diagnostic plots for history matching with pilot test #2 under different diffusivity rates (Alfarge et al., 2018e).***

conditions is not the same as under field conditions. The response timing of shale-oil reservoirs suggests that the diffusivity rate at the field scale is much slower than at the lab scale. Simulation studies also match the low diffusivity solutions, as reported by Alfarge et al. (2018e).

4.6 Lessons learned from miscible gas injection pilots

- Unconventional reservoirs respond much faster to injected NG than to injected CO_2, as shown in Figs. 4.11 and 4.12 versus Fig. 4.13. For example, in the second pilot, which was conducted in the north central part of the Elm Coulee Field in 2009, the Bakken formation responded very slowly to the injected CO_2. This was reported by Evans and Hoffman and is shown in Fig. 4.10. Approximately one year passed before production increased when CO_2 was injected in the Elm Coulee Area, Bakken Formation, Montana, as shown in Fig. 4.11. However, the pilot tests showed that the Eagle Formation responded very quickly to the injected NG. This was also clear when Lloyd W. Helms (Executive Vice President-Exploration & Production) mentioned, in the EOG Earning Calls (2016), that "the production response occurs quickly, within the first 2–3 months, and holds steady for longer."

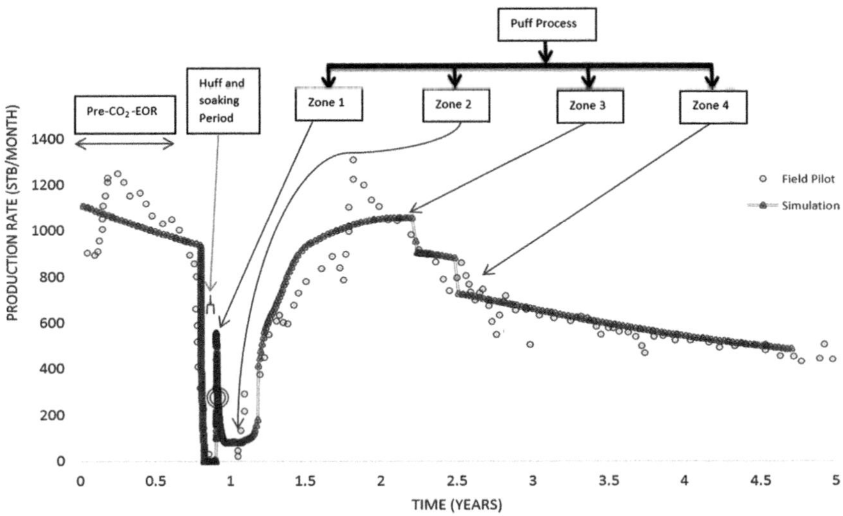

Figure 4.11 ***Typical plot for CO_2 -EOR huff-n-puff in shale -oil reservoirs (Alfarge et al., 2018e).***

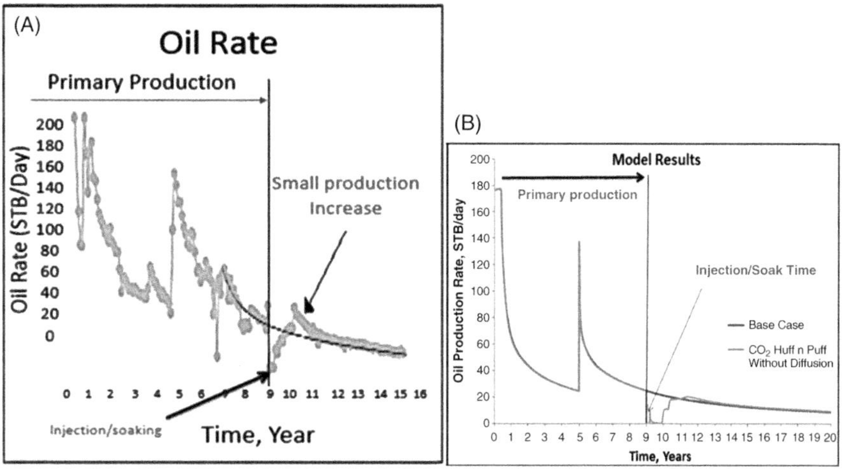

Figure 4.12 (A) CO_2 pilot test #2 (Hoffman & Evans, 2016). (B) History match from the simulated model (Alfarge, Wei, & Bai, 2017b).

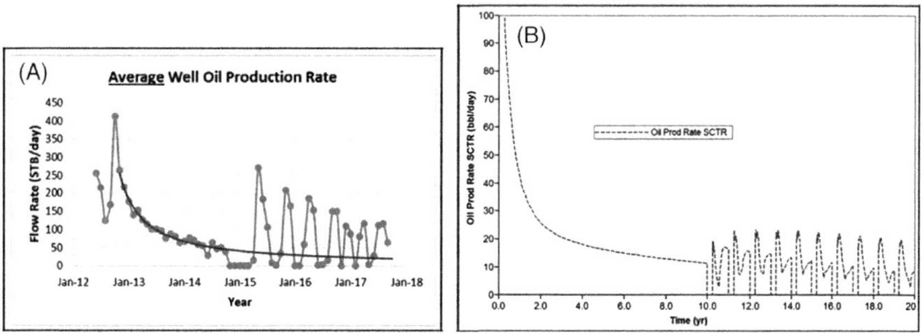

Figure 4.13 (A) NG-EOR pilot test #D in the Eagle Ford (Hoffman, 2018). (B) Simulation prediction for NG-EOR in the Bakken (Alfarge et al., 2017b).

- The number of cycles has a negative impact on CO_2-EOR but a positive impact on NG-EOR, as shown in Fig. 4.13 versus Fig. 4.14.
- The two previous points support the following statement by Alfarge et al. (2018d): "if the well or field conditions predict a low molar-diffusivity for the injected gases, the rich and lean gases would have a better feasibility than CO_2 in unconventional reservoirs."
- When the injected gas is produced back during the puff process, this hinders oil production due to the slippage effect. This causes bumps in the curve of the oil rate versus time in the puff process, as shown in the simulation model (Fig. 4.13B). However, these bumps do not appear in the pilot field data reported by TRC for the following two possible

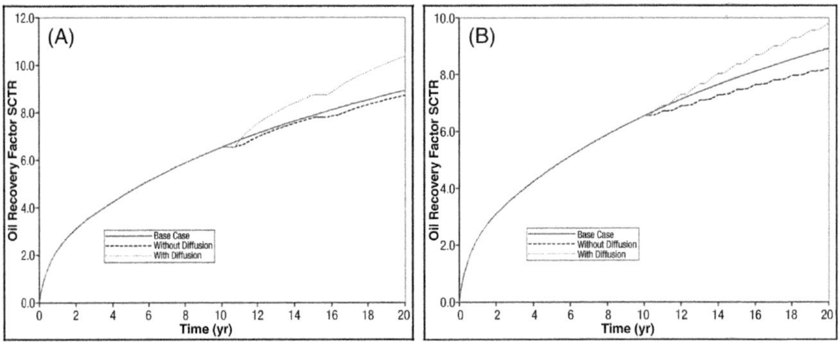

Figure 4.14 (A) Performance of CO_2-EOR (2-cycle scenario). (B) Performance of CO_2-EOR (10-cycle scenario) (Alfarge et al., 2017b).

reasons: (1) The volume of natural gas produced back during the puff process is reported incorrectly because the reported data, which are available through the TRC website, do not clearly indicate the quantity of the injected gas and whether or not it has been subtracted from the reported gas production data (Hoffman, 2018), (2) The operators generated some injection-induced fractures while injecting these natural gases into the Eagle Ford formation.

4.7 SUMMARY

1. Although most of the previous studies in this area of research recommended CO_2 as the best EOR technique to improve oil recovery in these formations, pilot tests clearly showed that natural gas performance exceeds CO_2 performance at the field scale. In addition, a clear gap exists between the results of experimental studies and pilot tests for the applicability of CO_2-EOR in shale-oil reservoirs.
2. Most of the previous experimental studies relied on the CO_2 molecular diffusion mechanism to predict the potential success of CO_2-EOR in shale reservoirs. However, the direct upscaling for this mechanism to the field scale via simulation methods, using the same lab-obtained CO_2 diffusion rate is misleading. The previous simulation studies also suffer from a variety of other drawbacks.
3. At the lab scale, the design of experiments reported that the TOC and exposure time are the two main parameters controlling CO_2-EOR success in shale reservoirs. The sensitivity of the oil recovery factor obtained by CO_2 to the TOC might suggest that the CO_2-EOR mechanisms in shale-oil reservoirs are different from those in conventional reservoirs.

4. The simulation results proved that the molecular diffusion mechanism plays a significant role in EOR by gas injection in the Bakken formation. However, CO_2 requires good molar diffusivity into the formation oil in order to enhance oil production in these shale reservoirs. The success of lean and rich gases requires less molar diffusivity compared with CO_2.
5. Some of the CO_2 pilot tests showed a good match with the simulated cases, which had low molecular diffusivity between the formation oil and the injected CO_2.
6. If the well or field conditions predict low molar diffusivity for the injected gases, then the rich and lean gases will have better feasibility than CO_2.

References

Alfarge, D., Wei, M., & Bai, B. (2017b). Numerical simulation study on miscible-EOR techniques for improving oil recovery in shale-oil reservoirs. *Journal of Petroleum Exploration and Production Technology*, *8*, 901–916. https://doi.org/10.1007/s13202-017-0382-7.

Alfarge, D., Wei, M., & Bai, B. (2017c). *IOR methods in unconventional reservoirs of North America: Comprehensive review*. Society of Petroleum Engineers. https://doi.org/10.2118/185640-MS.

Alfarge, D., Wei, M., Bai, B., & Almansour, A. (2017g). *Optimizing injector-producer spacing for CO_2 injection in unconventional reservoirs of North America*. Society of Petroleum Engineers. https://doi.org/10.2118/188002-MS.

Alfarge, D., Wei, M., Bai, B., & Alsaba, M. (2017i). *Analysis of IOR pilots in Bakken formation by using numerical simulation*. Society of Petroleum Engineers. https://doi.org/10.2118/188633-MS.

Alfarge, D., Wei, M., Bai, B., & Alsaba, M. (2017). *Selection criteria for miscible-gases to enhance oil recovery in unconventional reservoirs of North America*. Society of Petroleum Engineers. https://doi.org/10.2118/187576-MS.

Alfarge, D., Wei, M., & Bai, B. (2018c). A review of improved-oil-recovery methods in North American unconventional reservoirs. In C. Carpenter (Ed.). *Journal of Petroleum Technology*, *70*(1).

Alfarge, D., Wei, M., & Bai, B. (2018d). *A Parametric study to compare different miscible gases to enhance oil recovery in unconventional liquids rich reservoirs*. Society of Petroleum Engineers. https://doi.org/10.2118/189785-MS.

Alfarge, D., Wei, M., & Bai, B. (2018e). CO_2-EOR mechanisms in huff-n-puff operations in shale oil reservoirs based on history matching results. *Fuel*, *226*, 112–120. https://doi.org/10.1016/j.fuel.2018.04.012.

Alfarge, D., Wei, M., & Bai, B. (2018f). Data analysis for CO_2-EOR in shale-oil reservoirs based on a laboratory database. *Journal of Petroleum Science and Engineering*, *162*, 697–711. https://doi.org/10.1016/j.petrol.2017.10.087.

Alharthy, N., Teklu, T., Kazemi, H., et al. (2015). *Enhanced oil recovery in liquid-rich shale reservoirs: Laboratory to field*. Society of Petroleum Engineers. https://doi.org/10.2118/175034MS.

Chen, C., Balhoff, M. T., & Mohanty, K. K. (2014). *Effect of reservoir heterogeneity on primary recovery and CO_2 huff-n-puff recovery in shale-oil reservoirs*. Society of Petroleum Engineers. https://doi.org/10.2118/164553-PA.

Dong, C., & Hoffman, B. T. (2013). *Modeling gas injection into shale oil reservoirs in the Sanish Field, North Dakota*. Society of Petroleum Engineers. https://doi.org/10.1190/URTEC2013-185.

Du, F., & Nojabaei, B. (2019). A review of gas injection in shale reservoirs: enhanced oil/gas recovery approaches and greenhouse gas control. *Energies, 12*, 2355. 2019.

Fai-Yengo, V., Rahnema, H., & Alfi, M. (2014). *Impact of light component stripping during CO_2 injection in Bakken formation*. Unconventional Resources Technology Conference. https://doi.org/10.15530/URTEC-2014-1922932.

Hoffman, B. T. (2012). *Comparison of various gases for enhanced recovery from shale oil reservoirs*. Society of Petroleum Engineers. https://doi.org/10.2118/154329-MS.

Hoffman, B. T. (2018). *Huff-n-puff gas injection pilot projects in the Eagle Ford*. Society of Petroleum Engineers. https://doi.org/10.2118/189816-MS.

Hoffman, B. T., & Evans, J. (2016). *Improved oil recovery IOR pilot projects in the Bakken formation*. SPE-180270-MS paper presented at the SPE Low Perm Symposium held in Denver, Colorado, USA, 5–6 May 2016.

Jia, B., Tsau, J.-S., & Barati, R. (2017). *Role of molecular diffusion in heterogeneous shale reservoirs during CO_2 huff-n-puff*. Society of Petroleum Engineers. https://doi.org/10.2118/185797-MS.

Jin, L., Sorensen, J.A., Hawthorne, S.B., Smith, S.A., Bosshart, N.W., Burton Kelly, M.E., Miller, D.J., Grabanski, C.B., & Harju, J.A. (2016). Improving oil transportability using CO_2 in the Bakken System—A laboratorial investigation. In *Proceedings of the SPE International Conference & Exhibition on Formation Damage Control, SPE 178948, Lafayette, Louisiana, February 24–26*.

Kurtoglu, B. (2013). *Integrated reservoir characterization and modeling in support of enhanced oil recovery for Bakken* (Ph.D. dissertation). Colorado School of Mines.

Mohammad et al. (2018). *Carbon dioxide minimum miscibility pressure with nanopore confinement in tight oil reservoirs*. IOP Conf. Ser.: Earth Environ. Sci.167 012030. https://iopscience.iop.org/article/10.1088/1755-1315/167/1/012030/pdf.

Pu, W., & Hoffman, B.T. (2014). *EOS modeling and reservoir simulation study of Bakken gas injection improved oil recovery in the Elm Coulee Field, Montana*. Society of Petroleum Engineers. https://doi.org/10.15530/urtec-2014-1922538.

Ratner, M., & Tiemann, M. An overview of unconventional oil and natural gas: resources and federal actions. Congressional Research Service Report, 2015. Available online: https://fas.org/sgp/crs/misc/R43148.pdf.

Sanchez, R. D. (2014). Reservoir simulation and optimization of CO_2 huff-and-puff operations in the Bakken shale (MS thesis). University of Texas at Austin.

Schmidt, M., & Sekar, B. K. (2014). *Innovative unconventional 2EOR-A light EOR an unconventional tertiary recovery approach to an unconventional Bakken reservoir in Southeast Saskatchewan*. World Petroleum Congress.

Shoaib, S., & Hoffman, B. T. (2009). *CO_2 flooding the Elm Coulee Field*. Society of Petroleum Engineers. https://doi.org/10.2118/123176-MS.

Thomas, W. R., Helms, L. W., Driggers, T. K., Trice, D. W., & Thomas, G. L. (2016). *EOG Resources (EOG) Earnings Call, May 06, 2016*. https://seekingalpha.com/article/3972422-eog-resources-eog-william-r-thomas-on-q1-2016-results-earnings-call-transcript.

Vega, B., O'Brien, W. J., & Kovscek, A. R. (2010). *Experimental investigation of oil recovery from siliceous shale by miscible CO_2 injection*. Society of Petroleum Engineers. https://doi.org/10.2118/135627-MS.

Wan, T., Yu, Y., & Sheng, J. J. (2015). Experimental and numerical study of the EOR potential in liquid-rich shales by cyclic gas injection. *Journal of Unconventional Oil and Gas Resources, 12*, 56–67.

Wan, T., Sheng, J. J., & Soliman, M. Y. (2013). *Evaluate EOR potential in fractured shale oil reservoirs by cyclic gas injection*. Society of Petroleum Engineers. https://doi.org/10.1190/URTEC2013-187.

Wang, L., et al. (2017). Advances in improved/enhanced oil recovery technologies for tight and shale reservoirs. *Fuel*, *210*, 425–445. https://doi.org/10.1016/j.fuel.2017.08.095.

Verma, M. (2015). Fundamentals of carbon dioxide-enhanced oil recovery (CO_2-EOR)—A supporting document of the assessment methodology for hydrocarbon recovery using CO_2-EOR associated with carbon sequestration. https://pubs.usgs.gov/of/2015/1071/pdf/ofr2015-1071.pdf.

Xiong, Y. (2015). Development of a compositional model fully coupled with geomechanics and its application to tight oil reservoir simulation (Ph.D. thesis). Colorado School of Mines.

Yu, W., Lashgari, H., & Sepehrnoori, K. (2014). *Simulation study of CO_2 huff-n-puff process in Bakken tight oil reservoirs*. Society of Petroleum Engineers. https://doi.org/10.2118/169575-MS.

Yu, Y., Li, L., & Sheng, J. J. (2017). A comparative experimental study of gas injection in shale plugs by flooding and huff-n-puff processes. *Journal of Natural Gas Science and Engineering*, *38*, 195–202.

CHAPTER 5

Air injection in shale and tight oil reservoirs

Contents

5.1 Introduction 87
5.2 Types of air combustion 90
5.2.1 Dry forward combustion 90
5.2.2 Wet forward combustion 93
5.2.3 Reverse combustion 94
5.3 Modes of combustion reactions 95
5.3.1 Oxygen addition 95
5.3.2 Low-temperature oxidation or bond scission 95
5.3.3 Spontaneous combustion or high-temperature oxidation 96
5.4 Air injection EOR modeling 97
5.5 Performance of air injection in conventional reservoirs 98
5.6 Advantages and disadvantages of air injection 100
5.7 Performance of air injection in unconventional reservoirs 102
5.8 Air injection pilots in tight oil reservoirs 103
5.9 Pros and cons of applying air injection in shale-oil reservoirs 104
5.10 Numerical simulation for air injection in unconventional reservoirs 105
5.11 Summary 108
References 108

Abstract

This chapter discusses the viability of air injection for conventional reservoirs as an enhanced oil recovery (EOR) technology by discussing its EOR mechanisms, major impact factors, and successful projects. Then, the viability and feasibility of its application for EOR in unconventional reservoirs are presented. Some reservoir simulation study results are presented to validate the observations. This discussion of viability is only from a technical perspective and does not consider economics.

Keywords: air injection; shale; oil; tight oil; combustion; dry; wet

5.1 Introduction

The use of air injection as an EOR technique has become very popular in the oil industry since its discovery in the early 1900s (Ursenbach et al., 2010). In the 1950s and 1960s, air injection was successfully deployed in many heavy oil fields in California (Chu, 1977; Sarathi and Olsen, 1995). At that time,

Fundamentals of Enhanced Oil Recovery Methods for Unconventional Oil Reservoirs
http://dx.doi.org/10.1016/B978-0-12-818343-4.00005-X

Canada also contained giant fields of very heavy oil, thus creating the need for a technique to produce such oils. Due to the successes in California, some types of thermal methods were extended to Canadian heavy oils and bitumen from the 1960s through the 1980s (Ursenbach et al., 2010). However, because the Canadian oils are much heavier, in-situ combustion and/or fire-flooding has been used. Although some success was achieved in the pilot tests performed initially, many others failed. Therefore, Ursenbach et al. (2010) posited that air injection in heavy oils never reached its theoretical potential.

Air injection is a thermal recovery method whereby air is injected into the reservoir at a high pressure. Many physical and chemical processes take place in this EOR method, which depend on several factors such as the reservoir characteristics, nature of the oil, pressure/temperature of the injection process, and so on. However, for most instances, the injected air reacts with some of the oil, which may or may not initiate spontaneous ignition (Youtsos and Mastorakos, 2013). The ignition depends primarily on the temperature of the reservoir and whether the oil in the reservoir is heavy or light. Generally, spontaneous ignitions are achieved for light oil, while some form of external ignition mechanism is required for heavy oil, such as electric gas burners, pyrophoric agents, and other mechanisms. The achieved temperature also dictates which mode of reaction will occur in the reservoir—oxygen addition, low-temperature oxidation (LTO), or high-temperature oxidation (HTO). While LTO may have some advantages, the most sought-after reaction scheme is HTO because of the large amount of oil that can be recovered from the reservoir in this mode.

Upon the ignition of the oil in the reservoir, heat energy is released, and a fire front is created, as shown in Fig. 5.1. This fire front causes the release of CO_2 in the flue gas and increases the temperature, which results in

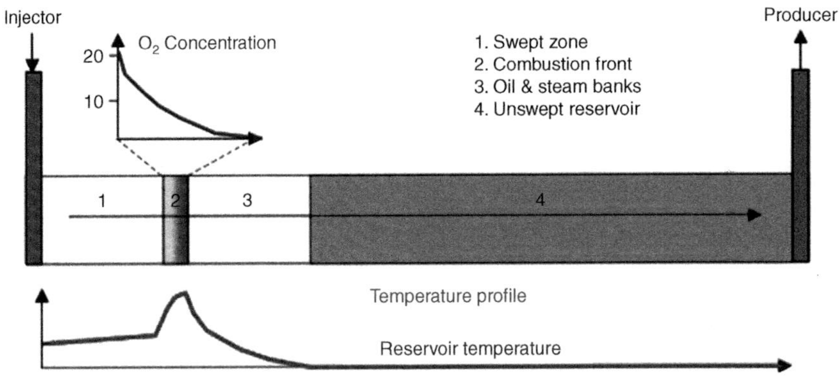

Figure 5.1 ***Air injection process schematic.*** *(Modified from Hughes & Sarma, 2006a).*

immiscible gas displacement, improved miscibility, reduction in interfacial tension, oil swelling, and reservoir re-pressurization. In addition, heat from the fire reduces oil viscosity and helps vaporize reservoir water into steam. The steam, hot water, and flue gas all act to drive oil in front of the fire toward the production wells. However, a number of factors come into play when considering the recovery mechanisms for air injection. Most EOR mechanisms are associated with the generated thermal effects; however, some additional recovery is also attributable to the pressure maintenance and oil swelling effects. The main air injection recovery mechanisms are summarized as follows:

- When oil is burned due to the generation of flue gases and heat generation, as well as the conversion of water to steam if any water is present, then these gases aid in the sweep efficiency, enhancing mobilization (Jia et al., 2012a).
- Due to thermal effects, as well as the quick vaporization of some of the lighter hydrocarbons and the water, the reservoir re-pressurizes rapidly, aiding in the recovery of hydrocarbons (Jia, 2012).
- Light components are easily extracted for subsequent natural gas liquid (NGL) recovery because due to the additional heat generated, they get converted into gas and are produced, whereby they are again converted to NGL by the surface production facilities (Hughes and Sarma, 2006a).
- Oil swelling takes place via flue–gas dissolution, which assists in oil movement and production (Dutton et al., 2004).
- The potential of miscible flooding increases due to the gases generated through the process (Das, 2009).
- The creation of thermally generated microfractures in the reservoir increases due to the excessive heat generation. Given their enhanced permeability, the fractures aid the movement of the oil. This effect becomes more accentuated when considering low-permeability reservoirs such as shales (Jia et al., 2014).
- Crude oil viscosity reduction occurs via thermal effects, hence easing the movement of the oil. This effect is more prominent when the crude oil is ignited in the form of HTO, less so in LTO, and in the form of oxygen addition reactions, this mechanism may not play a substantial role (Jia et al., 2012b).
- Bulldozing or pore-blocking effects may occur, whereby the surrounding reservoir area's pores are blocked. This helps to re-direct the gas flow to improve sweep efficiency (Jia and Sheng, 2016).

- Due to the excessive heat and the gases and combustion products generated, the probability of relative permeability alteration exists under miscible/immiscible conditions.
- Wettability and capillary pressure alteration can further improve oil recovery.

In addition to air injection, another process for heavy oil, called in-situ combustion (ISC), has become very popular. Although the two methods are similar to some extent, some differences exist which need to be highlighted. All ISC projects are air injection projects, but not all air injection projects are ISC projects. Most ISC projects require some ignition strategy. However, the required ignition can be achieved naturally, without using an external ignition source in some projects, due to their nature of the reservoir, their high pressure and high temperature as well as their oil properties. Overall, both processes require the injection of air, regardless of auto-ignition or some source of external ignition, and hence they are classified under the umbrella of air injection or high-pressure air injection (HPAI). Fig. 5.1 shows a typical air injection project. The reservoir has been divided into four zones: the swept zone, combustion front, oil and steam banks, and unswept reservoir. Gradually, the combustion front moves from the injector to the production well, and the oxygen concentration is the greatest just before the point at which combustion takes place; after that, it decreases considerably.

5.2 Types of air combustion

The injection of air results in the occurrence of three main types of air combustion processes, as explained in the subsequent sections.

5.2.1 Dry forward combustion

In this type of combustion project, only dry air is injected into a reservoir. Furthermore, the burning front moves in the same direction as the injected air (Ahmed and Nathan Meehan, 2012). During the forward combustion process, several distinct zones form in an oil reservoir, as shown in the schematic view of Fig. 5.2. The oil displacement mechanism and temperature profile associated with each of these formed zones are shown in Fig. 5.3. This kind of combustion requires two types of operations. If the temperature of the reservoir is already high enough and light oil is present in the reservoir, the chances of achieving auto-ignition are very high. On the contrary, if the temperature of the reservoir is not high, then downhole gas burners, electrical ignition devices, or pyrophoric agents may be used to

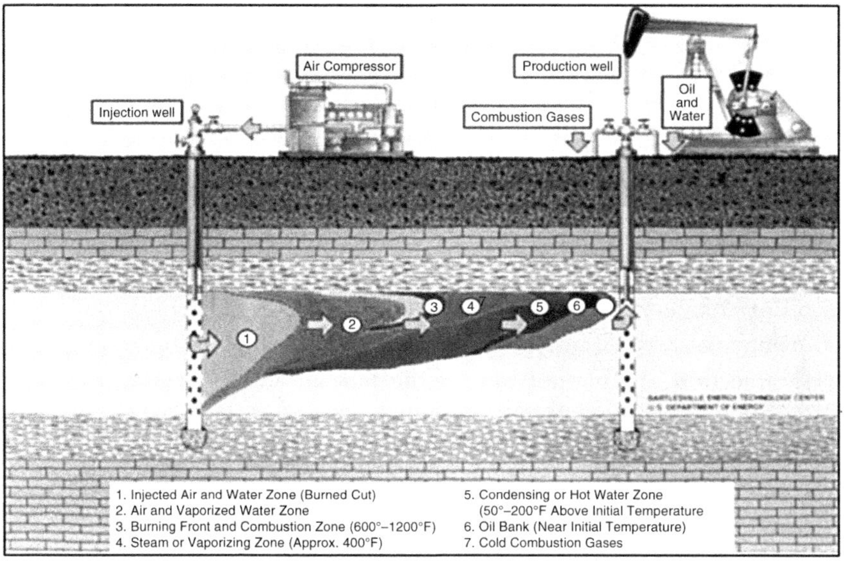

Figure 5.2 ***Dry forward combustion process schematic.*** *(Produced by Courtesy Bartlesville Energy Technology Center, US Department of Energy, Washington, DC and Modified from Ahmed and Nathan Meehan, 2012).*

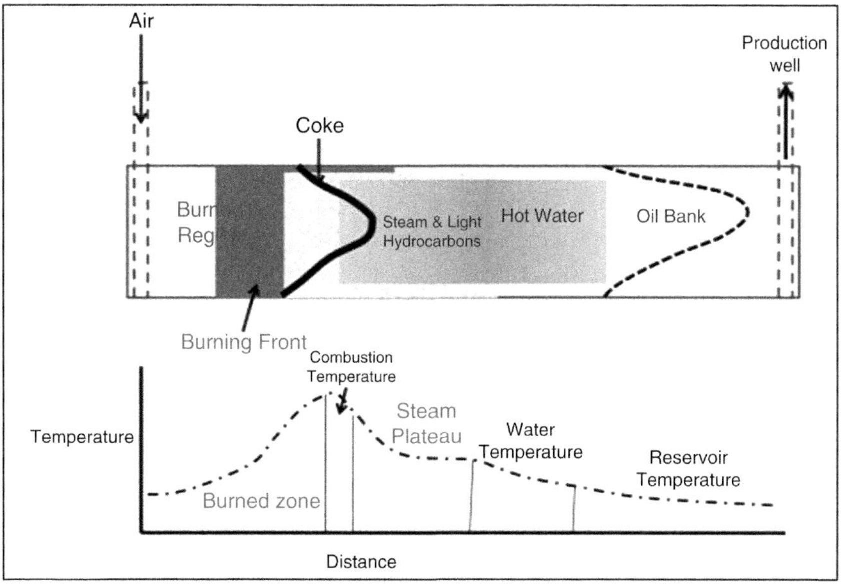

Figure 5.3 ***Schematic of temperature zones during dry forward combustion.*** *(Modified from Ahmed and Nathan Meehan, 2012).*

achieve ignition. After ignition has been achieved, the fire front moves from the injector to the production well via the continuous injection of air. Heat is generated along with flue gases, which improves oil recovery due to the advantages associated with ISC and the relevant thermal effects.

As shown in Figs 5.2 and 5.3, seven clear zones form during the forward combustion process (Ahmed and Nathan Meehan, 2012), which include:

- Burned zone: This is defined as the region, which is already burned. This zone is essentially composed of clean sand that is completely free of its oil or coke content, but it may be filled with air and contains a small quantity of residual unburned organic material. Due to the continuous air injection, the temperature of the burned zone increases from the injected air temperature at the injector to the temperature at the leading edge of the combustion.
- Combustion front zone: This zone is defined as the region situated ahead of the burned-out zone. Its temperature varies from 600°F to 1200°F. In this region, the oxygen combines with fuel, and high-temperature oxidation occurs.
- Coke zone: This zone is defined as the region situated just ahead of the combustion front zone. Due to the thermal cracking of the crude oil, the coke region forms where carbonaceous material has been deposited. The components filling the coke zone have a high molecular weight and boiling point temperatures. The fraction of the components left in the coke zone can represent up to 20% of the crude oil.
- Vaporizing zone: The vaporizing zone is located immediately ahead of the coke region and consists of vaporized light hydrocarbons, combustion products, and steam. It has a temperature variation ranging from the high temperature of combustion to the temperature necessary to vaporize the reservoir connate water.
- Condensing zone: The condensing zone is located just ahead of the vaporizing region. From this zone, oil is extracted by the following driving mechanisms: (1) the condensed light hydrocarbons displace reservoir oil miscibly, (2) condensed steam creates a hot water flood mechanism, and (3) the combustion gases provide additional oil recovery via gas drive. Temperatures in this zone have a typical range of 50°F–200°F above the initial reservoir temperature.
- Oil bank zone: The displaced oil from the condensing zone accumulates in the next zone, forming an oil bank. The temperature in this zone is slightly above the initial reservoir temperature, which leads to a minor improvement in the oil viscosity.

- Undisturbed reservoir: This zone consists of the undisturbed parts of the reservoir, which have not been affected by the combustion process. This zone lies further ahead of the oil bank.

5.2.2 Wet forward combustion

Wet forward combustion is similar to dry forward combustion, with the only difference being that water is injected with the air simultaneously or alternately (Schlumberger, 2020), as shown in Fig. 5.4. The advantage of injecting water with air is that water can effectively capture and advance more heat from the burned zone to the surroundings. It also serves to vaporize and assists in moving the flame front. The injected water recovers the heat from behind the burning front and transfers it to the oil bank ahead. Because of this additional energy, the oil displacement is more efficient and requires less air. In spite of these advantages, a wet combustion process cannot avoid liquid-blocking problems, and the use of wet combustion is limited by the oil viscosity. Wet combustion is also called in-situ steam generation or a combination of forward combustion and waterflooding, which is abbreviated as COFCAW (Schlumberger, 2020).

When conducted properly, there is evidence that water-assisted combustion will reduce the amount of fuel needed and hence will result in increased oil recovery and decreased air requirements to heat a given volume of reservoir. Up to 25% improvement in process efficiency can be achieved by injecting water with air (Ahmed and Nathan Meehan, 2012). However, determining the optimum water-to-air ratio is difficult because of reservoir

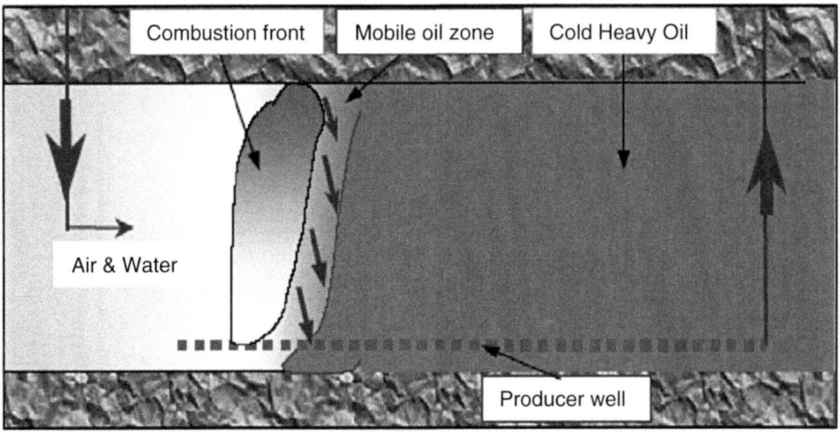

Figure 5.4 ***Schematic of wet forward combustion.*** *(Modified from Manahali and Salim, 2010).*

heterogeneities and gravity override that can affect fluid movement and saturation distributions. Injecting too much water can result in an inefficient fire front, thus losing the benefits of the process. Hence, care must be exercised while performing this type of combustion (Ahmed and Nathan Meehan, 2012).

5.2.3 Reverse combustion

In this process, the flame front moves opposite the direction of the air injection, as shown in Fig. 5.5. Initially, air is injected through the production well, and the fire is ignited. After the burning front propagates for some distance far away from the production well, air should be supplied only near the injection well (Schlumberger, 2020). While the burning front advances toward the injection well, the oil moves toward the production well. Reverse combustion actually means dry reverse combustion. This type of combustion can be used to recover extremely viscous oil or tar (Schlumberger, 2020). In reverse combustion, there is not much risk of liquid blocking because the hot zone is basically maintained near the production well. However, the reverse combustion process is not as efficient as dry forward combustion because lighter fractions of the oil are burned, and heavier fractions are left behind the burning front (Schlumberger, 2020).

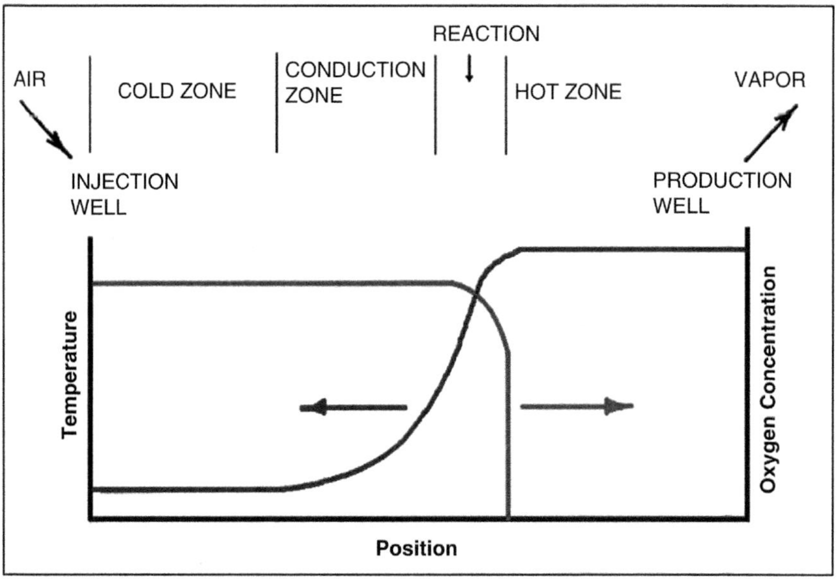

Figure 5.5 ***Schematic of reverse combustion.*** *(Modified from Berry and Parrish, 1960).*

Another drawback is that spontaneous ignition might occur near the injector well, which diverts the injected air to feed the combustion near the injector well instead of near the production well (Schlumberger, 2020). In practice, this has been difficult to achieve, and this method is not seen much in application.

5.3 Modes of combustion reactions

5.3.1 Oxygen addition

Oxygen additions are simply reactions in which oxygen is added into the reservoir oil. The reactions appear more so as bond scission than as heat. It is only an intermediate process in which partially oxygenated compounds and few or no carbon oxides or water are produced. Oxygen atoms can spontaneously bind into the molecular structure of the hydrocarbons, generating various oxygenated compounds, such as hydroperoxides, aldehydes, ketones, and acids, as well as a small number of carbon oxides, which will worsen the oil properties. Therefore, the biggest concern is whether the oxidation reaction can shift to the next mode of low-temperature oxidation.

5.3.2 Low-temperature oxidation or bond scission

Low-temperature oxidation (LTO) is also referred to as a bond scission reaction. The products of such reactions are water and partially oxygenated hydrocarbons, such as carboxylic acids, aldehydes, ketones, alcohols, and hydroperoxides (Burger, 1972). Because some heavy components are generated, the oil viscosity increases (Alexander et al., 1962), and the API gravity decreases (Mamora, 1995). LTO is ineffective at mobilizing oil (Gutierrez, Skoreyko, & Moore, 2009); hence, it is not a desirable type of oxidation reaction for reservoirs.

Various researchers have defined low-temperature oxidation reactions differently. Sarathi (1999) defined reactions occurring at temperatures below 204°C as LTO. Greaves et al. (2000) suggested that LTO occurs below 300°C, while Gutierrez, Taylor, and Kumar (2008) reported that it occurs below 250°C. Montes et al. (2010) proposed that LTO is the foremost reaction for light oils at temperatures below 150°C and for heavy oils below 300°C. Hence, it is evident that LTO takes place at temperatures ranging from approximately 150 to 300°C.

The exothermic intensity of LTO reactions is much milder than that of the high-temperature combustion reactions, which follow them and are the desired type of reactions in AIPs. The efficacy of LTO for enhanced oil recovery has always been questionable. Uncertainty has existed around whether LTO reactions can improve recovery efficiency and produce residual oil from reservoirs. Research has shown that the thermal effect of LTO can play a significant role in recovery performance because a higher temperature results in a higher oil recovery factor. A temperature increment of 40°C caused by LTO reactions can result in a 10% recovery factor increase (Huang et al., 2018). On the contrary, the effect of LTO on residual oil production remains unknown. LTO generates oxygenated compounds, which can increase the viscosity of the crude oil, thus resulting in recovery efficiency reduction. However, the benefits of the thermal effect of LTO should far outweigh its drawbacks in terms of the oil viscosity increment, leading to more successful projects. In this stage, a great amount of CO_2 and H_2O is produced, and heat is released that can result in coke generation. If the released heat raises the reservoir temperature to a certain degree, then it will switch to another mode, either spontaneous combustion or HTO reactions.

5.3.3 Spontaneous combustion or high-temperature oxidation

In spontaneous combustion reactions, some amount of heat is generated, producing visible results. However, it is still not the most effective method by which to increase recovery.

The most effective reaction for increasing recovery is considered to be HTO, the reaction between the oxygen in the injected air and the coke at temperatures above 343°C (Sarathi, 1999). Greaves et al. (2000) suggested that the temperature range for HTO is between 250°C and 340°C for light oil and between 400°C and 500°C for heavy oil. Gutiérrez et al. (2008) reported that these reactions occur between 250°C and 300°C for light oil and above 450°C for heavy oil. The main products are carbon dioxide, carbon monoxide, and water. These reactions generate a great deal of heat. HTO is extremely effective at mobilizing oil and is the desired state of operation for air injection. HTO is also considered a combustion reaction. Generally, in an LTO process, the temperature increases, but the viscosity reduction may not be significant. Also, the compounds generated by the LTO process, such as carboxylic acids, aldehydes, ketones, alcohols, and others (Burger, 1972), will increase the oil viscosity. For air injection to have significant advantages over non-oxidation gas injection, combustion must occur. Hence, the design and execution of the AIP must ensure that the said processes will proceed to the next level, HTO.

5.4 Air injection EOR modeling

Various modeling approaches for predicting the HPAI performance have been presented and proposed over the years. Sakthikumar et al. (1995) simulated the process of HPAI as an isothermal immiscible nitrogen flood to assess the feasibility of injecting air into a water-flooded light oil reservoir. AIP in the Maureen field was modeled as a miscible gas injection process by Fraim et al. (1997). Glandt, Pieterson, & Dombrowski (1999) designed their study to build a calibrated simulation model by matching Coral Creek's production history. They assessed the recovery potential of HPAI in this field while modeling the process as an isothermal flue gas flood with 10 components via an equation-of-state (EOS) model. Ren, Greaves, & Rathbone (2002) conducted a simulation study in three stages. First, the reaction models were formulated based on results obtained from the SBR (small batch reactor) experiments and matching oxidation tube experiments. Then, a 2-D cross-sectional base model was used to assess sensitivity. Finally, a 3-D small-section field simulation was conducted. Kuhlman (2000) evaluated the predictions obtained from different simulation approaches for a viscous-dominated (Horse Creek) and a gravity-stable (West Hackberry) reservoir. The study used a thermal combustion model, a flue gas EOS model, and injection of reservoir gas in a black oil model. Kuhlman (2000) concluded that a thermal model with at least six hydrocarbon components is preferable for modeling light oil air injection and that an EOS model can be used for hot, thick, very light oil reservoirs, but is not useful in cooler reservoirs. When low porosity is combined with low thickness or low API gravity, black oil models can predict the response to gas injection but should only be used as scouting tools.

The earlier-mentioned models were deployed at a time when awareness and knowledge of what was going on in the field were generally lacking. The reservoir simulation models discussed previously were evolving. Hence, conclusions were based on models, which used the same grid block sizes for each approach and included upscaled kinetics for thermal simulations. Comparisons of EOS and thermal simulations showed that the former could produce more oil. Later, thermal simulations were found to be more advanced but incorporated more uncertainties. Pascual et al. (2005) constructed a simulation model in a layer cake structure of 50 × 50 cells over a 56 m × 51 m area, with 13 vertical layers of approximately 5-m thickness, resulting in a total of 21,567 active cells. Properties were populated through the construction of maps, and the final geological model was exported in ECLIPSE format to build a black simulation model. In the compositional simulation study, the experimental PVT data were imported into CMG's EOS PVT software package,

WINPROP, to regress a pseudo-component compositional representation and generate a *K*-value compositional model. Then, the experimental laboratory combustion tube profiles were imported into CMG visualization software to be matched with the simulation. De Zwart et al. (2008) addressed the role that combustion plays in the incremental recovery of HPAI. In their study, numerical simulations were conducted in a 3-D model with real geological features, and a reservoir simulator with dynamic gridding capabilities was used to capture more realistically the physics of the combustion front. The investigated combustion and its associated temperature effects were based on isothermal simulations and multicomponent combustion runs. Such models result in a total desaturation of the air swept zones. Moreover, as the combustion front approaches the producer, medium-molecular-weight components (C7–C15) are transported and produced in the gas phase, leading to a significant increase in the peak production rates. Lacroix et al. (2003) presented an effective approach, based on a dynamic sub-gridding technique, which allowed the computational time to be reduced without any loss of precision in the IOR simulation process. Van Batenburg et al. (2011) presented a new robust semi-implicit method of nested dynamic local grid refinement, which can be successfully implemented in a general-purpose flow simulator, including simulation models with water flooding, chemical flooding, gas injection, and ISC. Their method incorporates an additional condition on the completion of a time step, which is based on how well the grid fits the current solution and which determines whether the grid needs to be adapted.

Clearly, simulation modeling and the associated assumptions have evolved over the years and have increased in their complexity, 3-D modeling, reaction dynamics, and modeling of the combustion front. Essentially, it is the modeling of the combustion front, as well as the most accurate depiction and approximation of the reaction kinematics for general oil compositions, which determine the accuracy and performance results for any simulation study. When the users and operators become familiar with the various air injection modeling approaches, the reservoirs being studied and the operational setting involved can generate a generalized set of results, which can be used for screening criteria or for predicting the performance of air injection in future candidate reservoirs.

5.5 Performance of air injection in conventional reservoirs

In the form of in-situ combustion (ISC) air injection, these projects have been in practice for a long time and have achieved promising results. This technology was patented in 1920 in the United States, and the first very short-term ISC field pilot (actually the first ignition operation) took place in the former Soviet Union in 1933–34. The true testing of an ISC process occurred in the United States during 1950–51. However, these initial tests were just the starting points with a short duration. The first extended field test began in 1963 on the Sloss field in Nebraska (Parrish et al., 1974a, 1974b). This project, carried out by Amoco, combined forward combustion and water flooding. The tested reservoir was a deep (6,200 ft), thin (11 ft), light-oil (38.8°API), and watered-out reservoir. This pilot recovered 83,992 bbl of oil, which was equivalent to 43% of the oil left in the five-spot pattern after water flooding. Hence, it was the HPAI process that was regarded as having delivered good and promising results. In 1967, the pilot was expanded from 80 acres to a 960-acre project, which recovered 527,000 bbl of incremental oil. The second application of HPAI was the West Heidelberg pressure-maintenance project in Mississippi, which started in 1971 as a secondary recovery project in the deep (11,400 ft) Cotton Valley sands (Huffman et al., 1983). The project continued to be a successful air injection project.

An important HPAI milestone was the implementation of commercial secondary HPAI projects in the North and South Dakota portions of the Williston basin, which started in 1979 and continued to be a technical and economic success. Air injection in the Buffalo field air injection units (located in northwest South Dakota) began in January 1979, and as of December 2007, approximately 240 Bscf of air had been injected into the field. A total of 17.2 million barrels of incremental oil was produced, equivalent to 9.4% of the original oil in place (OOIP).

The four original projects, three in the Buffalo Red River Unit and one in the Medicine Pole Hills Unit, were initiated by the Koch Exploration Company in 1987. Reservoirs in this area are typically deep, high-pressure, thin dolomites with permeability in the 10–20 md range. Water flooding in these reservoirs proved very challenging due to the poor injectivity associated with the low permeability and the large well spacing (De Zwart et al., 2008). The Total Minatome Corporation and its US subsidiary operated an HPAI project in the Horse Creek field, which also showed promising results (Watts, Hall, & Petri, 1997).

The previously mentioned projects indicate the promising aspects of air injection. However, it is necessary to remain mindful of certain considerations when assessing the air injection potential of a specific project for a conventional reservoir. First, having high pressure and temperature is beneficial in aiding auto-ignition, which ensures a higher project success rate than for projects in which external ignition is required.

Some researchers have reported effects of some rock and fluid properties, as well as certain operating factors, on the performance of air injection. The dip angle plays a role because having a higher dip angle ensures proper injection of the air. A reasonably high matrix porosity, permeability, and reservoir thickness ensure that the injection process is carried out and moves in the reservoir with sufficient penetration. Having a high and rich clay content with lower viscosity (<10 cP) at reservoir conditions ensures proper burning and movement of the fire front. For economical consideration, the minimum required value of 0.1 for reservoir porosity multiplied by oil saturation ($\phi So > 0.1$) is needed. Among these factors, one of the most important is the oxidation activity of crude oil in a reservoir, and the abovementioned factors are more or less associated with this aspect. Jia, Yuan, & Zhang (2012) summarized the selection criteria for HPAI, as shown in Table 5.1.

5.6 Advantages and disadvantages of air injection

Air injection is a good synergetic enhanced oil recovery (EOR) method because of some of its unique advantages. First, the greater availability and lower cost of air compared to other gases used for EOR purposes are very helpful factors. Other gases, such as methane and nitrogen, are marketable and have their own intrinsic values, too.

Steam injection has a depth limitation of 3000 m when the reservoir temperature and pressure reach steam's critical point. Air injection does not have a depth limitation, so the benefits of miscible displacement can be exploited at higher depths. Also, reports indicate that as the sand thickness decreases and the pressure increases, ISC becomes more favorable. ISC installations require a larger investment. However, most oil production facilities already possess a sizable gas compression capacity. In addition, ISC associated with flue gas injection provides a CO_2 disposal option, which may be financially beneficial in the case of a "CO_2 Tax" for environmental reasons (Tingas, 2000).

Despite the obvious advantages of air injection, some challenges and safety considerations must be kept in mind.

- As air injection involves the movement of a fire front and high-pressure air injection, the potential for an explosion is quite significant at both the surface and the wellbore.
- Due to the obvious presence of excessive quantities of air, corrosion of the surface equipment, wellhead tubulars, casings, and downhole tools can also be a big concern.

Table 5.1 Selection criteria for HPAI based on laboratory and field projects.

Screening criteria	**Basins**	**Carnarvon (Barrow Island) (Hughes & Sarma, 2006a)**	**Cooper-Eromanga (Kenmore) (Hughes & Sarma, 2006b)**	**Surat-Bowen (Moonie)**	**Middle- Eastern**	**Barnhart (Yannimaras and Spencer, 1999)**	**Ellenburger (Dembla Dhiraj, 2004)**
	Location	*Western Australia*	*South Australia/ Queensland*	*Queensland/New South Wales*	*Middle-Eastern*	*Texas*	*Texas*
Lithology	*Sandstone Carbonate*	*Sandstone Carbonate*	*Sandstone*	*Sandstone*	*Sandstone*	*Sandstone*	*Carbonate*
φSo	>0.1	Y 33–45	Y 44–50	Y 45	Y 30–36.6	/	/
API gravity Oil	20–50+	0.7–1.76	2.48	1.0	7.68–12.8	>25	40.7
viscosity (mPa s)	<10					2.91	/
Temperature °(C)	>75	46–99	80–141	66	89–98	>76	93
Depth (m)	>1000	350–1950	1300–2940	1730	1475–2469	>1676	2751–2779
Thickness (m)	1–20	50–130	5–30	20–60	/	/	3.96–5.79
Pressure (MPa)		<19.54	11.45	29.5	7.68–25.6	11	/
Permeability (mD)	0.1–1000	1–5000	4.4	300	/	1200	0.9–8.11
Sw (%)	<80	16–62	30–60	48	/	24.2	31–62

Source: Modified from Jia et al. (2012).

- For production wells, some of the problems that have been witnessed include gas interference, carbon dioxide corrosion, emulsions, and casing collapse.
- For injection wells, the primary problems have been near-well formation plugging and annular air leaks.

The corrosion issue can be controlled through passivation and an appropriate completion design in comparison with the explosion problem. Appropriate synthetic lubricants can help to mitigate air explosions in compressors and surface-injection lines. However, the backflow of hydrocarbons and unreacted oxygen entering a production well can cause additional uncertainty, resulting in the mixed gases becoming potentially flammable.

5.7 Performance of air injection in unconventional reservoirs

Shale-oil reservoirs usually produce light crudes (Alfarge et al., 2018). The liquids of shale reservoirs are typically present in high-temperature, high-pressure environments. Light oils under such conditions give shale oil the potential to achieve spontaneous ignition when air injection is implemented. However, due to the ultralow matrix permeability in shale reservoirs, air injectivity will likely be very low compared to conventional reservoirs. Therefore, air injection will not reach all areas of the reservoir, especially the very tight matrix of nano-Darcy permeability. There also can be issues around the high compressive power requirements for injecting sufficient air quantities through such deep, ultra-tight reservoirs. Furthermore, the amount of injected air could be higher than that for conventional reservoirs, as reported by some researchers. To make up for these deficiencies, air huff-n-puff injection could be a good option, as this mode may not require as high permeability as the flooding mode. In the huff-n-puff mode, the different timings of the huff and puff cycles can be altered to meet the injectivity and soaking requirements, on a case-by-case basis. The entire EOR design can be optimized for a specific reservoir and its properties.

Light oil has a better chance of spontaneous ignition than heavy oil; nonetheless, some metrics must be defined to screen reservoirs for the air injection process (AIP). Huang and Sheng (2017) provided a comprehensive review of the screening criteria for AIP based on thermal experiments, which can be summarized as follows:

1. The crude oil should continuously release heat between the pre-combustion temperature and the ignition temperature.
2. Oil with a lower activation energy is more favorable for AIP.

3. Oil with a lower ignition temperature is more promising for AIP.
4. The longer and more stable the exothermic activity region, the better the AIP.
5. Oil that consumes more oxygen and generates more flue gas during the LTO stage is more favorable for AIP.

These screening criteria can be summarized in one sentence: a more reactive oil is more feasible for AIP.

Their conclusions indicate that the crude oil's activation energy and frequency factor need to be less than 60 kJ/mole and higher than 2 per second, respectively, in order to make spontaneous ignition possible. Hence, if a reservoir generally follows these guidelines, there will be a possibility of spontaneous ignition.

5.8 Air injection pilots in tight oil reservoirs

An AIP has been conducted in South Dakota for the Williston Basin (Kumar et al., 2007). The field has a porosity of 11%–19% and oil viscosity of less than 2 cP. The cores showed very low oil saturation (the exact value is not available). Three HPAI projects were conducted in the Buffalo Field, one of which was the West Buffalo Red River Unit (WBRRU) project (Kumar et al., 2007). WBRRU produces from the Red River reservoirs, which are deep (~8,400 ft), thin (~15 ft), hot (215°F), low-permeability (~10 md) carbonates. The unit produces from the Red River carbonate reservoir of Ordovician age that occurs at an average depth of 8,400 ft. The primary recovery mechanism was liquid expansion and rock compaction, and the ultimate primary recovery was estimated to be 6.5% of the original oil in place. The unit contains 4,560 acres with 15 producing wells, which were designed in July 1987 and began undergoing air injection operations four months later. In this field, more than half of the production could be attributed to the thermal EOR effects in the Red River Zone in the West Buffalo (Kumar et al., 2007). Before unitization in this field, the oil production was declining steadily, but the response to air injection was very successful, with production increasing significantly over its historical decline. The unit production in July 1987 was approximately 200 BOPD, and a peak production rate of 500 BOPD was achieved in January 1990 with 15 producing wells after 2 BSCF of air was injected. As of December 2006, approximately 23 BSCF of air had been injected in this unit, resulting in a cumulative oil production of 2.3 MMSTB since the beginning of the injection. This project showed positive, promising results for the application of air injection for unconventional reservoirs. For another set of projects, the pilot tests executed in the Holt Sands Unit, the reservoir temperature increased from 230°C to 250°C (Fassihi et al., 2016). Of 25 surveyed projects, 5 projects achieved spontaneous ignition. Other successful light oil air injection projects include the Buffalo, Medicine Pole Hills Unit (MPHU), Horse Creek, Coral Creek, and Hackberry fields. These projects show promising increased oil recovery results, confirming the applicability of air injection for light oils, too. During these projects, field experiences indicated that spontaneous LTO can lead to ISC.

5.9 Pros and cons of applying air injection in shale-oil reservoirs

According to a review conducted on field experiences and laboratory studies, some key characteristics of reservoirs may control the success of an air injection project. Shale-oil reservoirs with high temperature and high pressure can accelerate light crude oil oxidation and do not require any ignition source. Clay-rich tight shale with a high specific surface area is favorable for performing catalytic oxidation and mitigating early gas breakthrough. A dip-angled shale reservoir is helpful for gravity drainage in air injection. The thermal effects generated in the AIP have the potential to dry shale and create thermal-induced microfractures in the reservoir to enhance air injectivity and to improve fluid flow from the oil-saturated matrix into a fracture pathway. This process often occurs along with rapid re-pressurization of the reservoir, which could result in both shrinkage fractures and overpressure fractures. The crude oil oxidation process in air injection has the potential to extract more light hydrocarbons. It may be an alternate strategy for an in-situ, upgrading, oil-bearing shale to produce hydrocarbon gases with the aid of air injection or air huff-n-puff injection.

Research has indicated that the confinement of the pore spaces influences the success of air injection projects in several ways (Huang and Sheng, 2018b). The confinement effect inhibits the endothermic distillation process, increases the exothermic activity of LTO, and leads to better fuel deposition. Smaller pore sizes also result in a lower HTO exothermic peak temperature. The aforementioned factors positively effect the performance of air injection. However, as indicated, they are more beneficial when the HTO stage has been achieved. The activation energy is the energy required for initiating the combustion reaction. The pore sizes affect the activation energy, with small pore sizes proving helpful in initiating air injection combustion reactions. Hence, confined pore spaces generally have a positive impact on the success of air injection projects.

Certain factors can make a reservoir unfavorable for air injection. The low air injectivity of shale-oil reservoirs might increase the oxygen concentration in production wells above 10% due to the ultra-tight formations, which results in safety hazards (Zhang et al., 2020). Furthermore, the tight matrix of unconventional reservoirs would make the heat loss more serious than in conventional reservoirs. Excessive mature shale oil has a low content of unsaturates, mainly aromatics and asphaltenes. This is adverse for fuel loaded during the AIP. The very thick zone needs a large amount of air injected into it to guarantee a stable oxidation reaction and a stable thermal front. The nano-Darcy matrix permeability and limited availability of fractures are barriers to achieving a high injection rate. Pore walls with a diameter of nanometers can have a drastic effect on the arrangement of molecules inside the pores. This confinement effect can significantly change the phase behavior in shale-oil reservoirs, increasing the complexity of understanding crude oil oxidation, evaporation, cracking, and combustion in the nanopores. More hydraulic and thermal-induced fractures are needed to diminish the pore confinement effect and to improve air injection performance in shale-oil reservoirs.

5.10 Numerical simulation for air injection in unconventional reservoirs

Building an air injection model for EOR requires obtaining the kinetic data from thermal experiments (Huang and Sheng, 2017). Generally, Huang and Sheng (2017) practiced thermogravimetry analysis (TGA) experiments and forms of the Arrhenius equation to obtain a single set of kinetic data. Their kinetic data change with the temperature, oil composition, and other parameters. However, the obtained kinetic data do not exactly represent the actual physics. Their air injection model contains two parts. One part defines the fluid pseudo-components and their physical properties. The other part defines the oxidation reaction and the corresponding kinetic data. In order to validate an air injection model, both air purging TGA experiments and nitrogen purging TGA experiments are conducted. The fluid model (pseudo-components and their properties) is validated by matching the model results with the nitrogen purging TGA experiments. Then, the kinetic data are obtained by matching the model results with the air purging TGA experiments. The selection of the pseudo-components is validated with the nitrogen purging TGA experiments, and the oxidation reactions are validated with the air purging TGA experiments. Their result shows that the combination of isomerization reactions and oxygen addition reactions do a good job of simulating a light oil AIP in the low-temperature region. Their research has shown that an air injection model can be built using the kinetic data by analyzing TGA experiments and hence can be used for modeling the reactions and the AIP.

According to their study, reservoir fluid compositions and properties are very important in guiding the reaction scheme design for simulating air injection. Notably, the reaction scheme is often treated as the uncertainty factor for simulating air injection. Because numerous reactions occur simultaneously during the oxidation, cracking, and combustion processes, the design of a plausible reaction scheme for the oil is needed for a qualitative analysis of EOR mechanisms. The kinetics of the combustion and cracking reactions in the cracking zone has been discussed widely in the literature (Bousaid and Ramey, 1968; Le Thiez and Lemonnier, 1990). The design of the process involves experimental evaluation of the process variables in laboratory combustion tube experiments. The variables proven experimentally to be important for the design of the process typically include fuel availability, air requirements, oxygen utilization efficiency, combustion peak temperature, combustion front velocity, and the effect of the porous matrix on the kinetic parameters. Among other factors, it is the lack of a clear understanding of the basic kinetics of the ISC processes, which has led to difficulties with the interpretation of field ISC projects.

The main difference between the AIP and other gas injection processes is the reaction between the crude oil and oxygen. The reaction can be classified as LTO, cracking, and HTO. Because there are hundreds of reactions based on the different oil components in the crude oil, it is impossible and unnecessary to build the reaction scheme with every component. Instead, the pseudo-components are used to build the reaction scheme. If the crude oil components are lumped into two or three pseudo-components, it perhaps cannot appropriately represent the oil properties. If the oil is divided into too many pseudo-components, the reaction scheme will become complex. Table 5.2 presents the dominant reaction schemes and the kinetic parameters typically required for modeling an air injection project (Huang and Sheng, 2017). Different reactions with their required activation energy enthalpy and frequency are provided. Generally, the reaction schemes for simulation models are approximated.

Table 5.2 Dominant reaction schemes and kinetic parameters in the simulation model.

Reactions	Description	Formula	Frequency factor	Enthalpy, Btu/lbmole	Activation energy, Btu/ lbmole
Reaction 1	Heavy component cracking with CH_4 generation	$C_7–C_{31} \rightarrow 7.336\ CH_4 + 6.200$ Coke	127,400	0	38,000
Reaction 2	Heavy component cracking with C_2,$C6_6$ generation	$C_7–C_{31} \rightarrow 2.794\ C_2.C_6 + 2.866$ Coke	89,100	0	39,000
Reaction 3	Light component cracking	$C_2–_6 \rightarrow 2.688\ CH_4 + 1.120$ Coke	50,000	0	45,000
Reaction 4	Heavy component burning	$C_7–C_{31} + 38.525\ O_2 \rightarrow 27.525\ H_2O + 33.521\ CO$ + Energy	3.846E+10	10,294	12,400
Reaction 5	Light component burning	$C_2–C_6 + 7.625\ O_2 \rightarrow 7.115\ H_2O + 6.220\ CO$ + Energy	3.846E+10	2,420	14,200
Reaction 6	Coke burning	Coke $+ 1.515\ O_2 \rightarrow 1.003\ H_2O + CO_2$ + Energy	300,000	461	5,500
Reaction 7	CH_4 burning	$CH_4 + 1.500\ O_2 \rightarrow 2H_2O + CO$ + Energy	3.846E+10	1,463	15,000
Reaction 8	CO burning	$CO + 0.500\ O_2 \rightarrow CO_2$ + Energy	191,000	300,000	2,000

Source: Modified from Huang and Sheng (2017).

5.11 SUMMARY

- Air combustion modes have a significant effect on the performance of air injection in terms of enhancing oil recovery. Correctly predicting the combustion mode leads to much more successful air injection projects.
- Many obstacles exist that could impair the feasibility of air injection in shale-oil reservoirs. Heat loss and injectivity are two major concerns.
- According to some experimental reports, air huff-n-puff has been proven to be an efficient method of improving recovery and could have better results than the flooding mode. However, the huff-n-puff cycles must be properly optimized before implementation.
- Some rock and fluid properties and operating parameters have a significant impact on the performance of air injection, such as the dip direction, pressure, temperature, and permeability of the reservoirs.
- The confinement effect of shale-oil reservoirs can significantly change the phase behavior, which might make it difficult to understand the crude oil oxidation and combustion process through the nanopores. More hydraulic and thermal-induced fractures might be needed to mitigate the pore confinement effect; however, overcoming conformance problems will be critical in achieving the planned air injection performance.

References

Ahmed, T., & Nathan Meehan, D. (2012). Chapter 6—Introduction to enhanced oil recovery. *Advanced reservoir management and engineering* (2nd ed.), Gulf Professional Publishing (pp. 541–585). https://doi.org/10.1016/B978-0-12-385548-0.00006-3.

Alexander, J. D., Martin, W. L., & Dew, J. N. (1962). *Factors affecting fuel availability and composition during in situ combustion*. Society of Petroleum Engineers. doi: 10.2118/296-PA.

Alfarge, D., Wei, M., & Bai, B. (2018). *Data analysis for CO2-EOR in shale-oil reservoirs based on a laboratory database*. Journal of Petroleum Science and Engineering. https://doi.org/10.1016/j.petrol.2017.10.087.

Berry, V. J., & Parrish, D. R. (1960). Secondary recovery and pressure maintenance—a theoretical analysis of heat flow in reverse combustion. *Transactions of AMIE*, *219*, 124–131.

Bousaid, I. S., & Ramey, J. (1968). Oxidation of crude oil in porous media. *Society of Petroleum Engineers of AIME Transaction*, *243*(1968.).

Burger, J. G. (1972). *Chemical aspects of in-situ combustion - heat of combustion and kinetics*. Society of Petroleum Engineers. doi: 10.2118/3599-PA.

Chu, C. (1977). A study of fireflood field projects. *Journal of Petroleum Exploration and Production Technology*, *29*(2), 111–120. doi: 10.2118/5821-PA. SPE-5821-PA.

Das, S. A. (2009). Study of oxidation reaction kinetics during an air injection process. Msc. Thesis. The University of Adelaide.

Dutton, S. P., Kim, E. M., Broadhead, R. F. (2004). Play analysis and digital portfolio of major oil reservoirs in the permian basin application and transfer of advanced geological and engineering technologies for incremental production opportunities. Technical

Report DE-FC26-02NT15131. National Energy Technology Laboratory (NETL) for the United States Department of Energy.

De Zwart, A. H., van Batenburg, D. W., Blom, C. P. A., Tsolakidis, A., Glandt, C. A., & Boerrigter, P. (2008). *The modeling challenge of high pressure air injection*. Society of Petroleum Engineers. doi: 10.2118/113917-MS.

Fraim, M. L., Moffitt, P. D., Yannimaras, D. V. (1997). *Laboratory testing and simulation results for high pressure air injection in a waterflooded North Sea oil reservoir.* In: SPE-38905-MS.41, Presented at the SPE Annual Technical Conference and Exhibition. San Antonio, Texas, October 5-8.

Fassihi, M. R., Moore, R. G., Mehta, S. A., & Ursenbach, M. G. (2016). *Safety considerations for high-pressure air injection into light-oil reservoirs and performance of the holt sand unit project.* Society of Petroleum Engineers. . doi:10.2118/169168-PA.

Greaves, M., Ren, S. R., Rathbone, R. R., Fishlock, T., & Ireland, R. (2000). *Improved residual light oil recovery by air injection (LTO process)*. Petroleum Society of Canada. doi: 10.2118/00-01-05.

Glandt, C. A., Pieterson, R., Dombrowski, A., et al. (1999). *Coral creek field study a comprehensive assessment of the potential of high-pressure air injection in a mature waterflood project.* In: SPE-52198-MS, Presented at the SPE Mid-Continent Operations Symposium. Oklahoma City, Oklahoma, March 28-31, http://dx.doi.org/10.2118/52198-MS.

Gutiérrez, D., Miller, R. J., Taylor, A. R., et al. (2009). Buffalo field high-pressure air injection projects 1977 to 2007 technical performance and operational challenges. *SPE Reservoir Evaluation and Engineering, 12*(4), 542–549.

Gutiérrez, D., Skoreyko, F., Moore, R. G., et al. (2009a). The challenge of predicting field performance of air injection projects based on laboratory and numerical modelling. *Journal of Canadian Petroleum Technology, 48*(04), 23–33.

Gutiérrez, D., Taylor, A. R., Kumar, V., et al. (2008). Recovery factors in high pressure air injection projects revisited. *SPE Reservoir Evaluation and Engineering, 11*(06), 1097–1106.

Hughes, B., & Sarma, H. K. (2006a). Burning reserves for greater recovery? Air injection potential in Australian light oil reservoirs. *Society of Petroleum Engineers*. https://doi.org/10.2118/101099-MS.

Jia, H., & Sheng, J. (2016). Numerical modeling on air injection in a light oil reservoir: Recovery mechanism and scheme optimization. *Fuel* 172, 70–80. ISSN 0016-2361. https://doi.org/10.1016/j.fuel.2015.12.063.

Jia, H., Zhao, J. Z., & Yang, H. J. (2014). Numerical simulation of mechanism of high-pressure air injection (HPAI) in light oil reservoirs. *Petroleum Exploration and Development, 41*(02), 235–243.

Jia, H., Yuan, C. D., Zhang, Y., et al. (2012a). *Recent progress of high pressure air injection process (HPAI) in light oil reservoir: laboratory investigation and field application.* Presented at the SPE Heavy Oil Conference Canada, Calgary, Alberta (June 2012), pp. 12–14. SPE-156974-MS.

Jia, H., Zhao, J. Z., & Pu, W. F. (2012b). Numerical analysis on the influence of thermal effects on oil flow characteristic in high-pressure air injection (HPAI) process. *Journal of Applied Mathematics, 2012*(2012), 1–17.

Jia, H. (2012). Investigation into the oxidation mechanism of crude oil and gas channeling prevention for application of high-pressure air injection (HPAI) process, PhD Dissertation. Southwest Petroleum University (2012).

Kuhlman, M. I. (2000). *Simulation of light-oil air injection into viscous-dominated and gravity-stable reservoirs*. Society of Petroleum Engineers. doi: 10.2118/59331-MS.

Kumar, V. K., Gutierrez, D., Moore, R. G., & Mehta, S. A. 2007. *Case history and appraisal of the West Buffalo Red River unit high-pressure air injection project*. Paper SPE107715 presented the Hydrocarbon Economics and Evaluation Symposium, Dallas, 1–3 April. . doi: 10.2118/ 107715-MS.

Lacroix, S., Renard, G., Lemonnier, P., & Taieb, C. (2003). Enhanced numerical simulations of IOR processes through dynamic sub-gridding. *Petroleum Society of Canada*. doi: 10.2118/2003-087.

Le Thiez, P. A., & Lemonnier, A. (1990). *An in-situ combustion reservoir simulator with a new representation of chemical reactions*. Society of Petroleum Engineers. doi: 10.2118/17416-PA.

Mamora, D. D. (1995). *New findings in low-temperature oxidation of crude oil*. Society of Petroleum Engineers. doi: 10.2118/29324-MS.

Montes, A. R., Gutierrez, D., Moore, R. G., Mehta, S. A., & Ursenbach, M. G. (2010). *Is high-pressure air injection (HPAI) simply a flue-gas flood?* Society of Petroleum Engineers. doi: 10.2118/133206-PA.

Manahali, A., & Salim, M. O. (2010). Top-down in-situ combustion in heavy oil reservoirs have strong bottom aquifer support. Available from: https://www.semanticscholar.org/paper/Top-down-in-situ-combustion-in-heavy-oil-reservoirs-Manahali-Salim/9b243dc21c87b8c954e3f9419e10cb9ec8e2b6de.

Parrish, D. R., Pollock, C. B., & raig, F. F., Jr. (1974a). Evaluation of COFCAW as a tertiary recovery method, Sloss field, Nebraska. *Journal of Petroleum Technology*, *26*(6), 676–686. http://dx.doi.org/10.2118/3777-PA. SPE-3777-PA.

Parrish, D. R., Pollock, C. B., Ness, N. L., & Craig, F. F. (1974b). *A tertiary COFCAW pilot test in the Sloss field, Nebraska*. Society of Petroleum Engineers. doi: 10.2118/3839-PA.

Pascual, M., Crosta, D., Lacentre, P., & Coombe, D. (2005). Air injection into a mature water-flooded light oil reservoir—laboratory and simulation results for Barrancas Field, Argentina, SPE94092. https://doi.org/10.3997/2214-4609-pdb.1.D030.

Ren, S. R., Greaves, M., & Rathbone, R. R. (2002). Air injection LTO process: An IOR technique for light-oil reservoirs. *SPE Journal*, 7(1), 90–99. SPE-57005-PA.

Huang, S., & Sheng, J. J. (2018). Feasibility of spontaneous ignition during air injection in light oil reservoirs. https://doi.org/10.1016/j.fuel.2018.04.052.

Huang, S., & Sheng, J.J. (2018). Effect of nanopore confinement on crude oil thermal-oxidative behavior. https://doi.org/10.1021/acs.energyfuels.8b02177.

Huang, et al. (2018). Experimental investigation of enhanced oil recovery mechanisms of air injection under a low-temperature oxidation process: thermal effect and residual oil recovery efficiency. *Energy Fuels*, *32*(6), 6774–6781. https://doi.org/10.1021/acs.energyfuels.8b01314.

Huang, S., & Sheng, J. J. (2017). A practical method to obtain kinetic data from TGA (thermogravimetric analysis) experiments to build an air injection model for enhanced oil recovery. https://doi.org/10.1016/j.fuel.2017.06.019.

Huffman, G. A., Benton, J. P., El-Messidi, A. E., et al. (1983). Pressure maintenance by in-situ combustion, West Heidelberg Unit, Jasper County, Mississippi. *Journal of Petroleum Technology*, *35*(10), 1877–1883. SPE-10247-PA. http://dx.doi.org/10.2118/10247-PA.

Hughes, B.L., & Sarma, H. K. (2006). *Burning reserves for greater recovery? Air injection potential in australian light oil reservoirs.* Presented at the Asia Pacific Oil & Gas Conference and Exhibition, Adelaide, Australia (11–13 September, 2006) SPE-101099-MS.

Sakthikumar, S., Madaoui, K., & Chastang, J. (1995). *An investigation of the feasibility of air injection into a waterflooded light oil reservoir*. Society of Petroleum Engineers. doi: 10.2118/29806-MS.

Sarathi, P. S. (1999). In situ combustion handbook-principles and practices. National Petroleum Technology Office, U.S. Department of Energy, Tulsa, Oklahoma. Available from: https://scholar.google.com/scholar?q=sarathi+1999+combustion&hl=en&as_sdt=0%2C5&as_ylo=1999&as_yhi=1999.

Sarathi, P.S., & Olsen, D. (1995). Field application of in situcombustion-past performance/future application. Technical Report NIPER/BDM-0086& CONF-940450 (DE95000116), US DOE Bartlesville Project Office, Bartlesville, Oklahoma.

Schlumberger (2020). Chlumberger oilfield glossary. Available from: https://www.glossary.oilfield.slb.com/Terms/w/wet_combustion.aspx.

Ursenbach, M. G., Moore, R. G., & Mehta, S. A. (2010). *Air injection in heavy oil reservoirs - a process whose time has come (again)*. Society of Petroleum Engineers. doi: 10.2118/132487-PA.
Van Batenburg, D. W., De Zwart, A. H., Boerrigter, P. M., Bosch, M., & Vink, J. C. (2011). Application of dynamic gridding techniques to IOR/EOR processes. https://doi.org/10.3997/2214-4609.201404745.
Tingas, J. (2000). Numerical simulation of air injection processes in high pressure light & medium oil reservoirs. PhD thesis. https://researchportal.bath.ac.uk/en/studentTheses/numerical-simulation-of-air-injection-processes-in-high-pressure-.
Watts, B. C., Hall, T. F., Petri, D. J. (1997). Horse creek air-injection project: an overview presented at the SPE rocky mountain regional meeting, Casper, Wyoming, 18–21 May. SPE-38359-MS. http://dx.doi.org/10.2118/38359-MS.
Youtsos, M. S. K., & Mastorakos, E. (2013). Numerical simulation of thermal and reaction waves for in situ combustion in hydrocarbon reservoirs. *Fuel* 108, 780–792, ISSN 0016-2361, https://doi.org/10.1016/j.fuel.2013.02.034.
Zhang, Y., Huanga, S., Shenga, J. J., Qi, J. (2020). Experimental and analytical study of oxygen consumption during air injection in shale oil reservoirs. https://doi.org/10.1016/j.fuel.2019.116462.

CHAPTER 6

Water injection in unconventional reservoirs

Contents

6.1 Introduction 114
6.2 Oil recovery due to waterflooding 117
6.3 Factors impacting the performance of waterflooding 120
6.4 Waterflooding-derived techniques 121
6.4.1 Low-salinity water injection 121
6.4.2 Carbonated water injection 122
6.5 Water injection in unconventional reservoirs 124
6.6 EOR mechanisms for water injection in unconventional reservoirs 124
6.7 Smart water injection for shale-oil reservoirs 126
6.8 Comparison of water injection in huff-n-puff mode versus flooding mode in shale-oil reservoirs 129
6.9 Factors affecting the performance of water injection in unconventional reservoirs 130
6.10 Field applications for water injection in unconventional reservoirs 132
6.10.1 Case study I 132
6.10.2 Case study II 132
6.10.3 Case study III 134
6.10.4 Case study IV 134
6.10.5 Case study V 135
6.10.6 Case study VI 136
6.10.7 Case study VII 137
6.10.8 Case study VIII 137
6.11 Summary 137
References 138

Abstract

This chapter explains how the mechanisms of water injection are different in conventional reservoirs than in shale and tight oil reservoirs. In addition, it compares the performance of water huff-n-puff versus water flooding in shale and tight oil reservoirs. The effect of the chemical composition of the injected water on the performance of the cyclic water-based EOR technique for unconventional reservoirs is discussed. The field applications of water injection in shale formations are demonstrated.

Keywords: waterflooding in unconventional reservoirs; water huff-n-puff; water imbibition mechanisms; unconventional reservoir development; hydraulic fracturing; unconventional EOR methods; huff-n-puff versus flooding mode

Fundamentals of Enhanced Oil Recovery Methods for Unconventional Oil Reservoirs
http://dx.doi.org/10.1016/B978-0-12-818343-4.00006-1

6.1 Introduction

The waterflooding practice began in the 1920s when water was injected into porous and permeable subsurface formations, including the pay zones of oil reservoirs from which the oil and water originally had come (Warner, 2007). By the 1930s, reinjecting the produced water had become a common oilfield practice. This practice was first implemented systematically in the Bradford oil field of Pennsylvania. At that time, water was injected into a single well; then, when the water-swept area increased, the adjacent watered-out wells were used as new injection wells to further expand the area of water invasion. This kind of waterflooding pattern was known as the "circle-flood" approach. It was later replaced by a "line flood" pattern (Fettke, 1938). In the line flood approach, two rows of production wells were staggered on both sides of an equally spaced row of water injection wells, as shown in Fig. 6.1 (Warner, 2007). In the 1920s, in addition to the line flood pattern, a "five-spot" well layout was used, as shown in Fig. 6.2. Many waterflooding technologies and common practices were developed in the United States between 1940 and 1970. By the 1970s, most onshore oil fields in the United States, USSR, and China were being produced using this technology in various well-pattern arrangements. To achieve a reasonable oil recovery factor and improve profits, the owners and operators of some US offshore oilfields and oilfields elsewhere in the world were applying water injection as a logical practice. Since then, many large-scale water injection projects have been applied to oil reservoirs.

Waterflooding has been used successfully for more than 70 years in conventional reservoirs. The main reason for its success is its inexpensive process for maintaining reservoir pressure and displacing oil. Waterflooding

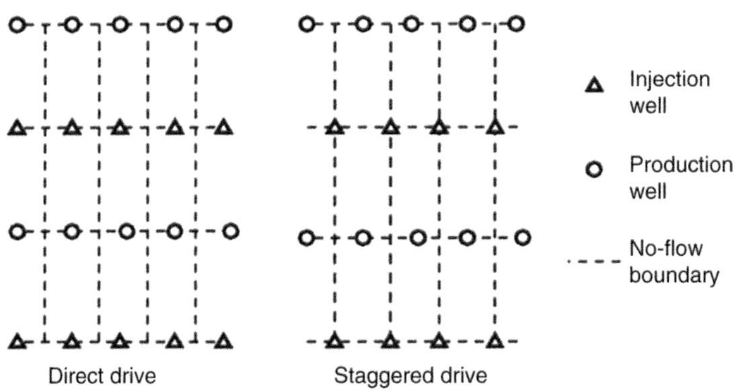

Figure 6.1 ***Line drive pattern for waterflooding in oil reservoirs.***

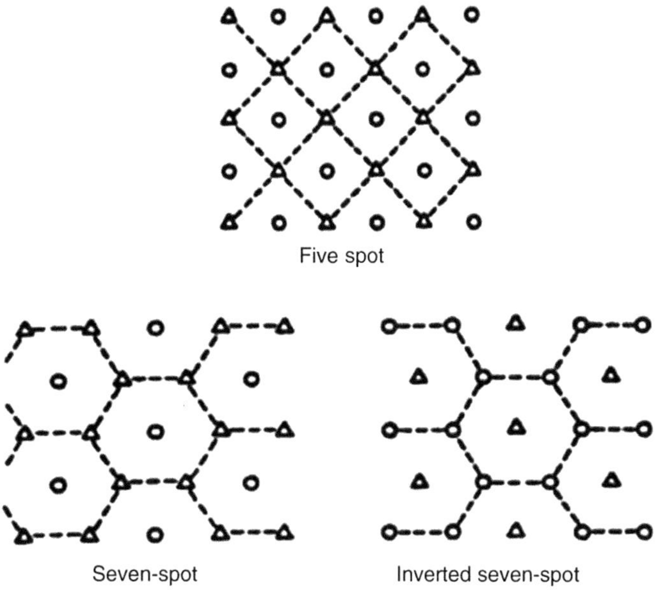

Figure 6.2 ***Five-spot and seven-spot drive pattern for waterflooding in oil reservoirs.***

is often utilized after the natural drive mechanisms have become ineffective. Currently, the term waterflooding also serves as an alternative name for secondary oil recovery. During waterflooding projects, water is injected into a reservoir through injection wells to drive the reservoir oil toward the production wells, as shown in Fig. 6.3. Sometimes, the injected water creates a bottom water drive on the oil zone, pushing the oil upward, as shown in Fig. 6.4. In earlier practices, water injection was conducted in the later phase of the reservoir's life, but now it is carried out in the earlier phase in order to avoid voidage and gas capping in the reservoir. Using water injection in the earlier phase enhances the improvement in the oil production much more significantly than using it after the bubble point pressure is achieved.

Figure 6.3 ***Waterflooding process in oil reservoirs.***

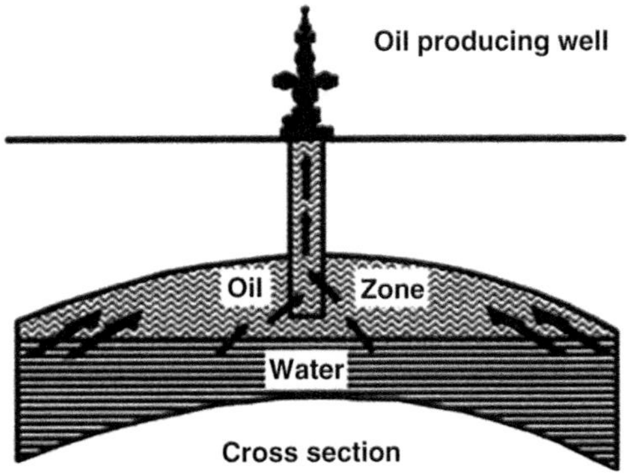

Figure 6.4 ***Schematic explaining bottom water drive (Slide share, 2020a).***

Improving the oil production using waterflooding becomes much less economical once the secondary gas cap forms. This occurs primarily because the injected water initially tends to compress the free gas cap and then later pushes the oil. Hence, much more injection water is required. Therefore, for better economy, water injection is carried out when the reservoir pressure is higher than the bubble point pressure. It is generally conducted when the reservoir is under the depletion drive mechanism or the water drive is weak.

The main factor, which determines the feasibility of waterflooding, is the mobility ratio between the displacing fluid (water) and the displaced fluid (oil), as shown in Fig. 6.5. When the mobility ratio (M) is less than 1, waterflooding is more effective. When the mobility ratio is greater than 1, problematic fingering becomes likely. For a traditional waterflooding project, the sensitivity of the reservoir rocks and the formation water to the injected water must be evaluated to ensure that no formation damage, fine migration, or other problems occurred during the flooding process. In waterflooding, either treated water (seawater, reservoir water) or fresh water (water from

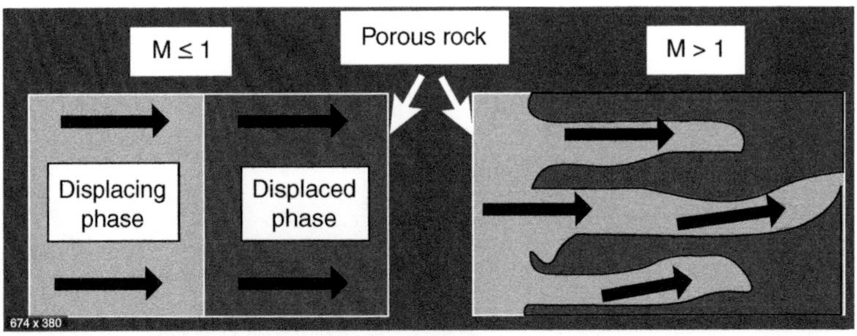

Figure 6.5 ***Effect of mobility ratio on sweep efficiency (Seright, 2005).*** *Seright, R. S. (2005, January 1). Clean Up of Oil Zones after a Gel Treatment. Society of Petroleum Engineers. doi:10.2118/92772-MS.*

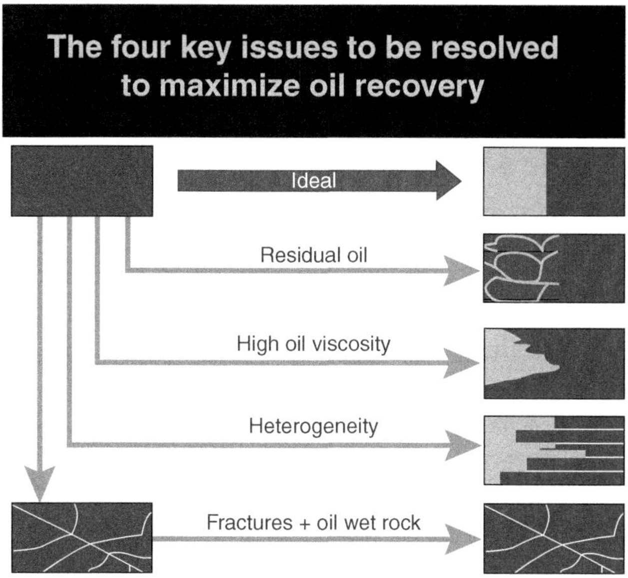

Figure 6.6 ***Main factors which reduce the oil recovery obtained by waterflooding (Slide share, 2020b).*** *Slide share. (2020b). Accessed in 2020 via the following link: https://slideplayer.com/slide/3848470/.*

lakes, rivers, shallow beds) is injected. Water injection, in general, suffers from two major problems that hinder its efficiency (Schumacher, 1978). First, it does not flush all the oil from the pore spaces as it moves through the reservoir rock because of the immiscibility of the fluids or the wettability conditions of the rock. Therefore, 25%–50% of the oil is left unrecovered during residual oil saturation, also called immobile oil saturation. Second, the waterfront can bypass part of the reservoir in terms of unswept volume due to the well placement, reservoir heterogeneity, mobility ratio, or the existence of fractures and vugs. To summarize, there are four factors, which reduce the oil recovery factor obtained by waterflooding, as shown in Fig. 6.6.

6.2 Oil recovery due to waterflooding

Oil recovery during waterflooding can be determined at any time in the life of any waterflood project if the following four factors are known:

1. Oil-in-place at the start of waterflooding: The oil-in-place at the time of initiating water injection is a function of the floodable pore volume and the oil saturation. The floodable pore volume is highly dependent on the selection and application of net pay discriminators, such as the permeability (and porosity) cutoffs. The oil-in-place at the start of the waterflooding can be calculated using the following equation:

$$S_{ox} = \left[1 - \frac{N_p}{N}\right]\left[\frac{B_o}{B_{oi}}\right](1 - SW_c) \tag{6.1}$$

where:
N = original oil-in-place, STB
N_p = primary oil production between the initial and current reservoir pressure, STB
B_o = oil formation volume factor at the current pressure, RB/STB
B_{oi} = oil formation volume factor at the initial pressure, RB/STB
SW_c = water saturation

2. Areal sweep efficiency: This is the fraction of the reservoir area that the injected water will contact, as shown in Fig. 6.7. It depends primarily upon the fractional flow properties of the oil and water, the injection-production well pattern used to flood the reservoir, the pressure distribution between the injection and production wells, and the directional permeability.
3. Vertical sweep efficiency:Vertical sweep refers to the fraction of a formation in the vertical plane which water will contact, as shown in Fig. 6.8. This will depend primarily upon the degree of vertical stratification existing in the reservoir.

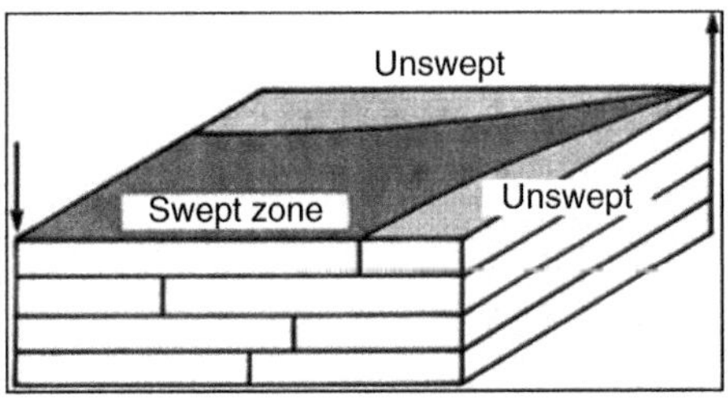

Figure 6.7 *Schematic explaining the concept of areal sweep efficiency (Green & Willhite, 1998).*

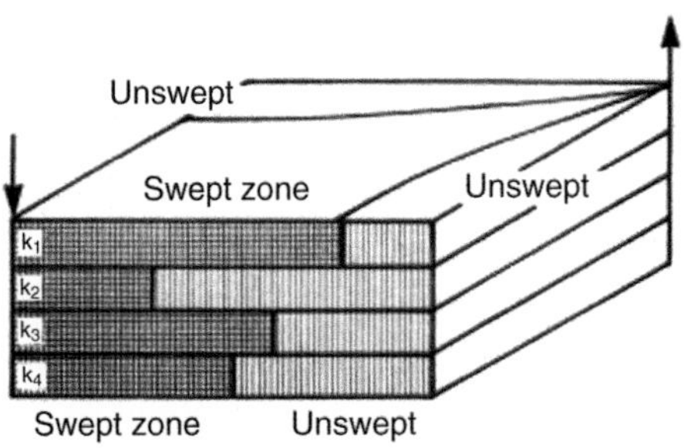

Figure 6.8 *Schematic explaining the concept of vertical sweep efficiency (Green & Willhite, 1998).*

The vertical sweep efficiency multiplied by the areal sweep efficiency gives the total sweep efficiency, which is termed the macroscopic efficiency and is shown in the following equation:

$$E_V = E_A \times E_I \tag{6.2}$$

where:
E_V: macroscopic efficiency
E_A: areal sweep efficiency
E_I: vertical sweep efficiency

4. Displacement efficiency: This represents the fraction of oil that has been recovered from a zone swept by a waterflood or other displacement process, as shown in Fig. 6.9. It is the fraction of the oil displaced from each pore, which is contacted by the encroaching driving fluids. Microscopic efficiency is another term for the displacement efficiency.

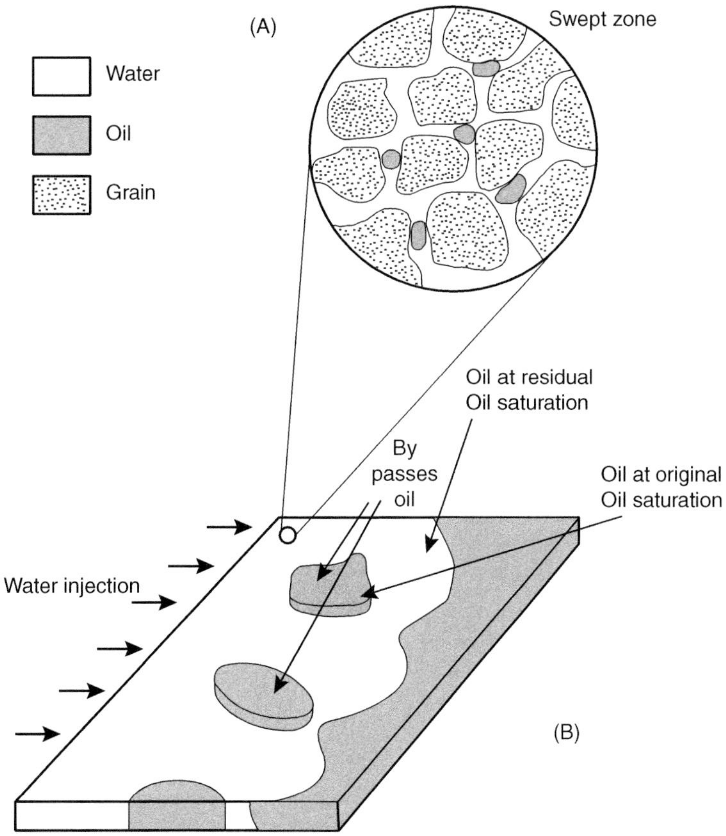

Figure 6.9 (A) Microscopic displacement and (B) residual oil remaining after a water flood. *(Modified from https://perminc.com).*

The displacement efficiency can be calculated using the following equation:

$$E_D = 1 - \left[\frac{\frac{S_{o2}}{B_{o2}}}{\frac{S_{o1}}{B_{o1}}} \right] \tag{6.3}$$

where:
E_D = displacement (microscopic) efficiency
S_{o1} = oil saturation at the start of the waterflooding
B_{o1} = oil formation volume factor at the reservoir pressure when the waterflooding begins
S_{o2} = current oil saturation
B_{o2} = oil formation volume factor at the current reservoir pressure
Multiplying the macroscopic efficiency by the microscopic efficiency gives the overall sweep efficiency, or the oil recovery factor (RF), which can be obtained by any waterflooding process.

$$RF = E = E_V \times E_D \tag{6.4}$$

where:
RF = oil recovery factor
E = overall efficiency
E_V = macroscopic efficiency
E_D = microscopic efficiency

6.3 Factors impacting the performance of waterflooding

Waterflooding recovery depends on a number of variables, and the ones that usually have the greatest impact are as follows:

- Oil saturation at the start of waterflooding
- Residual oil saturation
- Connate water saturation
- Free gas saturation at the start of water injection
- Water floodable pore volume
- Oil and water viscosity
- Effective permeability to the oil measured at the immobile connate water saturation (end point permeability)
- Relative permeability to water and oil
- Reservoir heterogeneity (Dykstra-Parsons coefficient, V)

- Waterflood pattern (symmetrical or irregular)
- Pressure distribution between injector and producer
- Injection rate
- Oil formation volume factor
- Economics

6.4 Waterflooding-derived techniques

As mentioned earlier in this chapter, waterflooding and injection techniques have been in practice for a long time in the oil and gas industry. However, waterflooding fails to recover much oil compared with the original oil-in-place. Hence, efforts were initiated to find superior ways and methods to increase oil recovery. Some of the new enhanced oil recovery (EOR) methods, which are considered water-injection–derived EOR methods include low-salinity water (LSW) injection and carbonated water injection.

6.4.1 Low-salinity water injection

One water-based EOR technique is low-salinity water (LSW) injection, also known as smart water injection. Many researchers have reported that an increment in the oil recovery can be observed if the salinity level of the injected water is reduced, as in the example shown in Fig. 6.10.

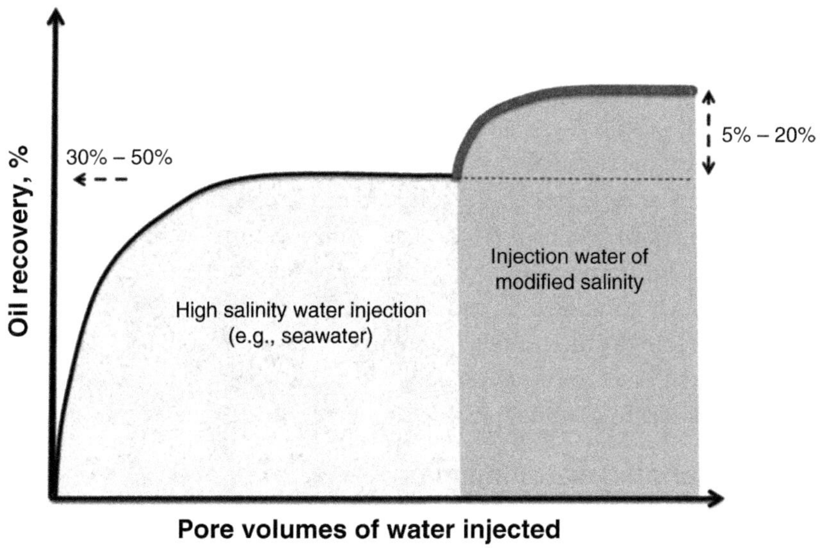

Figure 6.10 ***Schematic explaining the improved oil recovery due to the salinity modification of the injected water.*** *(Modified from: http://large.stanford.edu).*

Researchers have reported different mechanisms to explain the improved oil recovery. Some researchers have reported that this technique improves the microscopic displacement efficiency by modifying the reservoir's wettability (Muggeridge et al., 2014). Oil reservoir rocks might have a heterogeneous or "mixed" wettability. The low-salinity water makes the reservoir rocks slightly more (but not completely) water-wet as it progresses through the reservoir. This has the effect of mobilizing more of the oil behind the displacement front and increasing the oil recovery. The potential to use wettability alteration as the main recovery mechanism in an EOR process has been a major topic of research for many decades, since it was first recognized in 1959 by Wagner and Leach (1959) and tested in the field in 1962 (Wagner and Leach, 1959). They controlled the wettability by adjusting the pH and sodium content of the injected water. The recovery process involves injecting brine that has low salinity and is depleted of divalent cations (compared with the in-situ brine) into sandstone reservoirs. The salinity should be as low as possible without adversely affecting the flow performance. Pure water typically is not injected, as this can reduce oil recovery by causing swelling and deflocculation of some types of clay minerals, and subsequent blockage of the pore space. Since the first laboratory investigations of the effect of the water composition on the oil recovery in 1959, many tests have confirmed that lowering the brine salinity increases the oil recovery. Much of the core flooding and field evidence is consistent with the theory that low-salinity water injection progressively modifies the reservoir's wettability through multicomponent ion exchange (Lager, Webb, Black, Singleton, & Sorbie, 2008; Sorbie & Collins, 2010). One of the mechanisms causing areas of the pore walls to be oil-wet is the ion binding between the oil and mineral surface, mediated through multivalent cations such as Ca^{2+}, Mg^{2+}, and Fe^{2+}(Denekas & Mattax, 1959). Injecting lower-salinity water that has a reduced concentration of these divalent cations results in this ion-bound oil being released from the mineral surface (usually, but not exclusively, kaolinite) and such parts of rock surfaces becoming water-wet (Muggeridge et al., 2014). Note that this only has a slight effect on the bulk rock wettability; overall, it still has a mixed wettability. It is just slightly more water-wet. Some recent academic studies provide support for the proposed mechanism.

6.4.2 Carbonated water injection

Another water-based EOR technique is carbonated water injection. It has been observed that the direct injection of CO_2 might not result in an economically significant amount of additional oil recovery. From a storage perspective, storing CO_2 as a free phase in a reservoir can pose a significant leakage risk. An alternative CO_2 injection strategy that can eliminate many of the shortcomings of direct CO_2 injection is carbonated (CO_2-enriched) water injection (CWI). Compared to CO_2 injection, carbonated water has a much better sweep efficiency and a much slower CO_2 breakthrough because it is essentially as mobile as water. CWI also alleviates the adverse effect of the high-water saturation and the water shielding that results from mixing with the resident water, which facilitates CO_2 dissolution and the subsequent oil swelling. However, in direct CO_2 injection, the timescale for CO_2 diffusion into the oil can be several years due to the low sweep efficiency and gravity segregation (Solomon, 2007). Carbonated water injection (CWI) is considered a promising EOR technique because it enhances the sweep efficiency, allowing the dissolved CO_2 to transfer to the oil phase to improve the oil mobility and cause oil swelling (Esene et al., 2019). In addition to the EOR effectiveness, CWI provides a high CO_2 storage capacity in terms of geological storage, and a number of laboratory tests and field applications have confirmed the effectiveness of CWI (Esene et al., 2019).

According to some direct flow visualization experiments, CO_2 gradually leaves the injected carbonated water and dissolves into the oil during the injection process (Sohrabi et al., 2015). Such transportation processes then form some of the pore-scale interactions between the carbonated water and crude oil. Visualization experiments have shown that the EOR mechanisms of carbonated water are quite different from the well-known mechanisms observed in conventional CO_2 flooding. In addition to the usual CO_2-related mechanisms present during CWI, such as oil swelling and viscosity reduction, a new fluid phase forms within the oil (Sohrabi et al., 2015). According to some researchers, the formation of this new fluid phase is the major mechanism that significantly improves the performance of CWI and the amount of additional oil recovery achieved by CWI. CWI can significantly increase the oil recovery under both secondary mode (injected instead of conventional waterflooding) and tertiary mode (injected after conventional waterflooding); however, the performance of CWI could be significantly affected by the composition of the oil, including the amount of light and intermediate components (Sohrabi et al., 2015).

6.5 Water injection in unconventional reservoirs

Waterflooding is a mature recovery method in conventional reservoirs, but it has not been fully studied and understood in shale and tight oil reservoirs. The main concerns are the low permeability, low injectivity, and very weak connectivity of these reservoirs. Nonetheless, some research has been conducted to assess the viability of water injection in such types of reservoirs.

6.6 EOR mechanisms for water injection in unconventional reservoirs

Takahashi and Kovscek (2010) investigated the impact of different brine formulations in tight formations. They examined the impact of different brine formulations, ranging from acidic (pH of 3), to neutral, to alkaline (pH of 12), on siliceous shale samples. The highest oil recovery achieved during the spontaneous imbibition and forced core flooding was from the samples with high pH brine (alkaline), with an RF of 30% after the spontaneous imbibition and an RF of 95% after the forced core flooding. The authors attributed the improvement in the oil recovery from the high pH brine to the change in the wettability from intermediate water-wet to strongly water-wet conditions. The experiments conducted on the Bakken shale by Wang, Butler, Liu, & Ahmed (2011) demonstrated an increase in the shale permeability during brine core flooding. They attributed this enhancement to the mineral dissolution during the forced brine core flooding and to the shale cracking that resulted from clay swelling after the spontaneous imbibition. Fakcharoenphol, Charoenwongsa, Kazemi, & Wu (2013) pointed out that waterflooding changes the in-situ stresses of the formation due to the increase in the reservoir pressure and decrease in the reservoir temperature, which enhances the oil recovery of shale formations by reactivating existing natural fractures and/or creating new fractures, as shown in Fig. 6.11.

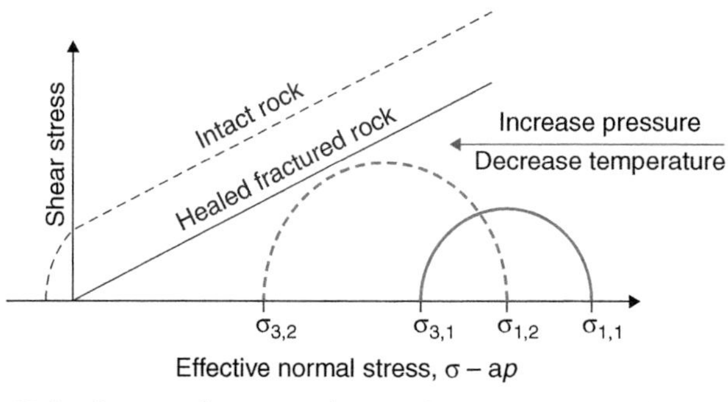

Figure 6.11 ***Mohr diagram for stress change during water injection.*** *(Modified from Fakcharoenphol et al., 2013).*

They proposed that the injected water significantly reduces the effective stress in shale reservoirs, especially with cold-water injection into deep or high-temperature formations such as the Bakken. The reduction in the effective stress could improve the hydrocarbon recovery from shale reservoirs in three ways. First, the effective stress reduction could result in maintaining or even improving the permeability of existing natural fractures. Second, it could induce new microfractures extending from the fracture surface into a tight matrix block, thus creating flow paths and helping to produce hydrocarbons from the matrix. Finally, the injected water could displace hydrocarbons through the induced microfractures. The process proposed by Fakcharoenphol et al. (2013) is shown in Fig. 6.12. Makhanov et al. (2012) demonstrated that imbibition could be a viable mechanism to transfer fluids from the fracture to the matrix in the Horn River shales. They also explained that the imbibition rate is higher along the bedding direction than across the bedding direction. (Morsy, Sheng, & Soliman, 2013c) studied

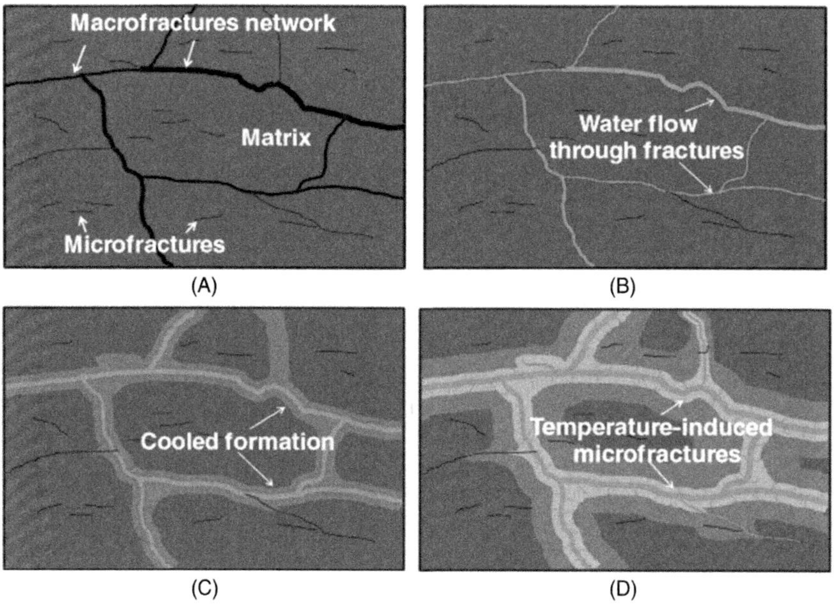

Figure 6.12 Enhanced hydrocarbon recovery sites resulting from cold-water injection in shale reservoirs. (A) initial condition, (B) early water injection period in which injected water flows only through the fracture network, (C) shale matrix block cooled, and (D) microfractures caused by temperature-induced stress. *(Modified from: Fakcharoenphol et al., 2013).*

the effects of water compositions (salinity, acid, and alkali) on the EOR potential of water imbibition.

The water huff-n-puff protocol was investigated, and its mechanism was found to be similar to that of CO_2 huff-n-puff. During the soaking period in CO_2 huff-n-puff, CO_2 diffuses into the shale matrix and extracts the oil via the countercurrent mechanism. However, during the soaking period in water huff-n-puff, the injected water imbibes into the rock, which results in extracting oils from the matrix to the cracks and channels. This occurs because oil and gas are mainly stored in the matrix, and cracks provide the main channel for fluid flow. The top priority of the oil-field development process in shale reservoirs should be crude oil extraction from the matrix. Because of the imbibition replacement between the matrix and the crack, fluid flow in tight oil reservoirs is different than in traditional reservoirs. Imbibition is a complex physical and chemical phenomenon, a process whereby the wetting phase enters the porous medium spontaneously and replaces the nonwetting phase through the capillary force action. In the process of imbibition, water first contacts the porous medium. Next, water enters the matrix through the capillary force and moves along the wall of the porous medium, replacing part of the movable oil. Imbibition oil recovery has become an interesting and widely investigated method for improving oil recovery in tight oil reservoirs. The effectiveness of the imbibition process depends on several parameters, including the size of the matrix block, the porosity and permeability of the rock, the viscosity of the fluid, the IFT, the wettability of the rock, and the contact area between the wetting fluid and the matrix. Tight oil reservoirs, which are rich with microcracks, are beneficial to imbibition oil recovery because natural fractures considerably increase the contact area between the fluid and the matrix.

6.7 Smart water injection for shale-oil reservoirs

Recently, quite a few studies have been conducted to investigate the effect of LSW injection on oil recovery enhancement in unconventional liquid-rich reservoirs. Different studies have reported that the maximum oil recovery can be achieved at an optimum concentration of salt in the brine injected into lab cores or in the field (simulation work). Wettability alteration and IFT may be the main mechanisms behind the incremental oil recovery in LSW injection. However, the underlying wettability alteration mechanisms remain controversial, although double-layer expansion and multicomponent ion exchange seem to be the main mechanisms. Unfortunately, most of the studies, which investigated the EOR mechanisms, focused on conventional reservoirs with a high permeability.

The LSW technique was applied recently in unconventional reservoirs by Morsy, Gomaa, & Sheng (2014). They conducted experimental and simulation studies to investigate the effect of LSW on the oil recovery using cores from the Eagle Ford formation. They compared the oil recovery from distilled water and 2% KCl, reporting a 19% RF from the samples placed in distilled water, and a 12% RF from those placed in 2% KCl. They concluded that the higher oil recovery from distilled water was due to shale cracking because of the clay swelling. Morsy and Sheng (2014) evaluated the change in the oil recovery for different shale rocks of Mancos, Marcellus, Barnett, and Eagle Ford when such cores interacted with NaCl and KCl brines with different concentrations from 0% to 30 wt%. They found that Mancos core plugs would crash into fragments when exposed to low-salinity solutions up to 15%, while Barnett core plugs showed consecutive cracks along bedding planes at low salinities. Only minor cracks were seen on Marcellus, while no visual cracks were noticed in Eagle Ford core plugs at low salinities.

Furthermore, their results indicated that 2%–15% of oil was recovered from oil-saturated shale plugs using water spontaneous imbibition. This is a wide RF range, with the actual RF in each situation depending upon the water salinity and rock mineralogy. Valluri et al. (2016) conducted another set of experimental tests, injecting different concentrations of sodium chloride and calcium chloride brines into sidewall cores of an ultra-tight, liquid-rich shale from South Texas. Their study confirmed the potential of low-salinity brines to change the wettability of shale rock and enhance the oil recovery more than just brine itself. Table 6.1 clearly summarizes the most significant studies conducted to investigate the feasibility of injecting LSW and water alone to improve the oil recovery in unconventional reservoirs. Table 6.2 summarizes the applicability, motivations, solutions, and obstacles around applying LSW and water injection methods in shale-oil reservoirs.

Table 6.1 The reported studies for low salinity water and brine flooding techniques in unconventional reservoirs.

S. No.	Authors (Year)	Paper number	Approach	Formation	IOR method	IOR mechanism
1	Wang et al. (2011)	SPE 138521	Experimental	Bakken	Brine	Shale cracking
2	Morsy et al. (2013)	SPE -167056-MS	Experimental/ simulation	Eagle Ford	LSW	Shale cracking
		SPE 167142	Experimental	Bakken	LSW	Wattability alteration and IFT reduction
3		URTeC: 1920887	Experimental	Niobrara	LSW	Shale cracking
4	Morsy and Sheng (2014)	CSA-5604-PA	Experimental	Mancos, Marcellus, Barnett, and Eagle Ford	LSW	Shale cracking, and water spontaneous imbibition
5		URTec2461736	Experimental/ simulation	Bakken	LSW	Osmosis
6	Valluri et al. (2016).	SPE-180274-MS	Experimental	Side wall Cores	LSW	Wettability alteration, and IFT reduction
7	Yu and Sheng (2016b)	SPE-180378-MS	Experimental	Eagle Ford	Cyclic water	Repressurization and fracturing

Table 6.2 Summary of the applicability of low-salinity water and brine flooding techniques in unconventional reservoirs.

Water injection	Applicability	Motivation	Solution provided	Obstacles
Water	Yes	Depletion and poor sweep efficiency	Pressure maintenance and oil displacement	Conformance control
LSW	Yes	Negative PC, poor sweep efficiency	Enhancing water imbibition by osmosis	Need-field tests

6.8 Comparison of water injection in huff-n-puff mode versus flooding mode in shale-oil reservoirs

Water injection via the flooding mode is the most common method used in the oil industry. However, because shale and tight oil reservoirs have ultralow permeability, injecting water in the flooding mode makes it very difficult to achieve any sweeping process.

The low injectivity in such reservoirs led the researchers to consider another mode by which water could be injected into such tight formations. Huff-n-puff is the most viable method by which to inject water and increase the oil production from shale reservoirs. Initial field trials and laboratory experiments have shown very promising results for the huff-n-puff mode. From the imbibition point of view, a water-wet formation is preferred for water injection in huff-n-puff mode. This conclusion is supported by experimental data obtained by Huang and Xiang (2004). The huff-n-puff experiments conducted by Yu and Sheng (2016a) indicated that the injection pressure can significantly affect the oil recovery. At the laboratory scale, increasing the soaking period could increase the oil recovery. Sheng and Chen's (2014) simulation results showed that the oil recovery from water huff-n-puff is only one-third of that from gas huff-n-puff. They also showed that the oil recovery resulting from water huff-n-puff is only 2% higher than that from waterflooding. The incremental oil RFs from waterflooding and water huff-n-puff are only 2%–3% higher than the primary depletion. Therefore, the EOR potential for water injection is limited. However, shale plays have huge oil-in-place (billions of oil barrels), so such an increment in the oil RF might result in billions of extra oil barrels as an incremental cumulative recovery.

The following findings summarize the researchers' conclusions about the comparison between waterflooding and water huff-n-puff:

1. For tight oil cores, the residual oil saturation of water huff-n-puff was found to be lower than that of conventional waterflooding.
2. To significantly improve the oil RF, using two to four cycles of water huff-n-puff is more suitable for tight oil reservoirs than waterflooding development.
3. One general mechanism for water huff-n-puff is that water preferentially invades the large pores and then imbibes into small pores to extract oil. Another important mechanism is that the invaded water and imbibed water increase the reservoir pressure and local pressure, boosting the drive energy. However, the main mechanisms of waterflooding are the pressure maintenance and sweeping process.
4. The injection pressure and soaking time can significantly affect the oil recovery obtained in the huff-n-puff mode.
5. The oil recovery mechanisms during water huff-n-puff share some similarities with the gas huff-n-puff mechanisms. However, the diffusion process is responsible for the oil recovery during gas huff-n-puff, while the imbibition process is the most dominant during water huff-n-puff. The oil recovery from water huff-n-puff was reported to be lower than that from gas huff-n-puff at the lab scale.
6. Some of the reported simulation studies showed that the oil recovery resulting from the water huff-n-puff process in unconventional reservoirs is 1%–2% higher than that from waterflooding.

6.9 Factors affecting the performance of water injection in unconventional reservoirs

One of the key issues impacting the success of water injection projects for shale is the permeability of the reservoir given the water imbibition and water–rock interaction. It is commonly accepted that water–rock interactions cause formation damage (reduce the formation permeability). However, many researchers have observed that the water may help to generate microfractures or open the preexisting microfractures in shale formations if no confining pressure is applied (Morsy et al., 2013a, 2013b, 2013c, 2014) . However, Behnsen and Faulkner (2011) and Faulkner and Rutter (2000) reported that with an isotropic confining pressure, a significant permeability reduction was observed when measured with water clay-bearing rocks or montmorillonite samples. Whether or not fractures can be induced to increase the permeability during water imbibition in shale cores under an isotropic compressive stress remains controversial. In contrast to conventional propped hydraulic fracture treatments, water fractures depend on the reactivation of natural fractures to induce a permanent shear-induced dilation, which enhances the reservoir's permeability (Moradian, Fathi, & Evans, 2016). Hydraulic fracturing is performed, in which shear failure is anticipated to dominate (Zoback et al., 2012) in shale under anisotropic stress. Several authors have reported that the hydration swelling due to water imbibition can weaken the mechanical strength of shale (even more than 60%) and reduce the shear-induced fracture conductivity. In terms of waterflooding performance, Sheng and Chen (2014) reported that the average reservoir pressure during water injection cannot be increased much higher than the pressure before water injection, as the high pressure is only near the injector. Their results also showed that the oil recovery from waterflooding is lower than that from gas flooding. However, the models did not consider water–rock interactions.

Per the simulations and experimental work carried out by some researchers, the following parameters must be considered for designing water-based EOR projects in shale-oil reservoirs:

1. The matrix permeability is the main parameter that causes low oil recovery from shale reservoirs. Designing a closer fracture spacing will positively influence shale-oil production.
2. In an ultralow porosity, ultralow permeability, and high oil viscosity shale-oil reservoir, injecting water in the flooding mode through a high conductivity fracture has less effect on improving oil recovery than gas injection.
3. Unlike miscible gas flooding that can reduce oil viscosity, the water injected in the flooding mode serves to maintain the reservoir pressure. Ultralow permeability worsens the sweep efficiency, leading to low productivity and low injectivity. However, the water injected in huff-n-puff mode with imbibition mechanisms would produce significantly different results.
4. The potential of cyclic water injection in shale-oil reservoirs is strongly a function of the imbibition mechanisms; a low oil RF may result if the imbibition process is very slow. Some of the experimental results showed that oil can be successfully extracted even in just the first two to four cycles.
5. Within a certain range, a longer soaking period causes a higher ultimate RF during the water huff-and-puff process. Once the maximum RF is

achieved, an extra-long soaking period might not contribute to additional recovery.

6. Increasing the injection pressure is beneficial in increasing the oil recovery. A high injection pressure can induce new microfractures in the matrix, which might help in increasing the flowability of the reservoir fluids.

6.10 Field applications for water injection in unconventional reservoirs

Quite a few water injection projects have been carried out in shale and tight oil reservoirs. Most of these have been implemented in the Bakken play and have not shown very promising results. Some of the projects were implemented in waterflooding mode (case I through case V), while other projects were conducted in huff-n-puff mode (case VI through case VIII).

6.10.1 Case study I

Crescent Point Energy began conducting water injection projects in the Bakken and Lower Shaunavon resources in 2006. The mode of water injection was waterflooding. The injection patterns were formed by situating horizontal injectors parallel to horizontal producers (Wood and Milne, 2011). A total of five patterns in the Bakken and three patterns in the Lower Shaunavon had been sanctioned by 2011 (Sheng, 2017). The first trial in the Bakken had a peak oil rate of 550 bbl/d from 50 to 100 bbl/d for four producers, while the second pattern had a lower oil production rate. The results of the other patterns were not reported at that time because it was too early to evaluate their performance. The porosity in the pattern area ranged from 14% to18%, and the permeability was less than 1 mD (Thomas et al., 2014).

6.10.2 Case study II

One waterflooding pilot was performed in the North Dakota section of the Bakken formation. This pilot had one injector in the center surrounded by four offset wells. Two of the producers, those situated to the east and the west, were 2300 ft away from the injector, while the other two, situated to the north and south, were 900 and 1200 ft away from the injector, respectively, as shown in Fig. 6.13. They injected 1350 bbl/day of water for 8 months, which raised the bottom-hole pressure to

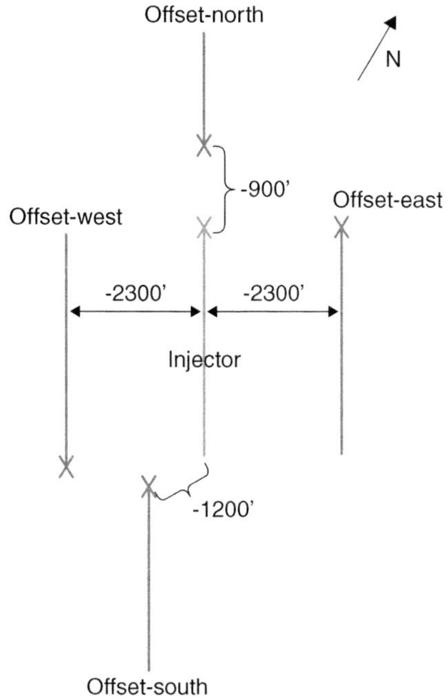

Figure 6.13 *General layout of pilot tests #4/7 (Hoffman & Evans, 2016).*

approximately 6000 psi. The offset wells to the east and west experienced a significant increment in the water production. The injected water achieved breakthrough in the producers in less than a month. However, there was no increment in the oil production. Interestingly, the quick breakthrough occurred in the wells located to the east and west of the injectors (2300 ft away from the injector) but not in the wells to the north and south of the injector (900 and 1200 ft, respectively, away from the injector). The reason for this behavior is that the injection process induced some fractures. These injection-induced fractures (IIFs) are commonly growing in the west–east direction in the United States (Baker et al., 2016). Moreover, according to the amount of time it took for the breakthrough to occur, these IIFs seem to have a length of more than half of the injector-producer spacing (Baker et al., 2016). For the second cycle of flooding, they reduced the injection flow rate to 380 bbl/day at a bottom-hole pressure of 5500 psi for 8 months. However, the results were similar to the first cycle, showing no increment in the oil production.

However, the water production in the offset wells decreased.

This provides further evidence that they created IIFs because IIFs would usually close when the injection rate gets lowered (Baker et al., 2016).

6.10.3 Case study III

Another test was carried out in 2014 in the Bakken formation in Montana. There was one injection well and several offset wells for producing and monitoring. The operator started injecting nearly 1700 STB/day of water for 3 months. After that, the injection rate was reduced to below 1000 STB/day for 5 months. Just like in the other pilots, some additional produced oil was observed in the offset well, with substantial water production due to conformance problems. These conformance problems may have been caused by the IIFs created during the injection process. Furthermore, some researchers argued that the observed additional produced oil did not necessarily result from the injection process. They thought that the additional oil could have resulted from the fracture hits going on in some of the close wells in that area of the field.

6.10.4 Case study IV

Sorensen and Hamling (2016) reported another pilot conducted in the Parshall Field well operated by EOG, the NDIC 16986. This well has been the subject of both produced water and field gas injection testing. A water injection process was conducted in the period from April 2012 to February 2014 in waterflooding mode. Fig. 6.14 shows the fluid production data for the injection well and the nearby offset well, NDIC 16461. Approximately 439,000 barrels of water were injected before the well was reopened for production in March 2014. In June 2014, EOG began injecting field gas mingled with the produced water. According to the information available, water was used to manage the effects of gas mobility in the fracture system, or if needed, to build system pressure with less gas volume (Sorensen & Hamling, 2016). Through August 20, 2014, a total of 88.729 MMcf of gas had been injected. Unfortunately, there are no data in the well file, suggesting that the operator considered the testing activities to be successful. However, changes in the fluid production rates of the two offset wells were observed, demonstrating that clear communication exists between wells.

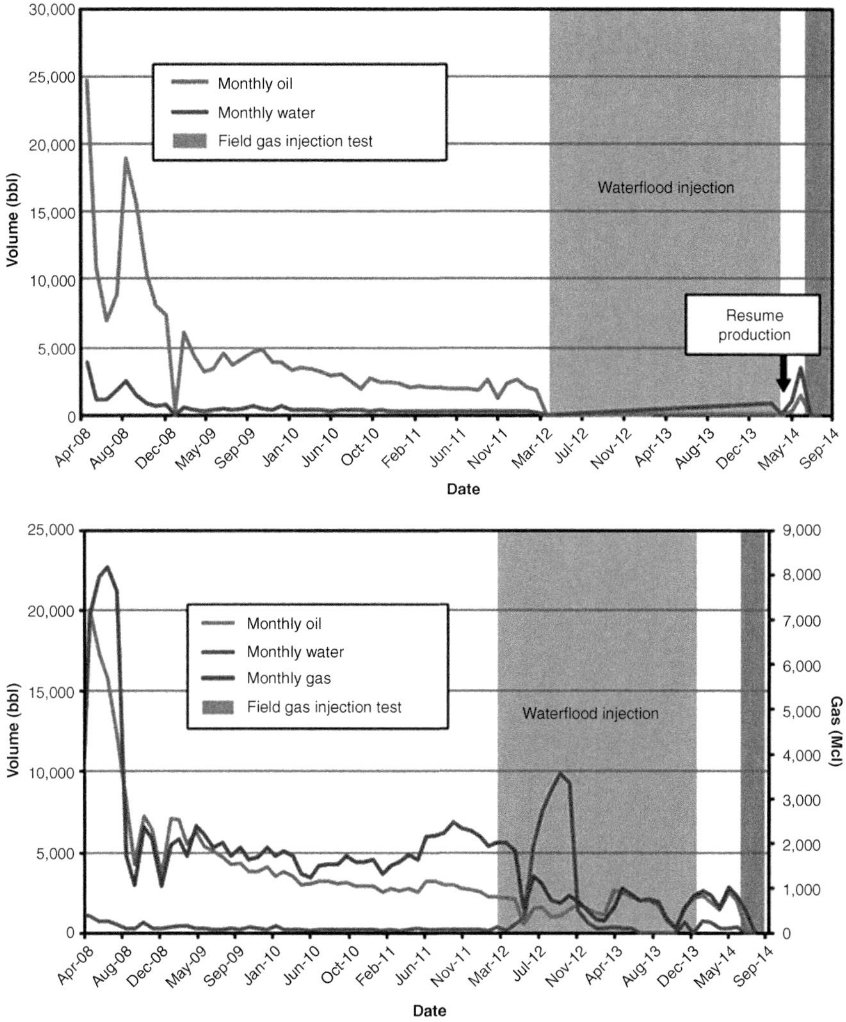

Figure 6.14 ***Fluid production data for the injection well and the nearby offset well, NDIC 16461 (Sorensen & Hamling, 2016).***

6.10.5 Case study V

Wood and Milne (2011) reported eight pilots conducted in the Canadian Bakken using waterflooding. Although most were in the early stages of execution, some already showed encouraging results. Interestingly, all of the pilot tests that have been performed in the Canadian Bakken have had approximately the same well pattern, the Toe-Heel pattern. This type of flooding pattern is useful for improving conformance control during EOR

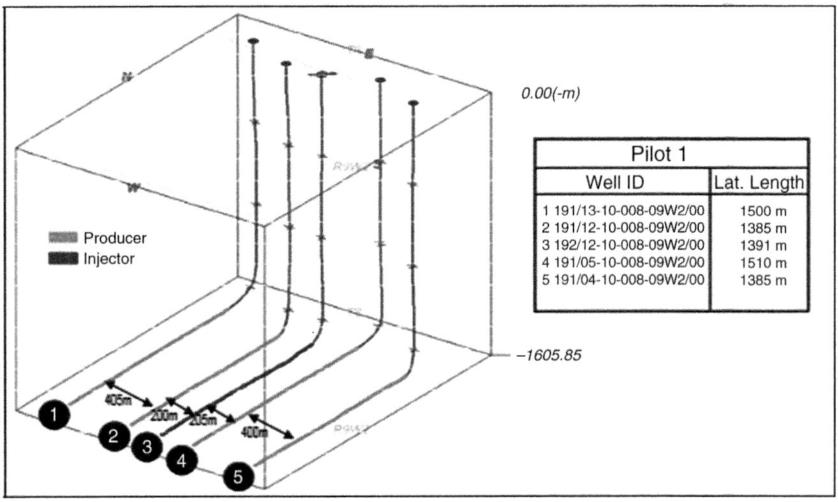

Figure 6.15 ***Well structure of Canadian Bakken pilot (Wood & Milne, 2011).***

injection. Also quite interesting is that the spacing between the injection wells and the production wells in the Canadian Bakken pilots is much shorter than the spacing between the injectors and producers in the US Bakken pilots, as short as 200 ft, although the porosity and permeability of the Canadian Bakken are much higher than those of the US Bakken. The short spacing might be one of the main reasons for the success of the Canadian Bakken pilots. The average lateral length for the production wells and injection wells, which were drilled horizontally in the Canadian Bakken area, is approximately 1 mile. Fig. 6.15 shows the Canadian Bakken pilot design structure with the number of wells, the spacing between them, and the lateral length for each one. The pilot had four producers and one injector in the center. The water injection rate scenarios and oil production rates are shown in Fig. 6.16. The results for this pilot test showed an increment in the oil production rate from around 75 to 550 bbl/day due to the water injection. Although the injection process was sporadic, as shown in Fig. 6.16, no injectivity problems in this project were reported.

6.10.6 Case study VI

One water huff-n-puff pilot test was conducted in the North Dakota part of the Bakken formation in 2012. The injection time was just over 1 month, the soaking time was 2 weeks, and the production time was 3–4 months. The injection rate was 1200 bbl/day. No water injectivity problems were observed, but little to no oil increment occurred (Hoffman & Evans, 2016).

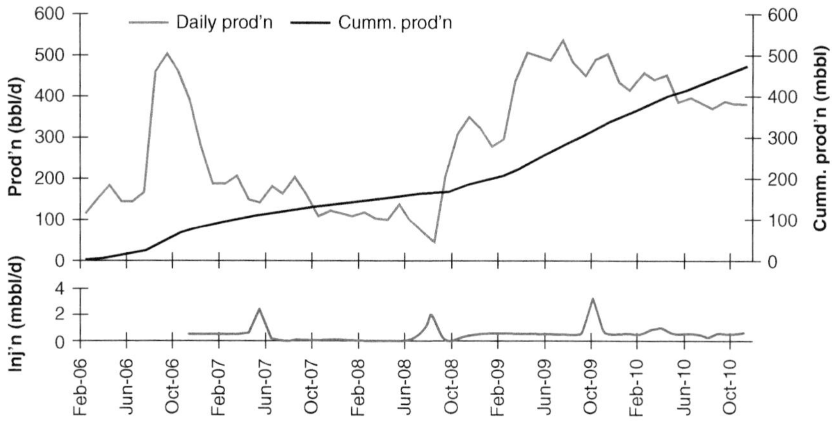

Figure 6.16 ***Injection and production rates of Canadian Bakken pilot (Wood & Milne, 2011).***

6.10.7 Case study VII

As early as 1994, water was injected by the Meridian Oil Company in NDIC 9660 in the Bicentennial Field in McKenzie County. Approximately 13,200 barrels of fresh water were injected into a horizontal well in the Upper Bakken Shale over 50 days. The well was shut in for two months before being opened for production. Unfortunately, the oil production rates remained below those obtained prior to the water injection. These low rates lasted for the remainder of the well's operational life (Sorensen & Hamling, 2016).

6.10.8 Case study VIII

EOG conducted a produced water injection test in a well in the Parshall Field, the NDIC 17,170. Injection began in the spring of 2012 with a plan to operate the well according to a huff-and-puff scheme with a 30-day injection and 10-day soak schedule. 10,000 barrels of water were injected in April, and 29,000 barrels were injected in May 2012. There was no observable increase in oil production.

6.11 SUMMARY

1. Although the waterflooding technique has been used in conventional reservoirs for many decades, water huff-n-puff in unconventional reservoirs is not yet mature; there is a critical need to develop and understand its concept.

2. The water imbibition mechanism is the main player that controls the success or failure of water huff-n-puff in shale-oil reservoirs.
3. Adding some chemical agents to the injection water and/or manipulating the composition of the water used in the huff-n-puff process is crucial to enhancing oil recovery in unconventional oil reservoirs, as reported in some experimental and pilot cases.
4. Although most of the lab and simulation results indicate very good performance of water injection in terms of enhancing oil recovery in shale-oil reservoirs, the performance of water huff-n-puff is poor in the pilot tests conducted in shale formations.
5. Conformance problems, as well as the imbibition rate, appear to be the biggest obstacles to the success of water huff-n-puff in unconventional reservoirs.

References

Baker, et al. (2016). *The myths of waterfloods, EOR floods and how to optimize real injection schemes.* SPE-179536-MS paper was prepared for presentation at the SPE Improved Oil Recovery Conference held in Tulsa, Oklahoma, USA, 11–13 April 2016.

Behnsen, J., & Faulkner, D. R. (2011). Water and argon permeability of phyllo-silicate powders under medium to high pressure. *Journal of Geophysical Research, 116*, B12203.

Denekas, M. O., & Mattax, C. C. (1959). Effects of crude oil components on rock wettability. *Transaction AIME, 216*, 330–333.

Esene et al. (2019). Comprehensive review of carbonated water injection for enhanced oil recovery. *Fuel* 237, 1086–1107. https://doi.org/10.1016/j.fuel.2018.08.106.

Fettke, C. R. (1938). The Bradford oil field, Pennsylvania and New York. Mineral Resources Report M21, Pennsylvania Geological Survey, Harrisburg, Pennsylvania, pp. 298–301.

Fakcharoenphol, P., Charoenwongsa, S., Kazemi, H., & Wu, Y. S. (2013). The effect of water-induced stress to enhance hydrocarbon recovery in shale reservoirs. *SPEJ, 18*(5), 897–909.

Faulkner, D. R., & Rutter, E. H. (2000). Comparisons of water and argon permeability in natural clay-bearing fault gouge under high pressure at 20°C. *Journal of Geophysical Research, 105*, 16415–16426.

Green and Willhite. (1998). *Enhanced Oil Recovery. SPE Textbook Series.* Vol. 6. Society of Petroleum Engineers. https://store.spe.org/Enhanced-Oil-Recovery-P21.aspx.

Hoffman, B.T., & Evans J., (2016). *Improved oil recovery ior pilot projects in the bakken formation.* SPE-180270-MS paper presented at the SPE Low Perm Symposium held in Denver, Colorado, USA, 5–6 May 2016. http://www.investorvillage.com/uploads/44821/files/CPGdundee.pdf (accessed October 27, 2013).

Huang, D. -Z., & Xiang, D. (2004). Mechanisms of huff-n-puff water injection to displace oil. *Petroleum Geology and Recovery Efficacy, 11*(5), 39–43.

Lager, A., Webb, K. J., Black, C. J. J., Singleton, M., & Sorbie, K. S. (2008). Low salinity oil recovery—An experimental investigation. *Petrophysics, 49*, 28–35.

Makhanov et al. (2012). *An experimental study of spontaneous imbibition in Horn River shales.* SPE-162650-MS presented at SPE Canadian Unconventional Resources Conference, Calgary, Alberta, Canada, October 30–November 1.

Moradian, Z., Fathi, A., Evans, B. (2016). *Shear reactivation of natural fractures in hydraulic fracturing.* American Rock Mechanics Association. https://www.onepetro.org/conference-paper/ARMA-2016-829.

Morsy et al. (2013a). *Potential of improved waterflooding in acid-hydraulically-fractured shale formations.* Paper SPE 166403 presented at the SPE Annual Technical Conference and Exhibition held in New Orleans, Louisiana, USA, September 30–October 2.

Morsy et al. (2013b). *Potential of waterflooding in shale formations.* Paper SPE 167510 presented at the Nigeria Annual International Conference and Exhibition held in Lagos, Nigeria, August 5–7.

Morsy, S., & Sheng, J. J. (2014). Effect of water salinity on shale reservoir productivity. *Advances in Petroleum Exploration and Development, 8*(1), 9–14. https://doi.org/10.3968/5604.

Morsy, S., Gomaa, A., Sheng, J.J. (2014). *Improvement of eagle ford shale formations water imbibition by mineral dissolution and wettability alteration.* In: Paper SPE 168985 Presented at the SPE Unconventional Resources Conference, 1–3 April, The Woodlands, Texas, USA.

Morsy, S., Sheng, J. J., & Soliman, M. Y. (2013c). *Waterflooding in the Eagle Ford shale formation: Experimental and simulation study.* Society of Petroleum Engineers. https://doi.org/10.2118/167056-MS.

Muggeridge et al. (2014). *Recovery rates, enhanced oil recovery and technological limits.* https://doi.org/10.1098/rsta.2012.0320.

Schumacher, M.M. (1978). *Enhanced oil recovery: Secondary and tertiary methods* (Chapter 2, p. 23). Park Ridge, NJ: Noyes Data Corp.

Seright, R.S. (2005). *Clean up of oil zones after a gel treatment.* Paper presented at the SPE International Symposium on Oilfield Chemistry.

Sheng, J., & Chen, K. (2014). Evaluation of the EOR potential of gas and water injection in shale oil reservoirs. *Journal of Unconventional Oil and Gas Resources, 5*, 1–9.

Sheng, J. (2017). *Critical review of field EOR projects in shale and tight reservoirs.* https://doi.org/10.1016/j.petrol.2017.09.022.

Slide share. (2020a). Accessed in 2020 via the following link: https://slideplayer.com/slide/10369263/.

Slide share (2020b). Accessed in 2020 via the following link: https://slideplayer.com/slide/3848470/.

Sohrabi, M., Emadi, A., Farzaneh, S.A., & Ireland, S. (2015, September 28). A Thorough Investigation of Mechanisms of Enhanced Oil Recovery by Carbonated Water Injection. Society of Petroleum Engineers. doi:10.2118/175159-MS.

Solomon, S. (2007). *The Bellona Foundation—Fact sheet: CO_2 storage.* Bellona report. Available from: http://www.bellona.org/factsheets/1191921304.33.

Sorbie, K.S., & Collins, I. (2010). A proposed pore-scale mechanism for how low salinity waterflooding works. *Proceedings of SPE Improved Oil Recovery Symposium.* (pp. 24–28). Tulsa, OK: Society of Petroleum Engineers.

Sorensen & Hamling (2016). https://www.aogr.com/magazine/cover-story/historical-bakken-test-data-provide-critical-insights-on-eor-in-tight-oil-p.

Takahashi, S., & Kovscek, A. R. (2010). Spontaneous countercurrent imbibition and forced displacement characteristics of low-permeability, siliceous shale rocks. *Journal of Petroleum Science and Engineering, 71*(1–2), 47–55. https://doi.org/10.1016/j.petrol.2010.01.003.

Thomas, A., Kumar, A., Rodrigues, K., Sinclair, R.I., Lackie, C., Galipeault, A., & Blair, M. (2014). *Understanding water flood response in tight oil formations: a case study of the lower Shaunavon.* Society of Petroleum Engineers. doi:10.2118/171671-MS.

Valluri et al. (2016). *Study of the rock/fluid interactions of sodium and calcium brines with ultra-tight rock surfaces and their impact on improving oil recovery by spontaneous imbibition.* SPE-180274-MS conference paper presented at the SPE Low Perm Symposium held in Denver, Colorado, USA, 5–6 May 2016.

Wagner, O. R., & Leach, R. O. (1959). Improving oil displacement efficiency by wettability adjustment. *Transaction AIME, 216*, 65–72.

Wang, D., Butler, R., Liu, H., & Ahmed, S. (2011). *Flow-rate behavior and imbibition in shale.* Society of Petroleum Engineers. https://doi.org/10.2118/138521-PA.

Warner Jr., H. R. (2007). *Petroleum Engineering Handbook. Volume V—Reservoir Engineering and Petrophysics* (pp. 1037–1102). https://petrowiki.spe.org/index.php?title=PEH:Waterflooding&mobileaction=toggle_view_mobile.

Wood, T., Milne, B. (2011). *Waterflood potential could unlock billions of barrels: Crescent point energy.* http://www.investorvillage.com/uploads/44821/files/CPGdundee.pdf.

Yu, Y., & Sheng, J. J. (2016a). *Experimental investigation of light oil recovery from fractured shale reservoirs by cyclic water injection.* Society of Petroleum Engineers. https://doi.org/10.2118/180378-MS.

Yu, Y., & Sheng, J. J. (2016). *Experimental investigation of light oil recovery from fractured shale reservoirs by cyclic water injection.* In: Paper SPE 180378 Presented at the SPE Western Regional Meeting Held in Anchorage, Alaska, USA, 23–26 May.

Zoback et al. (2012). *The importance of slow slip on faults during hydraulic fracturing stimulation of shale gas reservoirs.* SPE Americas unconventional reservoirs resources conferences.

CHAPTER 7

Chemical enhanced oil recovery methods for unconventional reservoirs

Contents

7.1 Introduction 141
7.2 Chemical EOR methods 145
7.3 Surfactant types 146
7.4 Surfactant huff-n-puff EOR 150
7.5 Surfactant injection during hydraulic fracturing process 151
7.6 Surfactant selection 154
7.6.1 Introduction of surfactant selection criteria 154
7.6.2 Principles of surfactant selection criteria 154
7.6.3 Data compilation 155
7.6.4 Functionality of oil recovery factor with CA, IFT, and capillary pressure 158
7.7 Limitations of surfactant EOR for unconventional reservoirs 160
7.8 Summary 160
References 161

Abstract

This chapter highlights the importance of chemical EOR methods in both conventional and unconventional reservoirs. Furthermore, it discusses the different types of chemical EOR methods, their EOR mechanisms, and the obstacles to their success. This chapter focuses on how to alter the oil affinity for shale reservoirs and increase the spontaneous imbibition using different chemical additives. The applicability of different types of surfactants in rocks with different compositions is introduced. The effect of the adsorption rate and imbibition penetration depth on surfactant performance is also discussed, as is the effect of the intensity of natural fractures on the imbibition rate.

Keywords: chemical EOR in unconventional reservoirs; surfactant huff-n-puff; water imbibition mechanisms; unconventional EOR methods; huff-n-puff versus flooding mode

7.1 Introduction

Although two-thirds of the original oil in place (OOIP) remains in conventional reservoirs after the primary and secondary oil recovery processes, the OOIP remaining trapped in unconventional reservoirs is much more than this. This happens because the oil is trapped by capillary forces (residual oil) and bypassed in some other way. Any mechanism that can increase the

Fundamentals of Enhanced Oil Recovery Methods for Unconventional Oil Reservoirs
http://dx.doi.org/10.1016/B978-0-12-818343-4.00007-3

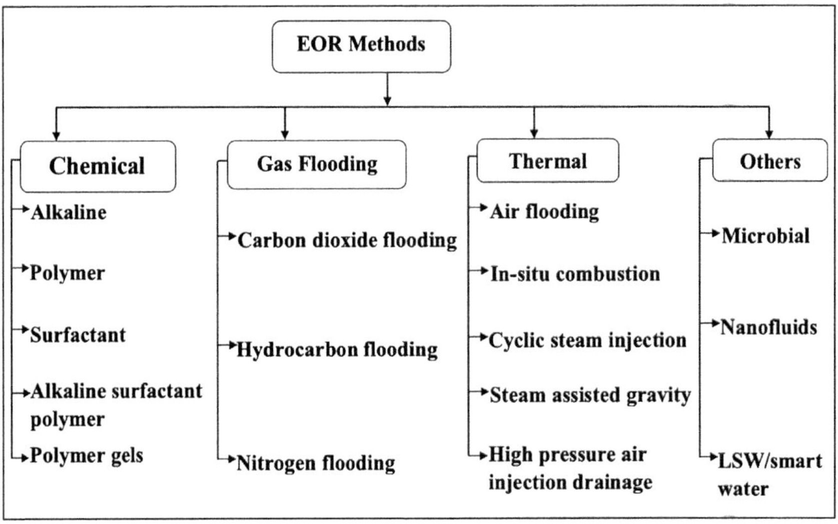

Figure 7.1 ***Classification of EOR methods.*** *(Modified from Sakthipriya, Doble, & Sangwai, 2015).*

oil recovery efficiency at either the micro scale, the macro scale, or both is beneficial for enhanced oil recovery (EOR) (Agi, Junin, Syamsul, Chong, & Gbadamosi, 2018; Gbadamosi, Kiwalabye, Junin, Augustine, 2018). Many EOR methods have been proposed to help in improving the total oil recovery efficiency, as shown in Fig. 7.1. Table 7.1 explains the factors that

Table 7.1 Most important factors impacting the microscopic recovery efficiency (E_D) and the macroscopic efficiency (F_V), on both the areal sweep efficiency (E_A) and the vertical sweep efficiency (E_I).

Factor (↑: increase)	Effect on E_A	Effect on E_I	Effect on E_D
Reservoir heterogeneity and anisotropy	++	++	N/A
Pore heterogeneity	N/A	N/A	+
Mobility	++	++	N/A
Well pattern (the correct pattern)	++	+	N/A
Capillary pressure	+	+	++
Reservoir thickness	+	++	N/A
Density ratio between displacing and displaced fluids	N/A	++	N/A
Distance between the injector and producer	+	++	N/A
Interfacial tension	+	+	++
Wettability (more water-wet)	+	+	++

Note: Strong effect ++; weak effect +; no effect N/A.

impact the microscopic recovery efficiency (E_D) and the macroscopic efficiency (E_V). Table 7.2 lists the main advantages and disadvantages of the most common EOR methods. Table 7.3 explains the solutions of different EOR methods to the most common reservoir problems, which lead to

Table 7.2 Main advantages and disadvantages of the most common EOR methods.

EOR method	Advantages	Disadvantages
Polymer flooding	• Excellent mobility control agent • Good improvement in sweep efficiency, with some debatable improvement in displacement efficiency • Requires easy mechanical considerations in terms of injection pressure (does not need high horsepower injection equipment	• Require high permeability (some studies indicated K > 50 md) • Not beneficial for reservoirs with high permeability contrast • Works at a specific range of reservoir temperature, and affected by divalent, dilutions, and salts
In-situ combustion	• Beneficial at very high viscous oil combined with gravity drainage process • Achieves a high recovery due to improving the oil mobility	• Field projects: a few successful and the majority disappointing • Oil blocking problems caused high fuel deposition and combustion front stalling • Difficult to control the fire front
Steam flooding	• Easy to control • Does not cause oil cracking • Does not produce environmentally objectionable gases • Less deposits built up in the wells	• Not economical when the distance between the injector and producer is long • Not economical when the reservoir experiences high heat loss • Difficult to perform in reservoirs with a depth > 5000 ft due to injection pressure problems

(*Continued*)

Table 7.2 Main advantages and disadvantages of the most common EOR methods. (*Cont.*)

EOR method	Advantages	Disadvantages
CSS	• Fast return on investment • Results in a high oil-production rate • Used when the interwell flow resistance is very strong	• Not very successful in thin reservoirs due to heat loss • The caprock should be sufficiently thick to tolerate a high steam injection pressure • Low recovery (generally <20%)
SAGD	• Requires lower injection pressure due to the lateral lengths of the horizontal well • Provides larger drainage area which results in RF > 50% • Overrides elimination and blind spot minimization	• Needs two horizontal wells (economic issues) • Requires high steam quantities • Limited by oil-well depths as imposed by steam's critical pressure
Gels	• Good method for no-crossflow reservoirs with high permeability contrast • Dual benefits in terms of EOR and water-cut reduction	• Not very beneficial for reservoirs of low permeability contrast with crossflow • Potential formation damage if bullhead injection is used
CO_2 flooding	• Benefits the environment when captured CO_2 is used • Easy to achieve miscible condition • Low T and low P to reach supercritical fluid condition	• Relatively low sweep efficiency • CO_2 resources
Surfactant	• Significantly reduces IFT • Modifies wettability	• Surfactant loss by adsorption • Emulsion

low oil recovery. Note that the information provided in Tables 7.1–7.3 are based on the authors' experience and might change over time according to knowledge advances and research findings.

Table 7.3 EOR solutions for the most common reservoir problems.

Reservoir problem	Solution 1	Solution 2	Solution 3
Mobility viscous fingering	Polymer: if oil viscosity is not very high and good permeability >50 md	WAG: if high oil viscosity, thin reservoir, low permeability	—
3-D water coining	Polymer gel with DPR property	Gel disk	—
2-D water coning	Water shut-off: gel treatment	Polymer gel	—
Casing leaks	Cement squeezing	Combination of mechanical isolation methods and gels	Combination of expandable tubulars and gels
High-K matrix layer with and without cross flow	Polymer: permeability ratio between layers not very big	Gel: permeability ratio between layers is big and no cross flow	Gel–polymer combination: permeability ratio between layers is big and cross flow exists
Depleted reservoir with low permeability	Miscible CO_2 flooding	Waterflooding: oil has low viscosity	—
Higher residual oil saturation	ASP: if it has good permeability >100 md	Surfactant: low adsorption rocks exist	Alkaline
Reservoirs heavy oil reservoirs	Cyclic steam or steam flooding: if the distance between injector and producer is short	In-situ combustion: if the oil is incredibly viscous	SAGD: if the reservoir is not very deep

7.2 Chemical EOR methods

Chemical EOR methods have been considered efficient oil recovery techniques for recovering bypassed oil and residual oil trapped in conventional and unconventional reservoirs because of their higher efficiency, technical and economic feasibilities, and reasonable capital cost (Gbadamosi et al., 2019). This category of EOR methods relies on the injection of chemical materials to enhance oil recovery. The application of chemical EOR methods became popular in the 1980s due to higher oil prices and technological advancements that enabled a clearer understanding of their mechanisms (Gbadamosi et al., 2019). Chemical EOR methods increase oil recovery via different mechanisms. The chemicals injected with the water slug can alter the fluid–fluid and/or fluid–rock interaction in reservoirs, including lowering the interfacial tension (IFT) between the driving fluid and the oil, changing the wettability from oil-wet to water-wet, and/or controlling flow conformance by increasing injectant fluid viscosity or reducing the preferential fluid flow through channels (Gbadamosi et al., 2019).

The most common chemical EOR methods are polymer flooding, surfactant, and alkaline flooding (Samanta, Bera, Ojha, & Mandal, 2012). Different combinations of chemical injectant modes have been designed, studied, and applied for EOR processes (Gbadamosi et al., 2019). These include the binary mix of alkali–surfactant (AS), surfactant/polymer (SP), and alkaline/polymer (AP), and the triple mix of alkaline/surfactant/polymer (ASP) slugs. The use of foam enhanced by surfactants and polymers for improving stability and mobility control has been reported and found to improve oil recovery (Kumar, Gaur, Mandal, 2017 Bera, Kumar, Ojha, & Mandal, 2013;Samin et al., 2017). Furthermore, nanofluids could be combined with surfactant for use as a potential EOR technique. Currently, some investigators also count smart water/low salinity waterflooding as a chemical EOR technique.

7.3 Surfactant types

Among chemical EOR methods, surfactant has been the most investigated and has shown the most promise for improving oil recovery in unconven tional oil reservoirs. In petroleum systems, wettability is defined as the affinity of either water or oil to spread onto a rock surface (Zhang et al., 2018). The fluid that has higher affinity to the rock is then called the wetting phase, and the nonwetting phase is the other fluid (Anderson 1986). Shale-oil reservoirs are well known as intermediate-wet to oil-wet; the rock affinity of reservoirs with this kind of wettability would prevent the aqueous phase from invading the matrix to displace the oil. Therefore, applying surfactant to change the rock wettability and enhance the water imbibition has become one major strategy to improve oil recovery in such complex formations (Sheng, 2015).

Conventional surfactant flooding has not been investigated as an EOR method in unconventional reservoirs because of the major concerns surrounding injectivity in such low permeability formations (Zhang et al., 2018). Most of the relevant studies in the literature investigated the potential of surfactants in huff-n-puff mode, where they demonstrated the ability to noticeably increase the early flowback production and make the oil production sustainable for a long time (He and Xu, 2015; He et al., 2015b; Rane and Xu, 2014, 2015a, 2015b). However, other researchers investigated the potential of surfactants to be used as a conventional EOR technique in shale-oil reservoirs (Shuler, Tang, Zayne, & Tang, 2001; Dawson, Nguyen, Champion, & Li, 2015).

Alvarez, Neog, Jais, and Schechter (2014) conducted experimental work to evaluate the potential of nonionic and anionic surfactants to alter wettability and improve oil recovery in unconventional, liquid-rich reservoirs. They used carbonate and siliceous preserved sidewall cores in their experiments. They found that surfactants can lower the contact angle (making the cores more water-wet) and improve the oil recovery, although the performance differs between nonionic and anionic surfactants. Dawson et al. (2015) conducted experimental work on how surfactants can be used in the Bakken formation to enhance oil recovery, and they upscaled the lab results to the field scale using numerical simulation methods. The oil recovery factor obtained with spontaneous imbibition enhancement in their cores ranged from 30% to 40%. This indicates great potential for surfactant imbibition to enhance oil recovery in unconventional oil reservoirs. Other researchers used oil-on-plate methodology in their experimental work, where they investigated surfactant penetration into oil molecules (Kumar et al., 2008). They concluded that the oil-removal efficiency of a surfactant is a function of its chemistry and the oil type. Alvarez and Schechter (2016b) investigated the effect of different types of surfactants on the IFT and contact angle (CA) using Permian basin cores. They found that all types of surfactants could change the wettability from oil-wet to water-wet. However, the anionic surfactant demonstrated better performance in reducing both the IFT and CA in unconventional, liquid-rich cores. Nguyen et al. (2014) investigated the effect of different types of surfactants, including nonionic, cationic, anionic, and amphoteric, for spontaneous imbibition into shale cores in both the Bakken and Eagle Ford. They found that surfactant imbibition generally increased oil recovery in Eagle Ford shale outcrop cores and Bakken reservoir cores more than when only brine was used. They also reported that wettability alteration is more important than IFT reduction using surfactants. Still other researchers found that using a surfactant with a stimulation fluid enhanced the penetration of the fracturing fluid into the tight formations much more than for cases in which surfactant was not used (Alvarez et al., 2014). Some researchers used computed tomography (CT) scanning to monitor the penetration depths for different cases in their experiments (Alvarez et al., 2014). They posited that the reason for the increase in the penetration depth of the fracturing fluid when using surfactant is due to the IFT reduction, which, in turn, would enhance the productivity index of these reservoirs.

In practice, many different types of surfactants exist, including nonionic, cationic, anionic, and amphoteric. Regardless of the surfactant type, surfactant molecules act as a bridge between the water and oil phases present in the pore system, and the two parts of the surfactant molecule (called the head and tail) have different affinities for the aqueous and oleic phases (Zhang et al., 2018). This means that the head of the surfactant molecule is hydrophilic, which has a higher affinity toward the aqueous phase, whereas the tail of the surfactant molecule is hydrophobic and therefore has a higher affinity to the oil molecule. It is known that amphiphilic surfactant molecules have the ability to decrease the water-oil IFT. In addition, proper surfactant molecules can alter the wettability of the rock surface from oil-wet to water-wet, resulting in the spontaneous imbibition of water into the pores (Zhang et al., 2018). Therefore, spontaneous imbibition is mainly a function of the capillary forces according to the Young-Laplace equation. Capillary forces are significant and complex in shale nanopores, so the main objective is to alter the CA and IFT simultaneously and significantly to enhance oil recovery in unconventional liquid reservoirs (ULR).

The other two types of chemical EOR, alkaline and polymer, are not as attractive as surfactant EOR. Injectivity problems constitute the main barrier in applying polymer in shale-oil reservoirs, although conformance problems are more dominant in the reported pilot tests (Hoffman and Evans, 2016). Also, injecting polymer into these reservoirs would plug the pores throats, which are very small in those plays. Regarding alkaline, its potential in unconventional oil reservoirs has not been investigated in any reported study, possibly because this chemical agent is not compatible with the complex mineral compositions of such reservoirs. Table 7.4 clearly summarizes some significant studies conducted to investigate the feasibility of chemical EOR methods to improve oil recovery in unconventional reservoirs. Table 7.5 summarizes the applicability, motivations, solutions, and obstacles surrounding the application of chemical EOR methods in shale-oil reservoirs.

Table 7.4 Reported chemical IOR methods in unconventional reservoirs.

SN	Authors, year	Paper number	Approach	Formation	IOR method	IOR mechanism
1	Wang et al. (2012)	SPE-153853-PA	Experimental	Bakken	Surfactant	Wettability alteration
2	Kathel and Mohanty (2013)	SPE-166281-MS	Experimental	X	Surfactant	Wettability alteration
3	Alvarez et al. (2014)	SPE-169001-MS	Experimental	Preserved sidewall core	Surfactant	Wettability alteration/ IFT reduction
4	Nguyen et al. (2014)	SPE-169085-MS	Experimental	Eagle Ford and Bakken	Surfactant	Wettability alteration and IFT R.
5	Dawson et al. (2015)	SPE-175937-MS	Experimental/ Simulation	Bakken	Surfactant	Wettability alteration
6	Xu, He, Rane, Yin, & Neeves (2015c)	SPE-175536-MS	Experimental	X	Surfactant	IFT reduction and deep penetration
7	Alvarez and Schechter (2016a)	SPE-177057-PA	Experimental	Permian Basin	Surfactant	Wettability alteration and IFT reduction
8	Alvarez and Schechter (2016b)	SPE-179688-MS	Experimental	Bakken	Surfactant & CNF	Wettability alteration and IFT reduction
9	Alvarez and Schechter (2016c)	URTeC 2461651	Experimental	Bakken, Eagle Ford, Wolfcamp, Barnet	Water	Wettability alteration
10	Wang et al. (2016)	SPE-178489-PA	Experimental/ simulation	Bakken	Surfactant	Wettability alteration
11	Shuler et al. (2001)	SPE-179667-MS	Experimental	oil-on-a-plate	Surfactant	X
12	Li et al. (2016)	URTec2461736	Experimental/ simulation	Bakken	Surfactant	Wettability alteration

Table 7.5 Summary of chemical IOR methods in unconventional reservoirs.

Chemical category	Applicability	Motivation	Solution provided	Obstacles
Surfactant	Yes	1. Oil-wet reservoirs 2. Reservoirs with high IFT 3. Reservoirs with negative Pc 4. All reported lab and simulation studies proved an increment in oil recovery	1. Wettability alteration 2. IFT reduction 3. Enhance water imbibition	1. Needs field tests 2. Cost of surfactant 3. Limited penetration depth for surfactant 4. Slow imbibition rate
Alkaline	Not investigated	1. Oil-wet reservoirs 2. Reservoirs with high IFT 3. Reservoirs with negative Pc	1. Wettability alteration 2. IFT 3. reduction 4. Enhance water imbibition	Needs to be investigated
Polymer	Unlikely	Reservoirs with high heterogeneity	Improve sweep efficiency	Small pore throat

7.4 Surfactant huff-n-puff EOR

According to the literature, most studies strongly recommended avoiding the use of the surfactant imbibition process during the continuous flooding mode in unconventional formations. The main concern is that the transport ability of surfactants is too low to allow the surfactant to migrate through the tight porous media if no connected fracture networks are present between the injection and production wells. It is not economical to rely on the diffusion process to displace oil from the injection to the production wells. Wang, Zhang, Butler, & Olatunji (2016) proved that surfactant imbibition cannot proceed more than a few meters into the ultralow-permeability formations of unconventional reservoirs. However, they reported that the surfactant can penetrate deeper when the reservoirs have a high incidence of natural fractures because this increases the contact area between the injected surfactant and the reservoir. They also suggested that the most productive wells during the primary recovery stage are the best candidates for surfactant applications due to the high chance that they have many existing natural fractures. However, the combined injection of low salinity brine with surfactant has been reported as a good method to increase the imbibition rate (Li, Teklu, Abass, & Cui, 2016). Wang et al. (2016) supported this method in an indirect way in which the oil recovery factor was observed to improve very quickly, after just 2 h of the combined injection in their cores, during their experimental work, which had been under a certain value of irreducible water saturation (swi). The other cores, which had swi equal to zero, took 24 h to achieve an improvement in the oil recovery factor using surfactant injection alone. Therefore, injecting low salinity water is a crucial way to enhance the surfactant imbibition rate. To sum up, the surfactant imbibition rate is a challenging problem in these unconventional, ultra-tight plays. However, the high incidence of natural fractures and the combination of LSW with surfactant might solve this problem. All of the aforementioned reasons led the investigators to recommend using surfactant in huff-n-puff mode for unconventional formations.

7.5 Surfactant injection during hydraulic fracturing process

Most of the reported studies investigating the applicability of surfactants in unconventional reservoirs have recommended that a surfactant or any complex nanofluid (CNF) should be used during the fracturing process or in huff-n-puff mode. During fracturing, a surfactant can effectively enhance the permeability of the reservoir oil or gases, assist flowback, and/or enhance oil and gas production from liquid-rich shale formations. Two primary benefits can be achieved by adding surfactant to the fracturing fluids, including the potential for surfactants to initiate microcracks near primary fractures, and enhancing the spontaneous imbibition farther into the formation rocks (Xu et al., 2015a). Kim, Konno, & Dauskardt (2009) explained that a surfactant in aqueous solutions can significantly impact the crack growth rate in organosilicate thin films. The surfactant transports into and begins interacting with the surface of nanoscale confined cracks. Practically, this involves a hydrophilic interaction between hydrophilic functional groups of the surfactant and hydroxide terminal groups of the crack surface. It also involves a hydrophobic interaction between hydrophobic surfactant chains and methyl groups on the crack surface. Furthermore, it may also involve an electrostatic interaction between ionic head groups of the surfactant and the charged crack surface, as illustrated in Fig. 7.2. Xu et al.

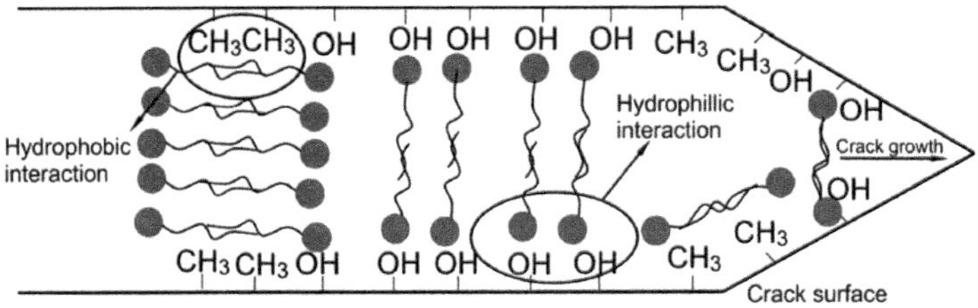

Figure 7.2 ***Interactions between surfactant molecules and organosilicate crack surfaces.*** *(Modified from Xu et al., 2015a).*

(2015a) explained the mechanism by which surfactant can create microcracks in shale-oil reservoirs. It is known that rock surfaces attain positive or negative charges in aqueous solutions. An ionic surfactant could more efficiently reverse the surface charge due to the long-range electrostatic interaction. As explained by Feng and Xu (2015), ionic surfactants appear to be more efficient in the pair interaction and wettability alteration in shale reservoirs. Another distinct operating mechanism for ionic surfactants is that they can gather near a rock surface that has an opposite charge, building up electrical double layers that overlap near the tip of the crack (Fig. 7.3). The overlapping double layers result in the production of a high concentration of charge species, such as ionic surfactants, to aggregate within the crack,

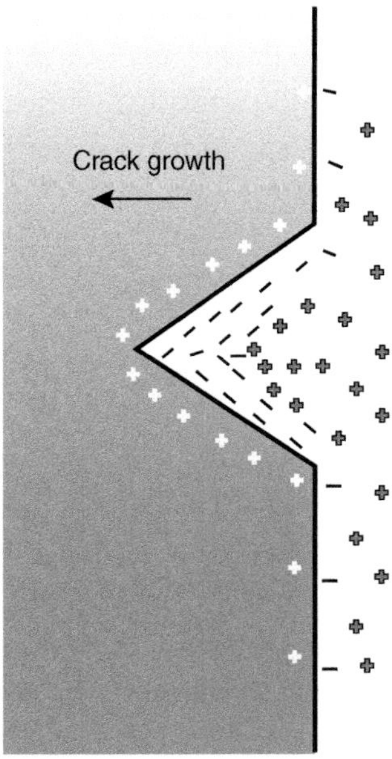

Figure 7.3 ***Double-layer condition inside a crack.*** *(Modified from Xu et al., 2015a).*

effectively lowering the chemical potential of aqueous solutions. The difference in chemical potential will likely result in a differential or osmotic pressure between the fluid in the crack and the rest in the bulk fluid. This ionic surfactant mechanism could activate and propagate more new cracks, particularly if the distance is very small between two crack surfaces, and the significant osmotic pressure can successfully create spontaneous cracking.

The second benefit of adding surfactant to the fracturing fluid is that it enhances the spontaneous imbibition farther into the formation rocks. Xu et al. (2015b) concluded that surfactant can enable spontaneous imbibition of fracturing fluids in liquid-rich shale plays, which could result in additional hydrocarbon recovery. As shown in the CT scans in Fig. 7.4, both anionic and nonionic surfactants were capable of penetrating deeper into the formation rocks than tap water alone (Xu et al., 2015a). There was about an order of magnitude difference, which is significant. The results of Xu et al. (2015b) appear to be consistent with the findings reported by Alvarez et al. (2014), where CT imaging was used to conclude that anionic and nonionic surfactants had a higher initial and total penetration magnitude than fracturing fluids without surfactants (Xu et al., 2015a).

As a result, the initial production can be enhanced, and production declines over time can be mitigated (Xu et al., 2015a). Using surfactant with a longer shut-in time after the fracturing process could enhance the post-fracture production performance (Li et al., 2016). A surfactant additive is

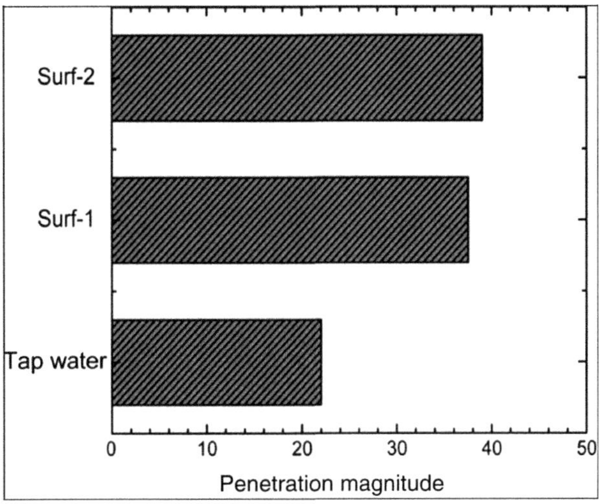

Figure 7.4 ***Penetration magnitude of different aqueous-based fluids with and without surfactant.*** *(Modified from Xu et al., 2015a).*

an important component to be included in fracturing fluids for ensuring higher productivity in unconventional oil and gas formations. Field production results from major liquid-rich shale plays in North America, such as the Barnett, Eagle Ford, Bakken, and Permian Basin, also suggest that the correct application of a surfactant can make a significant difference (He et al., 2015a). Surfactant injection increases oil recovery primarily by reducing the IFT between the aqueous and the oil phases. In addition, the exposure of oil-wet rock surfaces to the surfactant solution may change their wettability, which has proven effective in removing oil/water blocks in the hydraulic fractures and in the near-wellbore matrix (He et al., 2015a).

7.6 Surfactant selection

7.6.1 Introduction of surfactant selection criteria

Experimental results and field observations have indicated clearly that production can be enhanced by spontaneous imbibition by adding surfactants into completion fluids in unconventional liquid reservoirs (ULRs). During the hydraulic fracturing process, the surfactant molecules interact with the rock surface and alter its wettability and IFT (Zhang et al., 2018). The wettability alteration of shale reservoirs from oil-wet to water-wet enables the spontaneous imbibition of water into the matrix, which expels the oil out of the pore space toward the fractures. Understanding surfactant selection for the optimization of chemical EOR in ULRs is critical.

The capillary pressure plays a major role in the spontaneous imbibition process. Both the CA and IFT are also very important terms in the Young-Laplace capillary pressure equation, as well as in the spontaneous imbibition process. Some researchers have attempted to clarify the effect of the CA and IFT on the ultimate recovery by spontaneous imbibition in ULRs (Zhang et al., 2018). They compiled CA, IFT, and spontaneous imbibition experimental data from two of the most prolific shale reservoirs to provide insight into the relationship between the three variables. Based on the observed trends and correlations, a surfactant selection method can be developed to achieve optimal performance in ULRs.

7.6.2 Principles of surfactant selection criteria

Some surfactant selection methods depend primarily on the IFT value or the charge type of the surfactant's head group (Zhang et al., 2018). Having a lower IFT leads to better field performance due to the large imbibition rate observed in low IFT environments compared to high IFT systems

(Li and Horne, 2012). In addition, field tests have reported that low IFT systems can effectively improve oil recovery in conventional reservoirs (Zhang et al., 2018). Cationic surfactants or a mixture of cationic and nonionic surfactants were reported to provide the best oil recovery enhancement in carbonate reservoirs, and anionic surfactants are the most effective EOR agents in sandstone reservoirs (Negin, Ali, & Xie, 2017). However, because the properties of unconventional reservoirs are totally different from those of conventional reservoirs, the selection criteria methodology for surfactant EOR techniques in conventional reservoirs might not work in unconventional oil reservoirs. Therefore, a new selection criterion for the evaluation of surfactant performance in enhancing oil production from shale-oil reservoirs is needed.

Zhang et al. (2018) developed a selection criteria methodology for surfactant imbibition in unconventional reservoirs, compiling over 30 experiments with IFT, CA, and spontaneous imbibition data in their study database. Their data were mainly gathered from the Eagle Ford and Wolfcamp cores. The cores of the Eagle Ford formation are further divided into two groups, Eagle Ford A and Eagle Ford B. The summary of experimental data from all three systems is presented in Table 7.6. The procedure for their selection criteria methodology was as follows:

1. All IFT, CA, and imbibition production curve data were first presented and grouped based on the reservoir region.
2. With all data gathered, the analysis techniques for correlating the IFT and CA data with the recovery factor from the imbibition were presented.
3. The correlation trends provide suggestions on the surfactant selection criteria for achieving optimal production enhancement from the addition of a surfactant.

7.6.3 Data compilation

Fig. 7.5 presents both IFT (left column) and CA (right column) data. DW stands for distilled water, indicating the base cases in which no surfactant was added into the aqueous phase, and Surf indicates cases in which the aqueous phase contained surfactants (Zhang et al., 2018). For simplification, both DW and Surf cases are presented with the nomenclature of numbers (Zhang et al., 2018). Clearly, the addition of surfactant reduced the water-oil IFT. Most of the surfactants presented have IFT values larger than 1 mN/m, which is clearly higher than the surfactant system used for waterflooding in conventional reservoirs (Zhang et al., 2018). For the CA, adding surfactant

Table 7.6 Experimental data summary from all three systems.

		D	L	ϕ	CA	IFT	k	RF	Pc
	Fluid	**inch**	**inch**	**%**	**deg.**	**mN/m**	**nd**	**%**	**psi**
Wolfcamp	DW1	0.99	1.451	6.50	121	21.8	400	7.60	−286.25
	DW2	0.98	1.451	6.50	110.9	21.8	400	7.15	−198.27
	DW3	0.973	1.78	6.80	108.7	21.8	400	10.55	−182.26
	Surf1	0.982	1.702	6.30	62.6	9.8	400	18.37	113.20
	Surf2	0.98	1.593	6.40	57.4	0.4	400	33.95	5.45
	Surf3	0.982	1.536	6.50	56.6	9.8	400	19.73	137.54
	Surf4	0.974	1.96	6.50	55	0.4	400	24.30	5.85
	Surf5	0.974	1.887	6.50	50	3.9	400	32.60	63.91
	Surf6	0.99	2.037	6.80	47	0.9	400	18.90	16.01
	Surf7	0.989	1.777	6.80	39	8.9	400	47.30	180.36
	Surf8	0.979	1.481	6.40	32.4	0.4	400	28.51	8.54
EF-A	DW4	0.995	2.585	6.60	108.17	31.43	200	3.11	−406.48
	Surf9	0.992	2.350	6.60	85.90	15.32	200	6.89	45.43
	Surf10	1.001	2.393	6.60	69.05	10.25	200	19.77	151.99
	Surf11	0.999	2.170	6.60	57.32	7.17	200	16.64	160.56
	Surf12	1.002	2.639	6.60	51.53	5.85	200	24.06	150.93
	Surf13	0.994	2.154	6.60	34.67	17.14	200	29.70	584.63

Table 7.6 Experimental data summary from all three systems. (*Cont.*)

		D	L	ϕ	CA	IFT	k	RF	Pc
	Fluid	**inch**	**inch**	**%**	**deg.**	**mN/m**	**nd**	**%**	**psi**
EF-B	DW5	0.995	2.585	12.00	127	34.4	200	2.10	−1014.21
	DW6	0.997	2.288	12.26	94.7	34.03	200	1.97	−138.07
	DW7	0.999	1.165	12.00	93.2	34.03	200	2.90	−93.06
	Surf14	0.991	1.819	13.17	77.5	7.25	200	3.11	80.53
	Surf15	0.998	1.451	12.00	55	0.7	200	6.50	19.67
	Surf16	1.001	2.214	12.00	50	2.3	200	5.80	72.43
	Surf17	0.994	1.959	13.00	46	1.2	200	4.50	42.50
	Surf18	0.995	2.071	12.00	46.4	2.33	200	7.61	78.72
	Surf19	0.9935	1.953	13.17	41.2	4.48	200	3.60	173.00
	Surf20	0.998	1.453	12.00	40.3	0.22	200	6.84	8.22
	Surf21	0.993	2.028	12.26	39.9	1.76	200	9.32	66.86
	Surf22	0.955	2.0695	13.17	39.2	2.89	200	4.60	114.94
	Surf23	1.002	2.391	12.00	38	6.9	200	9.00	266.37
	Surf24	0.995	1.643	12.00	36.4	1.76	200	10.05	69.40
	Surf25	0.991	2.253	12.26	35.8	0.22	200	6.74	8.84
	Surf26	1.005	2.214	12.26	34.3	2.33	200	7.06	95.31
	Surf27	0.996	2.587	5.86	21.9	17.140	200	18.23	544.58
	Surf28	1.000	2.394	12.18	20.8	6.900	200	14.11	318.32

Source: Modified from Zhang et al. (2018).

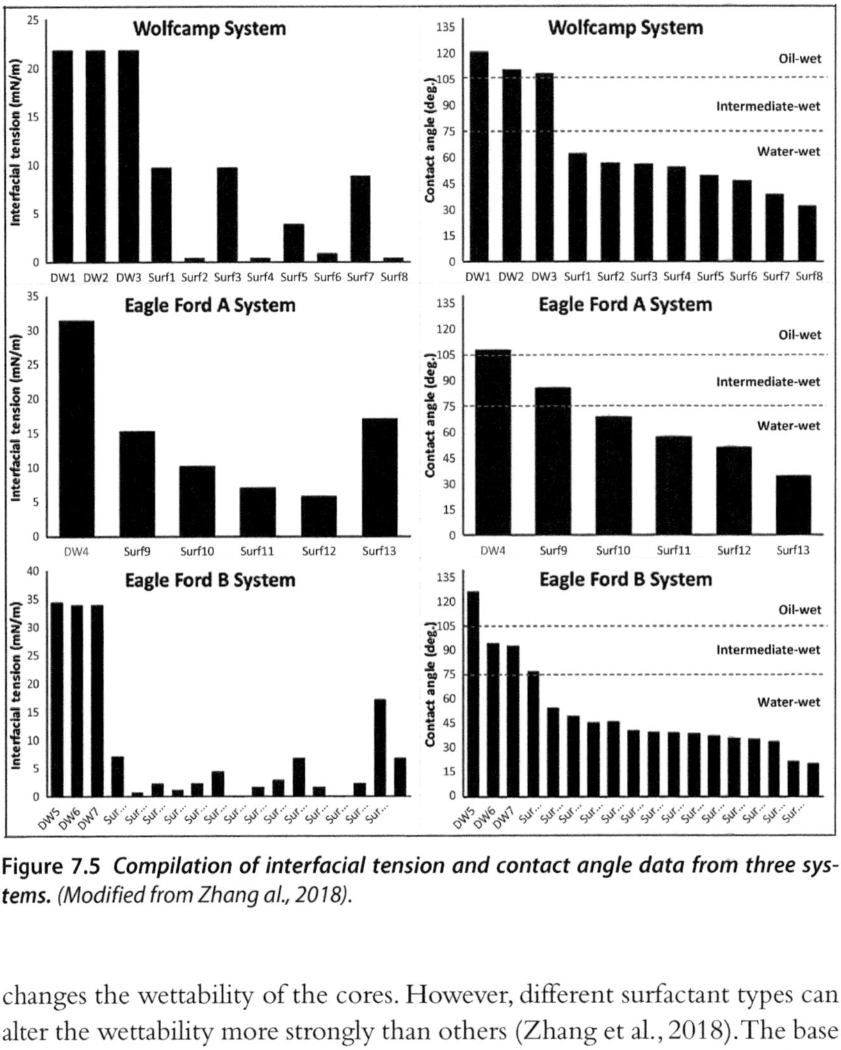

Figure 7.5 ***Compilation of interfacial tension and contact angle data from three systems.*** *(Modified from Zhang al., 2018).*

changes the wettability of the cores. However, different surfactant types can alter the wettability more strongly than others (Zhang et al., 2018). The base cases of shale cores have oil-wet to intermediate-wet wettability (Alfarge, Wei, & Bai, 2018). Therefore, adding surfactant could change their wettability to more water-wet and reduce the IFT, which makes the oil recovery more favorable, as shown in Fig. 7.6.

7.6.4 Functionality of oil recovery factor with CA, IFT, and capillary pressure

Fig. 7.7 illustrates a clear correlation between the CA and the recovery factor. Higher CA cores correlate well with lower oil recovery. The capillary pressure was reported as the driving force of oil production enhancement in

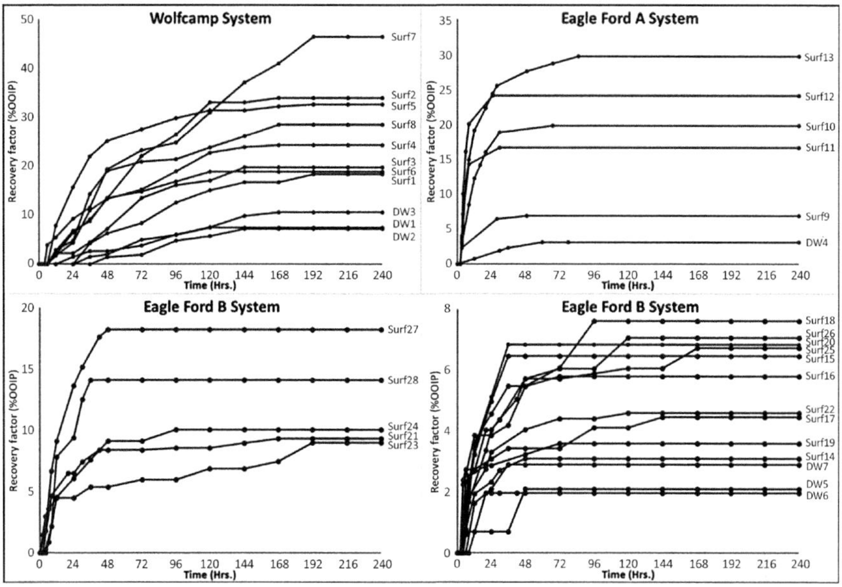

Figure 7.6 ***Production curve from spontaneous imbibition experiments.*** *(Modified from Zhang et al., 2018).*

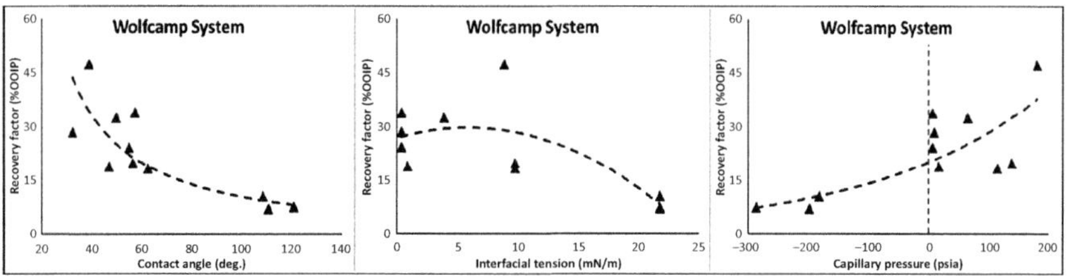

(A) Wolfcamp System Cores

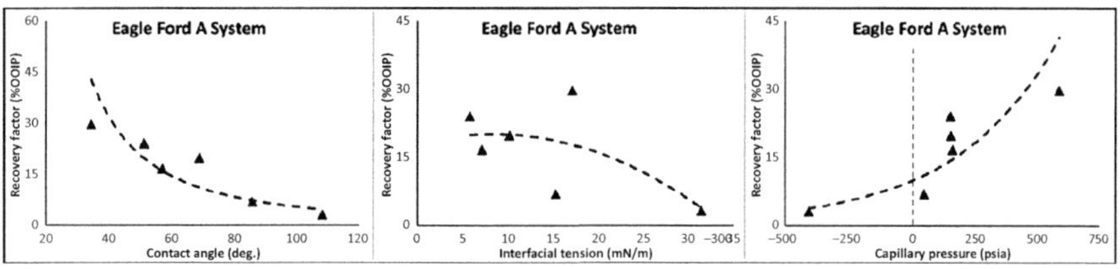

(B) Eagle Ford A System Cores

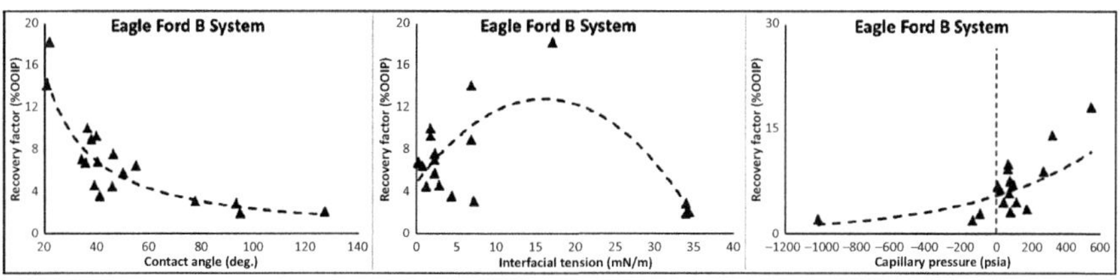

(C) Eagle Ford B System Cores

Figure 7.7 Plots of CA, IFT, and Pc versus recovery factor from (A) Wolfcamp system cores, (B) Eagle Ford A system, and (C) Eagle Ford B system. *(Modified from Zhang et al., 2018).*

the lab imbibition experiments of shale cores, so a positive correlation exists between the oil recovery factor and the Pc, as shown in Fig. 7.7 (Zhang et al., 2018). However, there is no significant trend between the IFT and the RF, as shown in Fig. 7.6 (Zhang et al., 2018).

To sum up, surfactant selection should depend on the factors, which significantly control oil recovery enhancement. For example, according to Zhang et al. (2018), the surfactants which significantly alter the wettability angle in shale cores are better than other surfactant types which reduce the IFT.

7.7 Limitations of surfactant EOR for unconventional reservoirs

Although surfactant imbibition has shown great performance in increasing oil recovery from shale cores, there are still some concerns surrounding the imbibition rate, which might be very slow at the field scale (Sheng, 2015). Furthermore, the surfactant adsorption density might be significantly high in such reservoirs, so these two factors must be considered sufficiently. Wang et al. (2016) investigated the imbibition rate of surfactant and the penetration depth into the matrix by upscaling their lab work to the field scale via numerical simulation methods. They concluded that surfactant cannot be very beneficial in increasing oil recovery if it is injected in reservoirs with only hydraulic fractures due to the limited penetration depth of surfactant into the matrix. However, when unconventional reservoirs have a high incidence of natural fractures, the surfactant technology can be quite promising in increasing oil recovery in such plays due to the increase in the contact area (Wang et al., 2016). However, other researchers have found that the high salinity in these reservoirs might be a big obstacle for applying surfactants because surfactant performance for reducing IFT and wettability alteration would be downgraded under high salinity conditions.

7.8 SUMMARY

1. Chemical EOR methods have been considered efficient oil recovery techniques to recover bypassed oil and residual oil trapped in conventional and unconventional reservoirs.
2. Among chemical EOR methods, surfactant shows the most promise for improving oil recovery in unconventional oil reservoirs.
3. Although reports indicate that surfactant imbibition performs well in increasing oil recovery from shale cores, some concerns remain surrounding the imbibition rate, which could be very slow at the field scale.

4. Surfactant should be used in the huff-n-puff mode during the stimulation process to increase the early flowback production and to make the oil production sustainable for a long time.
5. The main purpose of using surfactant during the fracturing process is to assist flowback, and/or enhance oil and gas production from liquid-rich shale formations.
6. Different types of surfactants require different operating conditions and could perform differently in shale-oil reservoirs based on the rock compositions and other parameters.

References

Agi, A., Junin, R., Syamsul, M. F., Chong, A. S., & Gbadamosi, A. (2018). *Intermittent and short duration ultrasound in a simulated porous medium*. Petroleum. doi: 10.1016/j.petlm.2018.03.012.

Alfarge, D., Wei, M., & Bai, B. (2018). *Data analysis for CO2-EOR in shale-oil reservoirs based on a laboratory database*. Journal of Petroleum Science and Engineering. doi: 10.1016/j.petrol.2017.10.087.

Alvarez, J. O., Neog, A., Jais, A., & Schechter, D. S. (2014). *Impact of surfactants for wettability alteration in stimulation fluids and the potential for surfactant EOR in unconventional liquid reservoirs*. Society of Petroleum Engineers. doi: 10.2118/169001-MS SPE-180270-MS15.

Alvarez, J. O., & Schechter, D. S. (2016a). *Wettability alteration and spontaneous imbibition in unconventional liquid reservoirs by surfactant additives*. Society of Petroleum Engineers. doi: 10.2118/177057-PA.

Alvarez, J. O., & Schechter, D. S. (2016b). *Altering wettability in Bakken shale by surfactant additives and potential of improving oil recovery during injection of completion fluids*. Society of Petroleum Engineers. doi: 10.2118/SPE-179688-MS.

Alvarez, J. O., & Schechter, D. S. (2016c). *Wettability, oil and rock characterization of the most important unconventional liquid reservoirs in the United States and the impact on oil recovery*. Unconventional Resources Technology Conference. doi: 10.15530/URTEC-2016-2461651.

Anderson, W. G. (1986). Wettability literature survey – Part 1: Rock/oil/brine interactions and the effects of core handling on wettability. *Journal of Petroleum Technology, 38*(10), 1125–1144. doi: 10.2118/13932-PA.

Bera, A., Kumar, T., Ojha, K., & Mandal, A. (2013). Adsorption of surfactants on sand surface in enhanced oil recovery: Isotherms, kinetics and thermodynamic studies. *Applied Surface Science, 284*, 87–99. doi: 10.1016/j.apsusc.2013.07.029.

Dawson, M., Nguyen, D., Champion, N., & Li, H. (2015). *Designing an optimized surfactant flood in the Bakken*. Society of Petroleum Engineers. doi: 10.2118/175937-MS.

Feng, L., & Xu, L. (2015). *Implications of shale oil compositions on surfactant efficacy for wettability alteration*. Presented at the SPE Middle East Unconventional Resources Conference and Exhibition, Muscat, Oman, January 26–28. SPE-172974-MS. http://dx.doi.org/10.2118/172974-MS.

Gbadamosi, A. O., Junin, R., Manan, M. A., et al. (2019). An overview of chemical enhanced oil recovery: Recent advances and prospects. *International Nano Letters, 9*, 171–202. doi: 10.1007/s40089-019-0272-8.

Gbadamosi, A. O., Kiwalabye, J., Junin, R., & Augustine, A. (2018). A review of gas enhanced oil recovery schemes used in the North Sea. *Journal of Petroleum Exploration and Production Technology, 5*, 1–15. doi: 10.1007/s13202-018-0451-6.

He et al. (2015a). Evaluation of surfactant performance in fracturing fluids for enhanced well productivity in unconventional reservoirs using rock-on-a-chip approach. http://dx.doi.org/10.1016/j.petrol.2015.10.008.

He, K., & Xu, L. (2015a). *Addressing operator concerns: will surfactant's performance be compromised under harsh conditions?* Presented at EUROPEC 2015, Madrid, Spain, 1–4 June. SPE-174361-MS. http://dx.doi.org/10.2118/174361-MS.

He, K., Yue, Z., Fan, C. et al. (2015b). *Minimizing surfactant adsorption using polyelectrolyte based sacrificial agent: a way to optimize surfactant performance in unconventional formations.* Presented at the SPE International Symposium on Oilfield Chemistry, The Woodlands, Texas, USA, 13–15 April. SPE-173750-MS. http://dx.doi.org/10.2118/173750-MS.

Hoffman, B.T., & Evans, J. (2016). *Improved oil recovery IOR pilot projects in the Bakken formation.* SPE-180270-MS paper presented at the SPE low Perm symposium held in Denver, CO, USA, May 5–6, 2016.

Kathel, P., & Mohanty, K. K. (2013). *EOR in tight oil reservoirs through wettability alteration.* Society of Petroleum Engineers. doi: 10.2118/166281-MS.

Kim, T. -S., Konno, T., & Dauskardt, R. H. (2009). Surfactant-controlled damage evolution during chemical mechanical planarization of nanoporous films. *Acta Materialia, 57,* 4687–4696.

Kumar, K., Dao, E. K., & Mohanty, K. K. (2008). *Atomic force microscopy study of wettabilityalteration by surfactants.* Society of Petroleum Engineers. doi: 10.2118/93009-PA.

Kumar, N., Gaur, T., & Mandal, A. (2017). Characterization of SPN pickering emulsions for application in enhanced oil recovery. *Journal of Industrial and Engineering Chemistry, 54,* 304–315. doi: 10.1016/j.jiec.2017.06.005.

Li, K., & Horne, R. N. (2013). Scaling of spontaneous imbibition in gas-liquid-rock systems. Presented at the SPE/DOE improved oil recovery symposium, Tulsa, OK. ttps://doi.org/10.2118/75167-MS.

Li, X., Teklu, W., Abass, A., Cui, Qi. (2016). *The impact of water salinity /surfactant on spontaneous imbibition through capillarity and osmosis for unconventional IOR.* URTec#2461736 presented at the unconventional resources technology conference held in San Antonio, TX, USA, August 1-3.

Negin, C., Ali, S., & Xie, Q. (2017). Most common surfactants employed in chemical enhanced oil recovery. *Petroleum, 3*(2), 197–211. doi: 10.1016/j.petlm 2016.11.007.

Nguyen, D., Wang, D., Oladapo, A., Zhang, J., Sickorez, J., Butler, R., & Mueller, B. (2014). *Evaluation of surfactants for oil recovery potential in shale reservoirs.* Paper SPE 169085 presented at the SPE improved oil recovery symposium, Tulsa, OK, April 12–16.

Rane, J. P. & Xu, L. (2014). *Monitoring residual surfactant in the flowback and produced water: A way forward to improve well productivity.* Presented at the SPE Saudi Arabia Section Technical Symposium and Exhibition, Al-Khobar, Saudi Arabia, 21–24 April. SPE-172190-MS. http://dx.doi.org/10.2118/172190-MS.

Rane, J.P. & Xu, L. (2015a). *Proper selection of surfactants for enhanced oil recovery applications based on interfacial rheology measurements of oil-water interfaces.* Presented at the SPE International Symposium on Oilfield Chemistry, The Woodlands, Texas, USA, 13–15 April. SPE-173767-MS. http://dx.doi.org/10.2118/173767-MS.

Rane, J. P. & Xu, L. (2015b). *New dynamic-surface-tension analysis yields improved residual surfactant measurements in flowback and produced waters.* SPE Prod & Oper Preprint. http://dx.doi.org/10.2118/172190-PA.

Samanta, A., Bera, A., Ojha, K., & Mandal, A. (2012). Comparative studies on enhanced oil recovery by alkali–surfactant and polymer flooding. *Journal of Petroleum Exploration and Production Technology, 2,* 67–74. doi: 10.1007/s13202-012-0021-2.

Samin, A. M., Manan, M. A., Idris, A. K., Yekeen, N., Said, M., & Alghol, A. (2017). Protein foam application for enhanced oil recovery. *Journal of Dispersion Science and Technology, 38,* 604–609. doi: 10.1080/01932691.2016.1185014.

Sakthipriya, N., Doble, M., & Sangwai, J. S. (2015). Enhanced oil recovery techniques for Indian reservoirs. In S. Mukherjee (Ed.), *Petroleum Geosciences: Indian Contexts*Springer, Cham: Springer Geology. doi: 10.1007/978-3-319-03119-4_11.

Sheng, J. J. (2015). Enhanced oil recovery in shale reservoirs by gas injection. *Journal of Natural Gas Science and Engineering*, *22*, 252–259. doi: 10.1016/j.jngse. 2014.12.002.

Shuler, P., Tang, H., Zayne, Lu., & Tang, Y. (2001). *Chemical processes for improved oil recovery from Bakken shale paper SPE 147531.* Presented at the Canadian unconventional research conference Calgary, Alberta, November 15–17.

Wang, D., Butler, R., Liu, H., et al. (2011). Flow rate behavior and imbibition in shale. *SPE Reservoir Evaluation & Engineering*, *14*(4), 505–512. doi: 10.2118/138521-PA, SPE-138521-PA.

Wang, D., Butler, R., Zhang, J., et al. (2012). Wettability survey in Bakken shale with surfactant-formulation imbibition. *SPE Reservoir Evaluation & Engineering*, *15*(6), 695–705. doi: 10.2118/153853-PA, SPE-153853-PA.

Wang, D., Zhang, J., Butler, R., & Olatunji, K. (2016). *Scaling laboratory-data surfactant-imbibition rates to the field in fractured-shale formations.* Society of Petroleum Engineers. doi: 10.2118/178489-PA.

Xu et al. (2015a). *Insights into surfactant containing fracturing fluids inducing microcracks and spontaneously imbibing in shale rocks.* SPE-175959-MS presented at the SPE/CSUR unconventional resources conference held in Calgary, Alberta, Canada, October 20–22, 2015.

Xu, L., He, K., Rane, J. P. et al. (2015b). *Spontaneously imbibed fluids for increasing contact area between hydraulic fracturing fluids and formation matrix in liquids-rich shale plays.* Presented at the SPE liquids-rich basins conference, Midland, TX, USA, September 2–3.

Xu, L., He, K., Rane, J. P., Yin, X., & Neeves, K. (2015c). Spontaneously imbibed fluids for increasing contact area between hydraulic fracturing fluids and formation matrix in liquids-rich shale plays. *Society of Petroleum Engineers.* doi: 10.2118/175536-MS.

Zaitoun, A., Makakou, P., Blin, N., Al-Maamari, R. S., Al-Hashmi, A. -A. R., & Abdel-Goad, M. (2012). Shear stability of EOR polymers SPE-141113-PA. *SPE Journal*, *17*, 335–339. doi: 10.2118/141113-pa.

Zhang et al. (2018). *Scaling for wettability alteration induced by the addition of surfactants in completion fluids: surfactant selection for optimum performance.* URTeC: 2889308 presented at the unconventional resources technology conference held in Houston, TX, USA, July 23–25.

CHAPTER 8

Selection criteria for miscible gases-based EOR in unconventional liquid-rich reservoirs (ULR)

Contents

8.1 Introduction 166
8.2 Significance of molecular diffusion 168
8.3 Numerical simulation setup 169
  8.3.1 Model selection 169
  8.3.2 Compositional model 171
  8.3.3 Systematic methodology for diffusivity level 172
8.4 Results and discussion 174
  8.4.1 Natural depletion 174
  8.4.2 Flow-type determination 175
  8.4.3 EOR stage for Bakken model 176
8.5 Closing remarks 182
8.6 Summary 182
References 183

Abstract

In this chapter, a two-step methodology is introduced to investigate the feasibility of using three different miscible gases in EOR from unconventional reservoirs: CO_2, lean gases, and rich gases. The integrated procedure presented in this chapter includes reservoir simulation and history matching of some miscible-gases used in EOR projects. The objective is to provide a proper evaluation of the molecular diffusivity of such gases in the field level of unconventional liquid-rich reservoirs (ULR). First, numerical simulation methods of compositional models are introduced, which are incorporated into the local grid refinement of hydraulic fractures. This chapter also presents an analysis of diffusion effects using an LS-LR-DK (logarithmically spaced, locally refined, and dual permeability) model. Second, the history matching process is conducted by incorporating different molar-diffusivity rates to find the correct diffusivity level in the field scale by matching the simulation models with the EOR pilot tests conducted in the Bakken formation of North Dakota, Montana, and South Saskatchewan. The main purpose of this chapter is to provide selection criteria for the proper type of miscible gases in terms of EOR in ULR.

Keywords: selection criteria for miscible gases-based EOR in ULR; natural gases-based EOR; CO_2-EOR; EOR methods simulation; shale oil reservoir simulation

Fundamentals of Enhanced Oil Recovery Methods for Unconventional Oil Reservoirs
http://dx.doi.org/10.1016/B978-0-12-818343-4.00008-5

8.1 Introduction

Miscible-gas injection has shown excellent results in conventional reservoirs with low permeability and light oils. Extending this approach to unconventional reservoirs, including shale oil reservoirs in North America has been extensively investigated over the last decade. The gases which have been investigated are CO_2, N_2, and natural gases. However, most studies have focused on CO_2 for different reasons. CO_2 can easily dissolve in shale oil, but it also causes the oil to swell and lowers its viscosity. Also, CO_2 has a lower miscibility pressure when used with shale oil rather than other gases such as N_2 and CH_4. Most experimental studies have reported that the molecular-diffusion mechanism for CO_2 is behind the increment in oil recovery obtained at lab scale (Alfarge, Wei, & Bai, 2017). Furthermore, most previous simulation studies have relied on the lab-diffusivity level for these miscible gases to predict the expected oil increment at the field scale (Alfarge et al., 2017).

Starting with lab-work tools, conducted an experimental investigation to compare results from injecting CO_2 and water in Bakken cores. They found that waterflooding enhanced oil recovery better than immiscible CO_2 in the huff-n-puff protocol. However, miscible and near-miscible CO_2 huff-n-puff tests performed better than waterflooding in enhancing oil recovery. Hawthorne et al. (2013) investigated the mechanisms behind increasing oil recovery by injecting CO_2 in Bakken cores. They proved that the diffusion mechanism is the main mechanism for CO_2 to increase oil recovery in these complex shale plays. However, extracting oil from a shale matrix by CO_2 requires long exposure times combined with large contact areas. Conducted experimental work on shale cores from the Mancos and Eagle Ford test sites to investigate the potential of CO_2 injection in these reservoirs. Their laboratory results indicated that cyclic CO_2 injection can improve oil recovery in shale oil cores by 33%–85% depending on the shale core type and other operating parameters. Compared the injection performances of different types of gases, such as CO_2, C_1-C_2 mixtures, and N_2, based on their experimental enhancement of oil recovery in Bakken cores. They concluded that injecting gas composed of C_1, C_2, C_3, and C_4, could produce nearly as much oil as a CO_2 injection, which was 90% in several Middle Bakken cores and nearly 40% in Lower Bakken cores. Also found that the counter-current mechanism is the main mechanism for these gases to recover oil in shale cores. Finally, Yu, Lashgari, Sepehrnoori (2014) experimentally investigated the N_2 flooding process on Eagle Ford core plugs saturated with dead oil. They examined the effect of different flooding durations and different injection pressures on N_2 flooding performance. They found that more oil was produced with a longer flooding time and higher injection pressure. To sum up, CO_2 showed a good potential to extract oil from shale cores in experimental works. Found miscible and near-miscible CO_2 huff-n-puff tests to be the best way to use CO_2 for EOR while and determined that CO_2 was the preferred gas for extracting oil from shale cores based in their experiments. However, finding the best methods and mechanisms are seemingly the key to getting the best results, and those results vary depending on the different gases used.

The numerical simulation studies of Shuaib and Wang with their coworkers (2009,2010) are among the earliest published studies in this area. Those models showed that 10%–20% of incremental oil could be recovered by continuous gas flooding while 5%–10% could be recovered by huff-n-puff gas protocol (Hoffman and Evans, 2016). Dong & Hoffman (2013) reported a numerical study evaluating CO_2 injection performance in the Bakken interval of a sector on Sanish Field in Mountrail County, North Dakota. They were able to increase the CO_2 injectivity in that field by drilling more horizontal injection wells. Their strategy predicted the possibility of injecting 5000 Mscf/day at a maximum injection pressure of 8000 psi. In their simulation study, they found that using the CO_2 injection method could increase oil recovery from 5% to 24% in that field. Xu and Hoffman (2013) evaluated the reservoir performance of the Elm Coulee field in Eastern Montana under CO_2 flooding with different hydraulic fracture orientations. They found that transverse fractures could produce a higher oil recovery factor, but these transverse fractures had a lower utilization value than longitudinal fractures due to breakthrough problems. Constructed a model in which gas could be injected into a hydraulic fracture along a horizontal well, and the production process could occur in an adjacent fracture, which intersected within the same well. They found a substantial improvement in oil recovery by injecting CO_2 in reservoirs with fluid flow from fracture to fracture. Introduced a new model, which considers capillarity and adsorption effect of the tiny pores for shale reservoirs. They found that using this model would properly simulate CO_2-EOR in unconventional reservoirs. Furthermore, capillarity consideration in the modeling process can predict higher oil recovery by CO_2 injection than in models, which do not include the capillarity property.

This chapter is dedicated to produce selection criteria for miscible gas enhanced oil recovery (EOR) in shale oil reservoirs based on numerical simulation methods. A detailed study for determining the level of CO_2 diffusivity in real-field conditions has also been conducted in this work. A comparison of CO_2 performance with lean gas and rich gas is presented based on the authors' investigations and aim to clarify the flow and recovery mechanisms of different gases in shale-reservoirs.

8.2 Significance of molecular diffusion

Gravity drainage, physical diffusion, viscous flow, and capillary forces are the common forces, which control fluid flow in porous media. However, one force might eliminate the contributions of others depending on the reservoir properties and operating conditions.

Molecular diffusion is defined as the movement of molecules caused by Brownian motion or composition gradients in a mixture of fluids (Mohebbinia & Wong, 2017). This type of flow would be the most dominant flow in fractured reservoirs with a low-permeability matrix when gravitational drainage is inefficient (Moortgat & Firoozabadi, 2013; Mohebbinia & Wong, 2017). Moortgat and Firoozabadi (2013) proved that gas injection is greatly affected by molecular diffusion, among other EOR processes. In simulation investigation, ignoring or specifying an incorrect diffusion rate can lead to an incorrect estimate of oil recovery from gas injection. This inaccuracy is not only due to the variations in the miscibility process between the injected gas and formation oil but also due to the change of flow path for the injected gas species from fractures to the formation matrix. This change in the fractures to matrix path can strengthen the penetration of injected gases into to the tight matrix, resulting in more gas diffused through shale oils. Measuring the matrix effect adds to the oil production. Concentrating solely on the injected gas species in fractures eliminates the ability to obtain an accurate oil recovery total.

The Péclet (*Pe*) number is a class of dimensionless numbers, which have been used to measure the relative importance of the molecular diffusion flow to the convection flow. This number can be calculated as shown in Eq. (8.1). If the *Pe* number is less than 1, diffusion is the dominant flow. However, if the *Pe* is greater than 50, convection is the dominant flow. The dispersion flow is dominant when *Pe* is in range of 1 to 50 (Hoteit & Firoozabadi, 2009). Fig. 8.1 explains the flow regimes according to *Pe* number cutoffs.

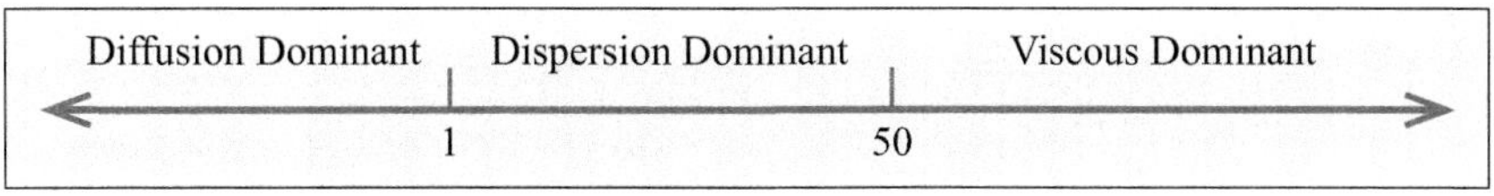

Figure 8.1 ***Flow regimes according to Péclet number cutoffs.***

$$\mathrm{Pe} = \frac{\text{Diffusion time}}{\text{Convection time}} = \frac{L^2/D}{L/v} = \frac{Lv}{D} \tag{8.1}$$

where v is the bulk velocity, L is a characteristic length, and D is the diffusion coefficient.

8.3 Numerical simulation setup

8.3.1 Model selection

To simulate the diffusion process, the majority of previous studies used a single-porosity model that requires a tremendous grid refinement to represent an intensely fractured shale-oil reservoir (Wan & Sheng, 2015). In our study, the logarithmically spaced, locally refined, and dual permeability (LS-LR-DK) model was used due to its ability to better capture the physics of the fluids flow in fractured tight reservoirs. An advanced general equation-of-state (EOS) compositional simulator was used to build an EOS model for Bakken oil and to simulate the compositional effects of reservoir fluid during the primary and EOR processes. This chapter also presents our simulation of the countercurrent mechanism of molecular diffusion for CO_2-EOR, which was introduced in the experimental work of Hawthorne et al. (2013). We tried to build a numerical model, which has the typical fluid and rock properties of a Bakken formation, one of the most productive unconventional EOR formations in the United States. In this model, three different EOR-miscible gases were injected, including CO_2, lean gas, and rich gas in separate steps of the huff-n-puff protocol through hydraulically fractured wells. All the mechanisms summarized in Table 2.1 have also been considered in simulation studies. In this field case study, the production well was stimulated with five hydraulic fractures with spacing between them of 200 ft. The simulation model includes two regions, the stimulated reservoir volume (SRV) and unstimulated reservoir volume (USRV), as shown in Fig. 8.2.

The dimensions of the reservoir model are 2000 ft × 2000 ft × 42 ft, which corresponds to length, width, and thickness, respectively. The

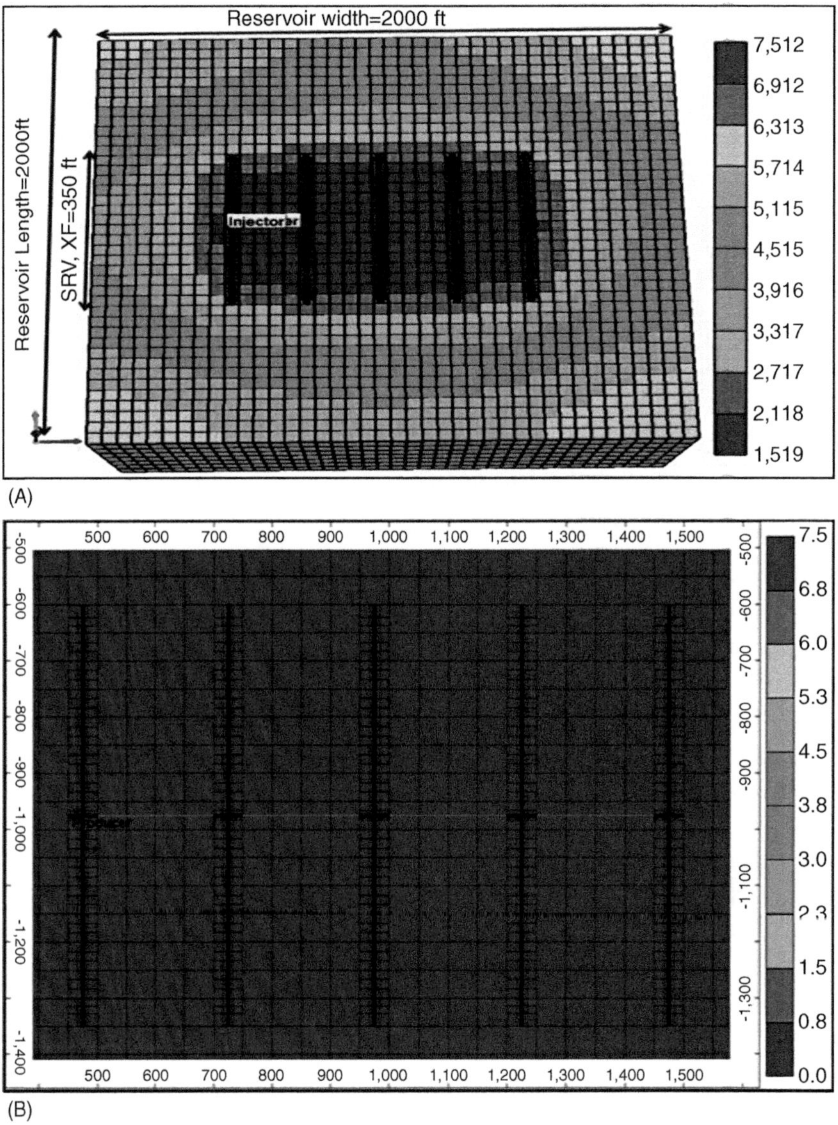

Figure 8.2 (A) Average pressure in a depleted well and (B) a closed view for stimulated reservoir volume.

dimensions of the hydraulically fractured region are five fractures with a half-length of 350 ft in the J direction, a width of 0.001 ft in the I direction, and a fracture height of 42 ft in the K direction. Fracture conductivity is 15 md ft. The other model input parameters are given in Table 8.1.

Table 8.1 Model input parameters for the base case.

Parameter	Value	Unit
Model dimensions	2000 × 2000 × 42	ft
Production time	20	year
Reservoir top	8000	ft
Reservoir temperature	240	°F
Reservoir pressure	7500	psi
Initial water saturation	0.3	value
Total compressibility	1×10^{-6}	psi^{-1}
Matrix permeability	0.005	mD
Matrix porosity	0.085	value
Horizontal well length	1000	ft
Total number of fractures	5	value
Fracture conductivity	15	mD-ft
Fracture half-length	250	ft
Fracture height	42	ft

8.3.2 Compositional model

The typical Bakken oil was simulated with 42 API°, 725 SCF/STB, and 1850 psi as oil gravity, gas oil ratio, and bubble point pressure, respectively. It is well known that compositional models are the most time-consuming models due to the number of components in a typical reservoir oil. This model has 34 components, so it would take a long time for the simulator to complete running one scenario. The common practice is to carefully lump the reservoir oil components into a short representative list of pseudo-components. These pseudo-components would be acceptable if they match the laboratory-measured phase behavior data. The supplied data for the reservoir oil must have a description of associated single carbon numbers and their fractions, saturation pressure test results, separator results, constant composition expansion test results, differential liberation test results, and swelling test results. All the available data can be used for tuning the EOS to match the actual fluid behavior.

The simulated Bakken oil was included in the seven pseudo-components using WinProp-CMG, as presented in Table 8.2. WinProp is an EOS-based fluid behavior and pressure-volume-temperature (PVT) modeling package. In WinProp, laboratory data for fluids can be imported and an EOS can be tuned to match its physical behavior. Fluid interactions can then be predicted, and a fluid model can be created. Table 8.3 presents the Peng–Robinson EOS fluid description and binary interaction coefficients

Table 8.2 Compositional data for the Peng–Robinson EOS in the model oil.

Component	Mole fraction	Critical pressure (atm)	Critical temperature (K)	Acentric factor	Molar weight (g/mol)
CO_2	0	7.28E + 01	3.04E + 02	0.225	4.40E + 01
N_2–CH_4	0.2704	4.52E + 01	1.90E + 02	0.0084	1.62E + 01
C_2H–NC_4	0.2563	4.35E + 01	4.12E + 02	0.1481	4.48E + 01
IC_5–CO_7	0.127	3.77E + 01	5.57E + 02	0.2486	8.35E + 01
CO_8–C_{12}	0.2215	3.10E + 01	6.68E + 02	0.3279	1.21E + 02
C_{13}–C_{19}	0.074	1.93E + 01	6.74E + 02	0.5672	2.20E + 02
C_{20}–C_{30}	0.0508	1.54E + 01	7.92E + 02	0.9422	3.22E + 02

Table 8.3 Binary interaction coefficients for Bakken oil.

Component	CO_2	N_2-CH_4	C_2H-NC_4	IC_5-CO_7	CO_8-C_{12}	C_{13}-C_{19}
CO_2						
N_2–CH_4	1.01E − 01					
C_2H–NC_4	1.32E − 01	1.30E − 02				
IC_5–CO_7	1.42E − 01	3.58E − 02	5.90E − 03			
CO_8–C_{12}	1.50E − 01	5.61E − 02	1.60E − 02	2.50E − 03		
C_{13}–C_{19}	1.50E − 01	9.76E − 02	4.24E − 02	1.72E − 02	6.70E − 03	
C_{20}–C_{30}	1.50E − 01	1.45E − 01	7.79E − 02	4.27E − 02	2.51E − 02	6.00E − 03

of the Bakken crude oil with different gases. Fig. 8.3 represents the two-phase envelope for Bakken oil, which was generated by WinProp-CMG.

8.3.3 Systematic methodology for diffusivity level

To determine the reasons causing the gap in the CO_2 performance in lab conditions versus what happened in the pilot tests and field scale conditions, we need to start with screening the parameters which we are sure of and the parameters which have some ambiguity. The parameters which are known in these pilot-tests are the following:

- PVT data, and oil composition properties
- Major wells and stimulation criteria
- Rock properties
- CO_2 injection operating parameters (rate, pressure, time)

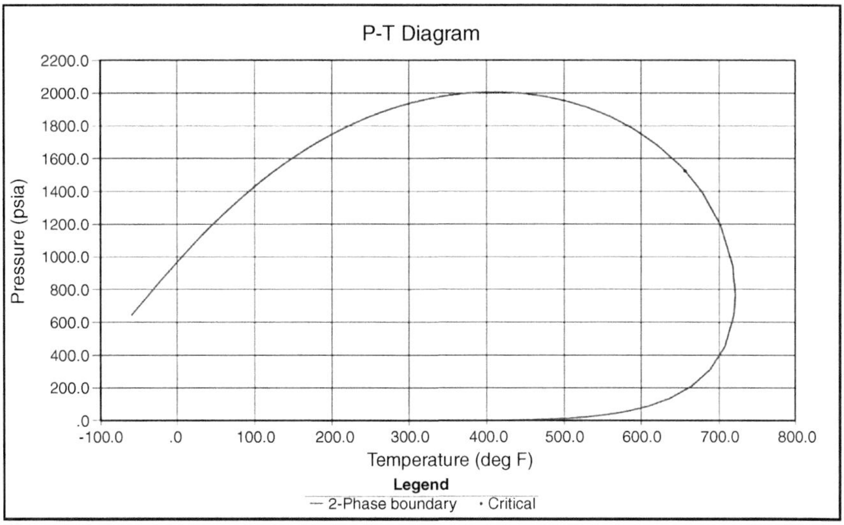

Figure 8.3 ***The two-phase envelope for Bakken oil.***

- Produced oil rate versus time
- CO_2 performance in lab conditions (most of the mechanisms and observations)

The most important parameters, which have an ambiguity in these pilot-tests, follow:

- CO_2 diffusivity level in the field scale
- Natural fracture intensity, porosity, and permeability in the field scale

According to this diagnosis, the systematic methodology for this work could be categorized under two reversed scenarios. First, a numerical simulation study was conducted to upscale the reported outcomes of experimental studies to the field conditions. As a result, different forward diagnostic plots were generated from different combinations of CO_2 physical mechanisms with different shale-reservoirs conditions. Second, different backward diagnostic plots were produced from a historical match with CO_2 performances in field pilot projects, which were performed in a Bakken formation, which spread across North Dakota and Montana. Finally, we fit the backward with the forward diagnostic plots to report and diagnose some findings regarding the performance of the injected CO_2 on the field scale. Fig. 8.4 shows the detailed methodology for this study. This study is the first numerical simulation study to integrate the two reversed approaches in this area of research. All of the previous numerical simulation studies either chose the direct upscale for the lab observation or built a merely conceptual model.

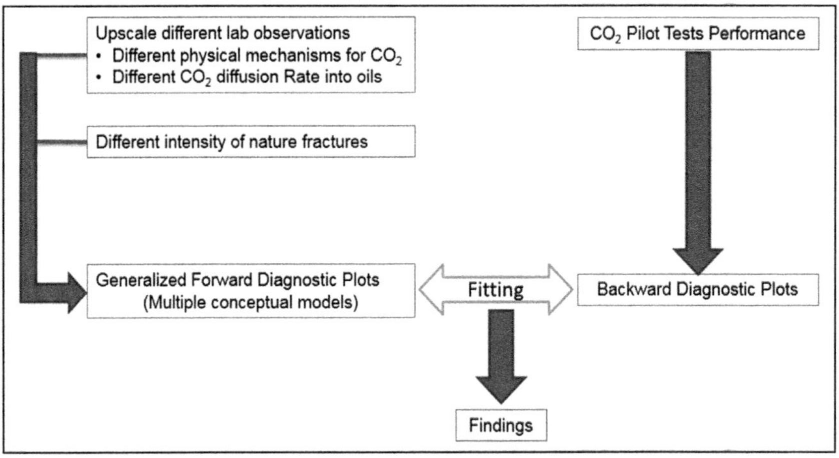

Figure 8.4 ***Systematic methodology to determining the diffusivity level.***

8.4 Results and discussion

8.4.1 Natural depletion

The reservoir model was initially run in a natural depletion mode for 7300 days (20 years). The production well, which was hydraulically fractured, was subjected to the minimum bottom-hole pressure of 1500 psi. The simulated Bakken well performance in natural depletion is shown in Fig. 8.5. In the natural depletion scenario, it was clear that the production well started with a high production rate initially. Then, it showed a steep

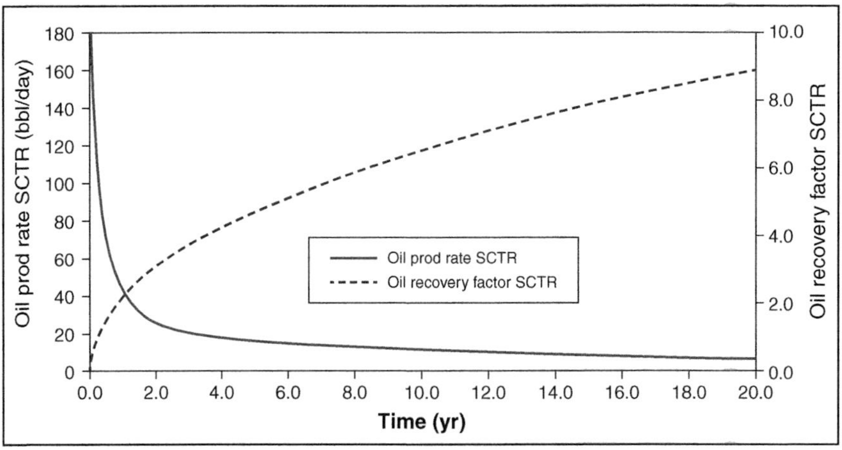

Figure 8.5 ***Reservoir performance under natural depletion conditions.***

decline rate until it got leveled off at a low rate. This is typical of what is happening in most, if not all ULRs of North America. If the reader investigates the pressure distribution in the reservoir model as shown in Fig. 8.1, the reader can find that the main reason for that fast reduction in production rate is due to the pressure depletion in areas that were closed to the production well. However, the reservoir pressure is still high in the areas that are far from the production well. This is due to the low-pressure maintenance capability of the low permeability formations during the production.

8.4.2 Flow-type determination

From the *Pe* number distribution in each grid, in the formation matrix areas, the results indicated that the *Pe* number is way below 1 for both the gas phase and oil phase, which means that the diffusion flow is the most dominant flow in the formation matrix, as shown in Fig. 8.6. However, in the hydraulic fracture parts, the viscous flow is clearly dominated where *Pe* is way above 100. In the natural fractures areas, the results indicated that the *Pe* number is significantly changeable where it is less than 1 in the areas that are far from hydraulic fractures; however, it is greater than 100 in the areas that are close to the hydraulic fractures, as shown in Fig. 8.7. According to the average value of the *Pe* number in natural fractures areas, the dispersion flow could be the most dominant.

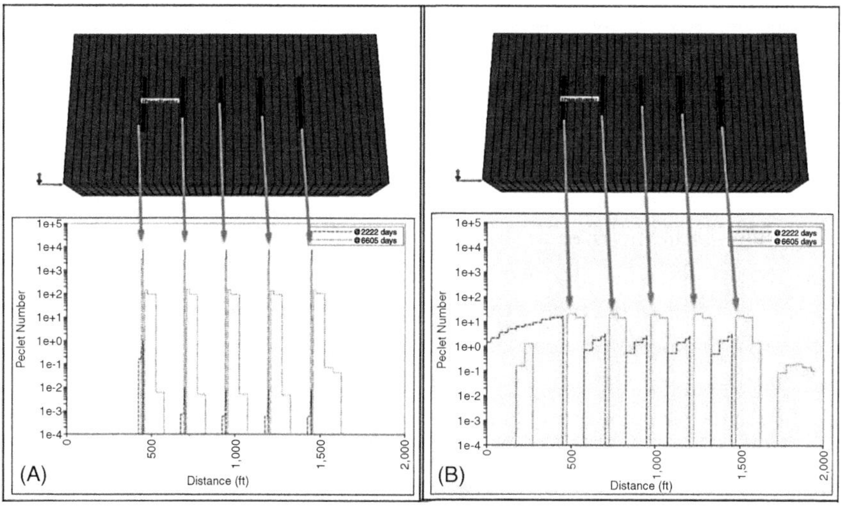

Figure 8.6 ***Péclet number distribution along the cross section in the matrix model.*** (A) Represents the gas phase, and (B) represents the oil phase.

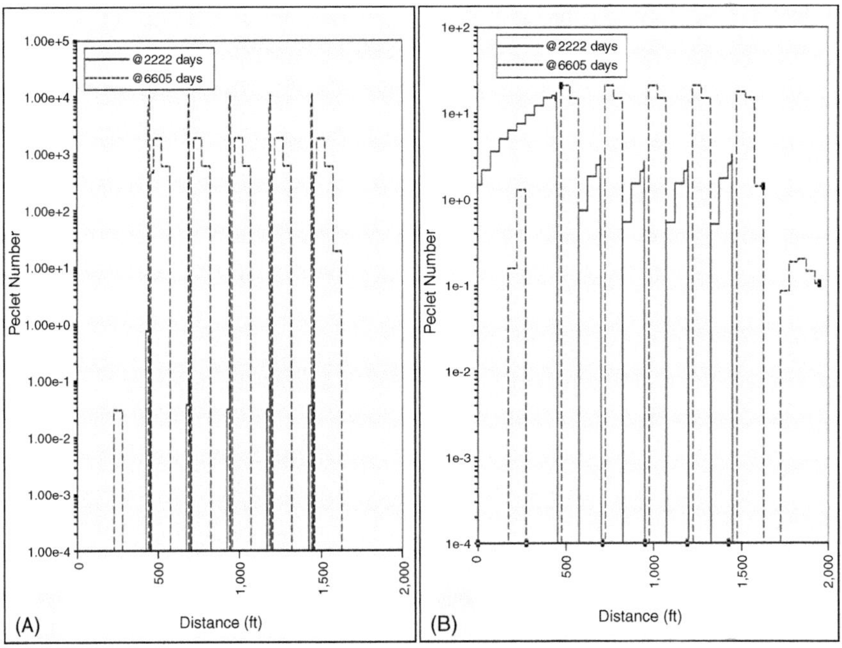

Figure 8.7 ***Péclet number distribution in the natural fracture model (A-gas phase while B-oil phase).***

8.4.3 EOR stage for Bakken model

In the EOR stage, we conducted a simulation investigation on injected CO_2, lean gas, and rich gas in a Bakken production well as a huff-n-puff protocol for eight cases, as listed in Table 8.4. The molecular-diffusion mechanism was switched on to attain some scenarios and switched off to others to investigate its effect for improving oil recovery in the simulated cases. The EOR stage started after 10 years of natural depletion. Lean gas is 90% C1 and 10% C2+, and rich gas is 65% C1 and 35% C2+.

Results indicate that as long as the molar-diffusion mechanism is switched on, the CO_2 performance will exceed the performance of both the lean gas and rich gas, as shown in Fig. 8.8. For example, in Case 1, we notice that the performance of miscible gases from the best to the worst as CO_2, lean gas, and rich gas, in that descending order. This happens due to the difference in the concentration gradient between the injected fluid and the formation fluid, according to Eq. (8.2). The concentration gradient is especially significant for CO_2; however, it is low for both lean gas and rich gas. This is particularly true for shale oil because the composition of shale oil usually contains high concentrations of light components (i.e., C1 and C2).

Table 8.4 The agenda and time breakdown for all simulated cases.

Case number	Number of cycles/10 years	Injected volumes/cycle	Soaking period	Diffusion mechanism
Case 1	10	Small (500 Mscf/day for 2 months)	1 month	ON
Case 2	10	Small (500 Mscf/day for 2 months)	1 month	OFF
Case 3	2	Small (500 Mscf/day for 6 months)	3 months	ON
Case 4	2	Small (500 Mscf/day for 6 months)	3 months	OFF
Case 5	2	Large (1500 Mscf/ day for 6 months)	3 months	OFF
Case 6	10	Large (1500 Mscf/ day for 2 months)	1 month	OFF
Case 7	2	Large (1500 Mscf/ day for 6 months)	3 months	ON
Case 8	10	Large (1500 Mscf/ day for 2 months)	1 month	ON

If we compare Case 1 and Case 3, we find that lean gas performance exceeds rich gas performance in Case 1, buy the reverse is true in Case 3, as shown in Fig. 8.8. This happens due to the difference in both molecular weight and concentration gradient between lean gas and rich gas. Because rich gas has a higher molecular weight than lean gas, it needs a longer

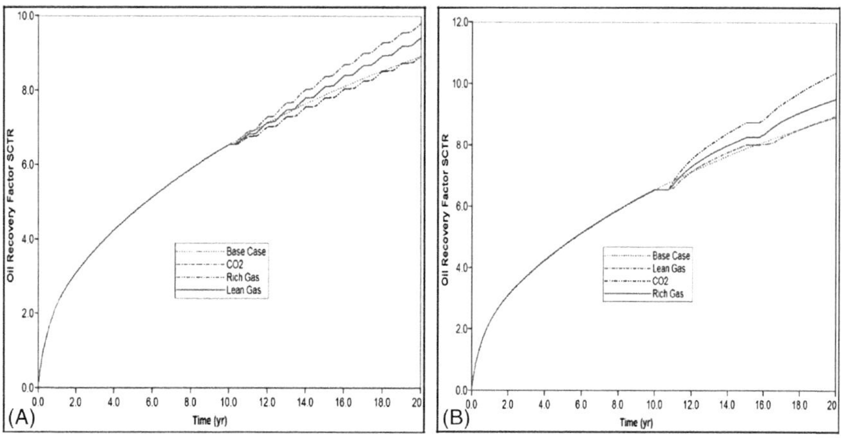

Figure 8.8 *Miscible gases performance (molar-diffusion mechanism is ON) (A represents Case1, and B represents Case 3).*

soaking period to invade the formation oil. These results are very consistent with the results reported by Hoteit and Firoozabadi (2009). In their model, which has been applied in conventional fractured reservoirs, they observed that methane would perform better than CO_2 in cases that do not consider the molecular-diffusion mechanism. However, the injected CO_2 would result in a higher increment for oil recovery in cases that use the molecular-diffusion mechanism, which is expressed as:

$$\text{Rate of diffusion} = \frac{C_D \times (C_1 - C_2) \times Ac}{tc} \tag{8.2}$$

where C_D is the molecular diffusion rate (0.0008–0.0004 cm^2/s was specified in this model); (C_1 − C_2) is the component concentration difference between the injected fluid and the target fluid; Ac is the contact area between the injected fluid and the target fluid, and tc is the separation distance between the injected fluid and the target fluid.

When the molar diffusion mechanism is switched off, the CO_2 performance is the worst compared with lean gas and rich gas. This happens mainly due to the larger molecules of CO_2 compared with those of lean gas and rich gas. CO_2 will not penetrate into the matrix far from hydraulic fractures if the molecular-diffusion rate is low, based on the principles guiding Eq. (8.2) and as shown in Fig. 8.9. However, the lean gas and rich gas penetrate deeper into matrix as compared to what happens during CO_2

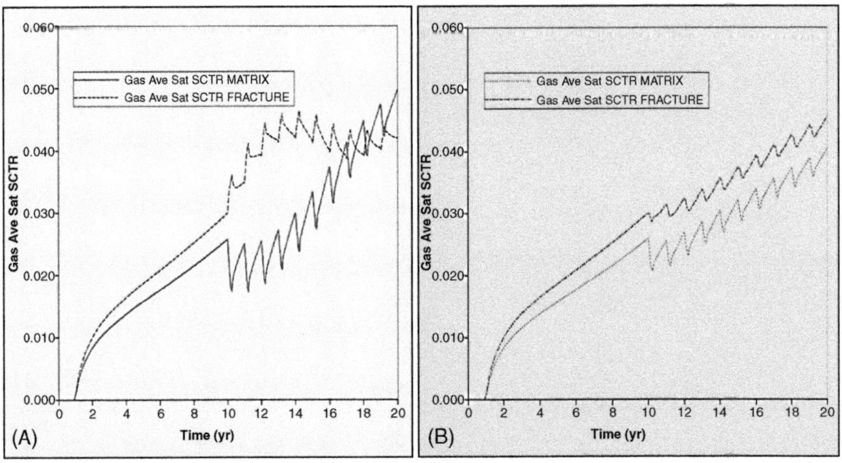

Figure 8.9 ***Gas saturation in matrix versus fracture (molar-diffusion is OFF).*** (A) Represents lean gas, and (B) represents CO_2.

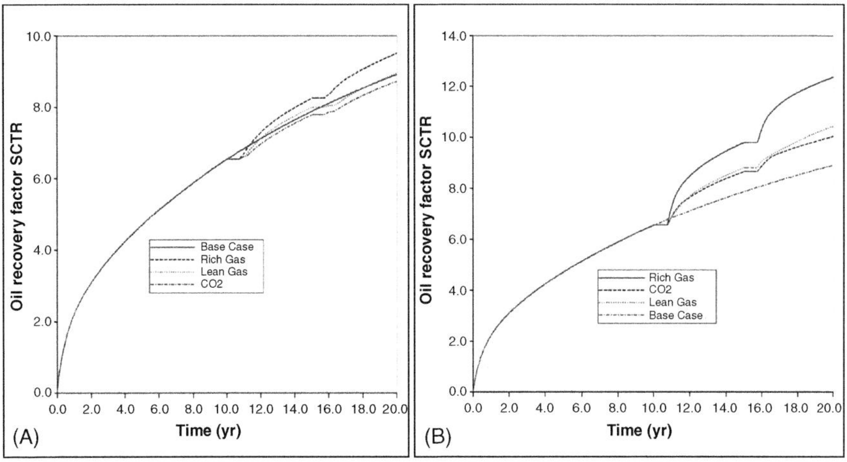

Figure 8.10 ***Miscible gas performance (molar-diffusion mechanism is OFF).*** (A) Represents Case 4, and (B) represents Case 5).

injection. This happens in Case 2, Case 4, Case 5, and Case 6. In the cases which have the molecular-diffusion mechanism switched off, lean gas and rich gas are alternatively in the lead for the best performance gases. The authors noticed that rich gas has a very strong soaking period function, whereas rich gas is in the lead for both Case 4 and Case 5 as shown in Fig. 8.10.

However, the authors noticed that lean gas has very strong injected volume function, whereas lean gas is in the lead, as shown in Case 6. This happens mainly in these shale reservoirs due to two main reasons. First, shale oil has always had a high content of light components making it more compatible with the rich gas performance for improving oil recovery as opposed to the lean gas performance. Second, rich gas has a larger molecular weight than lean gas; therefore, this enhances lean gas performance over rich gas performance. We assert that these two reasons are causing the rich gas to have a stronger function during the soaking period and the lean gas to have a stronger function of injected volumes. Table 8.5 lists the performance ranking for all three gases. Also, the performance functionality for each type of gas is shown in Fig. 8.11.

The authors used the typical fluid and rock properties of Bakken to build a model for that well. Different scenarios were run until the best relationship was obtained between the well model and the pilot test, as shown in Fig. 8.12. The model results and pilot test results are in agreement as shown in Fig. 8.12. However, one notable difference stands out

Table 8.5 The performance ranking for the three different gases.

Case number	Performance ranking
Case 1	CO_2 Lean gases Rich gases
Case 2	No difference in performance
Case 3	CO_2 Rich gases Lean gases
Case 4	Rich gases Lean gases CO_2
Case 5	Rich gases Lean gases CO_2
Case 6	Lean gases CO_2 Rich gases
Case 7	Rich gases CO_2 Lean gases
Case 8	CO_2 Lean Rich gases

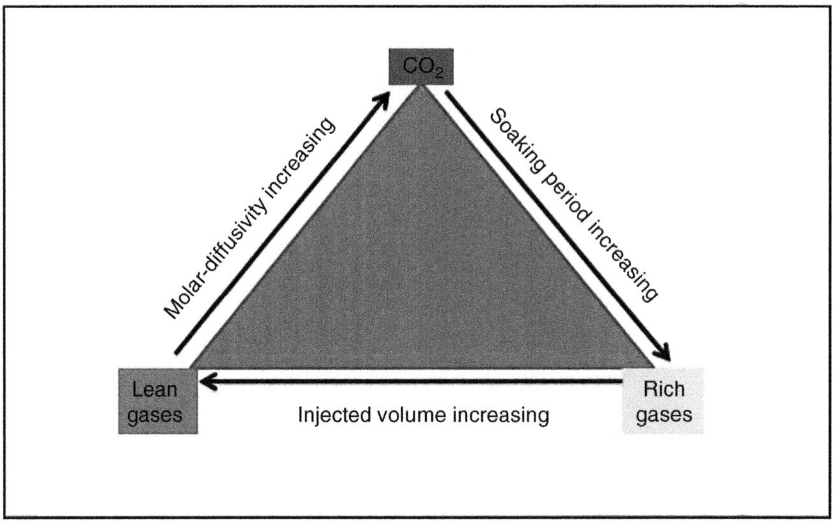

Figure 8.11 ***Applicability of miscible gases enhanced oil recovery in Bakken model.***

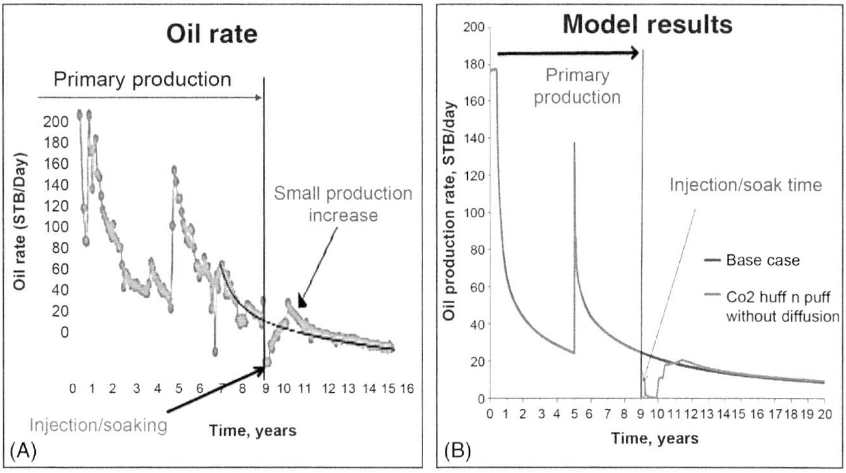

Figure 8.12 ***Pilot test performance versus the simulation match.*** (A) Represents field performance, and (B) represents simulation results).

wherein the oil production came quickly after the soaking period in the pilot test but took longer in the model. The authors attribute this difference to reported conformance problems caused by injection-induced fractures in the pilot projects where CO_2 was produced in the offset wells (Alfarge et al., 2017). Therefore, the produced-back CO_2 volumes in the producer were small, resulting in less of a hold-up effect on the produced oil. Notably, the authors have not induced injection fractures in the model featured herein. Therefore, large CO_2 production in large volumes occurred in the early huff-and-puff process of this model.

After investigating several EOR scenarios, the authors determined the best pilot test performance-simulation match can be obtained in a dual-permeability model with a low CO_2 molecular diffusivity. Two of the problems considered were: (1) Either the diffusion rate for CO_2 in reservoir conditions was too low or (2) the kinetics of the oil recovery process in the production areas exceeded the CO_2 diffusivity. The first possibility focused on the low diffusivity of CO_2 under shale reservoir conditions. This low diffusivity can be explained in two ways: (1) the contact area between the injected CO_2 and formation oil is small, and (2) the exposure time between the injected CO_2 and the formation oil is short. The contact area between CO_2 and formation oil is mainly the function of the natural fracture intensities in shale oil reservoirs. Although these types of reservoirs have a high intensity of natural fractures, the dual-permeability model was able to match the conducted pilot test results, including those with a low intensity of natural fractures. This indicated that either the natural fractures were not active or they were not connected to good pathways associated with hydraulic fractures.

8.5 Closing remarks

Most of the experimental studies reported that the CO_2 diffusion mechanism is behind the incremental increase in oil recovery obtained in lab conditions. These incremental oil recovery rates and/or the diffusion rates observed under lab conditions had been upscaled directly to the field scale by several researchers noted in the literature search, who used numerical simulation methods. This direct upscaling methodology is more optimistic than natural conditions dictate due to the fact that lab cores have higher contact area and longer exposure time to CO_2 than that found in established commercial reservoirs. Therefore, both simulation studies and experimental works were prone to predict a quick improvement in oil recovery by injecting CO_2 in these ULRs. This might explain why the results from the pilot tests using CO_2 as an injectant were disappointing and the results from the pilot tests using natural gases were encouraging (Alfarge et al., 2017). To sum up, the diffusion mechanism for CO_2 in the pilot tests had not been well recognized; therefore, CO_2 did not enhance oil production rate in those wells. The reasons behind the low-diffusion rate for CO_2 in pilot tests is due to either the kinetics of the oil recovery process in the production areas of these reservoirs being too fast or the CO_2 diffusion rate under field conditions being too slow (Alfarge et al., 2017). To sum up, the success of CO_2 in shale reservoirs mainly depends on understanding its main mechanisms, which are totally different from its mechanisms in conventional reservoirs. Although most unconventional EOR studies have investigated the applicability of CO_2, they have not properly investigated its principle mechanism on a field scale.

8.6 SUMMARY

- The *Pe* number calculations report a significant flow-type heterogeneity in shale reservoirs. However, diffusion flow is the most dominant.
- CO_2 needs a good molecular diffusivity information oil, so it can enhance oil production in these shale reservoirs.
- Some of CO_2 pilot tests showed a good match with the simulated cases that have a low diffusivity between formation oil and the injected CO_2.

- If the well or field conditions predict a low molecular diffusivity for the injected gases, the rich and lean gases will be more feasible than CO_2. However, rich gases need longer soaking periods and lean gases need larger volumes to be injected for more successful results.
- Generally, the number of cycles has no effect on injected gas performance. However, the duration of the soaking period has a significant effect.
- The timing of CO_2 injection performance is important. The earlier the CO_2-EOR, the better the enhancement in oil recovery.

References

Alfarge, D., Wei, M., & Bai, B. (2017). *IOR methods in unconventional reservoirs of North America: Comprehensive review.* SPE-185640-MS prepared for presentation at the SPE Western Regional Meeting held in Bakersfield, CA, USA, April 23–27, 2017.

Dong, C., & Hoffman, B. T. (2013). *Modeling gas injection into shale oil reservoirs in the Sanish Field, North Dakota*. Society of Petroleum Engineers. https://doi.org/10.1190/URTEC2013-185.

Hawthorne, S. B., Gorecki, C. D., Sorensen, J. A., Steadman, E. N., Harju, J. A., Melzer, S. (2013). *Hydrocarbon mobilization mechanisms from upper, middle, and lower Bakken reservoir rocks exposed to CO.* Society of Petroleum Engineers. https://doi.org/10.2118/167200-MS.

Hoffman, B. T., & Evans J. (2016). *Improved oil recovery IOR pilot projects in the Bakken formation*. SPE-180270-MS paper presented at the SPE low perm symposium held in Denver, CO, USA, May 5–6, 2016.

Hoteit, H., & Firoozabadi, A. (2009). Numerical modeling of diffusion in fractured media for gas-injection and recycling schemes. *SPE Journal, 14*(02), 323–337. SPE-103292-PA. http://dx.doi.org/10.2118/103292-PA.

Mohebbinia S., & Wong, T. (2017). *Molecular diffusion calculations in simulation of gasfloods in fractured reservoirs.* SPE-182594-MS, paper presented at the SPE Reservoir Simulation Conf., Montgomery, TX, USA, February 20–22, 2017, doi:10.2118/182594-MS.

Moortgat, J., & Firoozabadi, A. (2013). Fickian diffusion in discrete-fractured media from chemical potential gradients and comparison to experiment. *Energy Fuel, 27*(10), 5793–5805. http://dx.doi.org/10.1021/ef401141q.

Sheng, J. J. (2015). Enhanced oil recovery in shale reservoirs by gas injection. *Journal of Natural Gas Science and Engineering, 22*, 252–259. http://dx.doi.org/10.1016/j.jngse.2014.12.002.

Wan, T., & Sheng, J. (2015). *Compositional modelling of the diffusion effect on EOR process in fractured shale-oil reservoirs by gasflooding*. Society of Petroleum Engineers. https://doi.org/10.2118/2014-1891403-PA.

Xu, T., & Hoffman, T. (2013). *Hydraulic fracture orientation for miscible gas injection EOR in unconventional oil reservoirs*. Society of Petroleum Engineers. https://doi.org/10.1190/URTEC2013-189.

Yu, W., Lashgari, H., & Sepehrnoori, K. (2014). *Simulation study of CO_2 Huff-n-Puff process in Bakken tight oil reservoirs*. Society of Petroleum Engineers. https://doi.org/10.2118/169575-MS.

CHAPTER 9

Other enhanced oil recovery methods for unconventional reservoirs

Contents

9.1 Thermal enhanced oil recovery (EOR) techniques 185
9.2 Electromagnetic EOR techniques for shale oil reservoirs 187
9.3 Microbial EOR 188
    9.3.1 Introduction of microbial EOR 188
    9.3.2 Mechanisms of microbial EOR (MEOR) 189
    9.3.3 Field applications of microbial EOR 191
    9.3.4 Pros and cons of microbial EOR 191
9.4 Air-foam injection 192
9.5 Energized fluids for EOR 192
9.6 Other EOR methods 195
    9.6.1 CO_2 huff-n-puff followed by surfactant assist spontaneous imbibition (SASI) technique 195
    9.6.2 Chemical blend-based EOR for tight and shale oil reservoirs 196
    9.6.3 Fracturing improved oil recovery (IOR) techniques 196
9.7 Summary 198
References 198

Abstract

This chapter discusses the feasibility of other enhanced oil recovery (EOR) methods in shale and tight oil reservoirs such as thermal recovery, microbial EOR, and foam air injection. It provides some details of field applications for some of these EOR methods, if present. This chapter also presents obstacles, which impair the performance of these EOR techniques. It summarizes the theories behind the featured EOR methods, as well as their performance in the lab and in simulation methods.

Keywords: unconventional reservoir EOR methods; thermal EOR methods; foam air injection; microbial EOR methods; huff-n-puff versus flooding mode

9.1 Thermal enhanced oil recovery (EOR) techniques

Professionals in the field can often give the best summation of the most common EOR techniques. Sino Australian Oil and Gas Ltd. (2013) provided an excellent description of thermal recovery methods, which they

Fundamentals of Enhanced Oil Recovery Methods for Unconventional Oil Reservoirs
http://dx.doi.org/10.1016/B978-0-12-818343-4.00009-7

define as designed to raise the temperature of reservoir regions "to heat the crude oil in the formation and reduce its viscosity and/or vaporize part of the oil and thereby decrease the mobility ratio."

Thermal enhanced oil recovery (EOR) methods have different aspects as shown in Fig. 9.1. Thermal EOR methods include, but are not limited to, the injection of hot water, steam, or other gas, or by conducting in-situ combustion of oil or gas. The increased reservoir heat reduces surface tension, increases the effective permeability of the oil, and reduces reservoir seepage. Heated oil can also be vaporized and then condensed to improve the oil. However, this approach requires a substantial investment in labor and special equipment. Oil recovery by thermal EOR techniques differs from one method to another (Thomas, 2008). Cyclic steam stimulation (CSS), more commonly known as steam huff-n-puff, could result in an oil recovery factor of 10%–40% from the original oil in place (OOIP). Steam flooding could result in 50%–60% recovery from OOIP while hot waterflooding, steam-assisted gravity drainage (SAGD), and in situ combustion could result in 70%–80% oil recovery from OOIP.

Viscosity reduction, thermal expansion, and imbibition are the most dominant EOR mechanisms associated with accessory solution gas drives and gravity drainage (Wang et al., 2017). Thermal EOR methods have been cited as the most dominant EOR techniques implemented worldwide for heavy oil development. This is due to thermal EOR contributing to 47.7% of the enhanced oil production from 1959 to 2010 and 67% in 2010

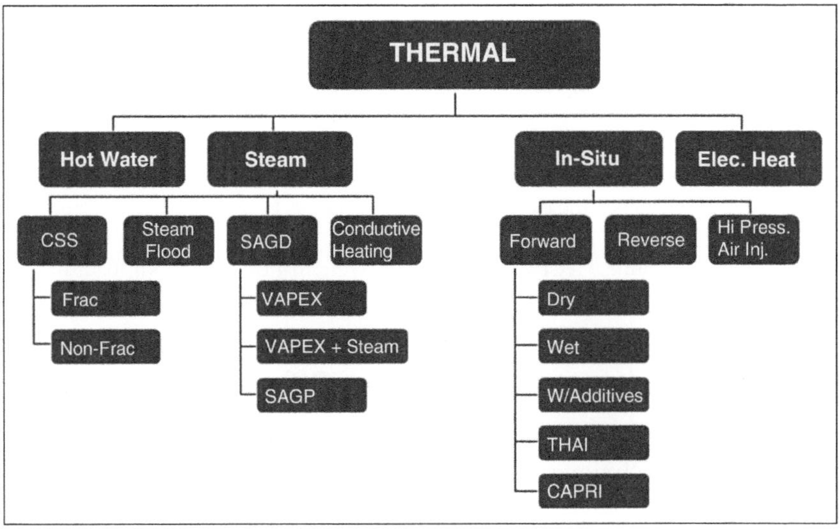

Figure 9.1 ***Thermal EOR methods.*** *Modified from: http://energy-cg.com.*

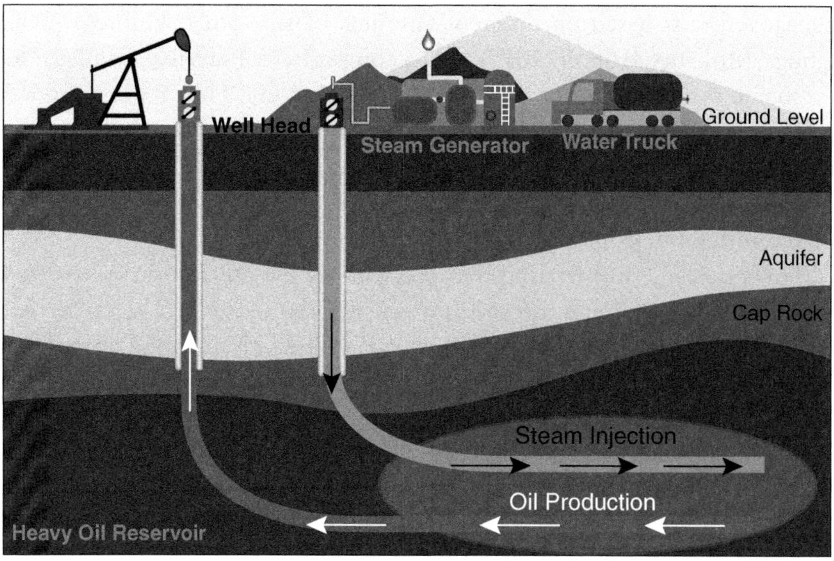

Figure 9.2 ***Schematic for SAGD.*** *Modified from She, Kong, Li, Hu, & Guo, 2019, an open access article distributed under Creative Commons Attribution License.*

(Koottungal, 2010). CSS or steam huff-n-puff injects steam for a period to heat the reservoir around the wellbore. Then, the steam is shut in the well for soaking time. After that, the well is opened for production. Steam flooding is similar to water flooding in that the steam is continuously injected to push the oil towards the producer. Fig. 9.2 illustrates the SAGD technique, which includes two parallel horizontal wells in a vertical plane, with the upper well serving as the injector and the lower well as the producer (Wang et al., 2017). SAGD could yield higher oil recovery and require less water and energy; however, as shown in Fig. 9.2, two wells are needed, and drilling two wells is not always economically feasible.

9.2 Electromagnetic EOR techniques for shale oil reservoirs

Increasing the temperature in a shale oil reservoir by using thermal methods could lead to in-situ heating at nearly 650°F, which may convert the kerogen to oil and gas. Although kerogen content in a shale oil rock is much lower than 10% of that found in typical oil shale, a shale oil formation may be heated so that the kerogen is converted to oil and gas. The shale formation permeability probably increases due to kerogen decomposition. Some researchers have investigated the amount of recovery via heating mechanism tests based on different amounts of kerogen in the reservoirs. Egboga, Mohanty, Balhoff (2017) used a compositional thermal simulator to investigate the feasibility of downhole in-situ heating. They recommended a high frequency electromagnetic (microwave) heating system be placed at the downhole wellbore to avoid the significant heat loss. In this case, microwave energy is introduced into a reservoir through a radiating element located in the horizontal section of the wellbore.

This process led to heating the reservoir via the adsorption of electromagnetic energy through the connate water in the reservoir. The same well was used for production and heating in a manner much like the CSS process where the well is not producing during the heating process. In the Egboga et al. (2017) simulation model, the properties of a Bakken oil were used. The heating energy accelerated the kerogen decomposition into oil and gas. This decomposition provided more pore space inside the rocks resulting in an increased porosity and permeability of the target formation. The porosity of the investigated model increased from 8% to 10%, and the permeability increased from 0.0015 to 0.002 mD. However, the effect of stimulation on the absolute value of the target formation by the heating process was not significant. Also, when the rock had 10% kerogen, the oil recovery by thermal stimulation was 1% higher than that from the thermal stimulation without kerogen. This indicates that the kerogen decomposition due to the heating process is not a dominant mechanism. This happens because the thermal heating energy and the resultant kerogen decomposition are concentrated locally near the heater. This observation is consistent with the permeability data. The simulation model of Egboga et al. (2017) shows that the oil recovery increased from 7.2% to 11.5% during the primary depletion within 1000 days of heating. Hence, it can be concluded that the presence of a high amount of kerogen may not play a major role in oil recovery process by thermal recovery methods for shale and unconventional reservoirs.

9.3 Microbial EOR

9.3.1 Introduction of microbial EOR

The microbial EOR (MEOR) process was first suggested by Beckmann in 1926 although it gained its recognition in the 1940s (Zobell, 1946). The first MEOR field implementation happened in the Lisbon fields in Union County, Louisiana, United States in 1954 (Lazar, Petrisor, & Yen, 2007; Ollivier & Magot, 2005). The acceptance of MEOR as an EOR technique faced worldwide resistance for decades (Nwidee, Theophilus, Barifcani,

Sarmadivaleh, & Iglauer, 2016). One reason for this resistance was that the microbiology age was less than 100 years old when this EOR technique was proposed (Nwidee et al., 2016). Therefore, the effectiveness of the MEOR process was viewed as a mere myth (Sen, 2008). Prior to the 1940s, university laboratories conducted MEOR research (Brown, 2010). These research efforts continued and eventually led to MEOR growth from extended laboratory work in the 1980s to field applications in the 1990s (Sen, 2008).

9.3.2 Mechanisms of microbial EOR (MEOR)

The MEOR technique can either inject microbial reaction products into the reservoir or inject microbes and nutrients into the reservoir so that microbial products can be generated as shown in Fig. 9.3 (She et al., 2019). MEOR techniques rely on two primary components: microbes and nutrients. Microbes (bacteria) are generated by either an exogenous or indigenous source. Exogenous bacteria are cultivated on the surface while indigenous bacteria are naturally present in the reservoir (Sheng, 2013). The indigenous bacteria are mostly found in reservoir rocks, utilize hydrocarbon,

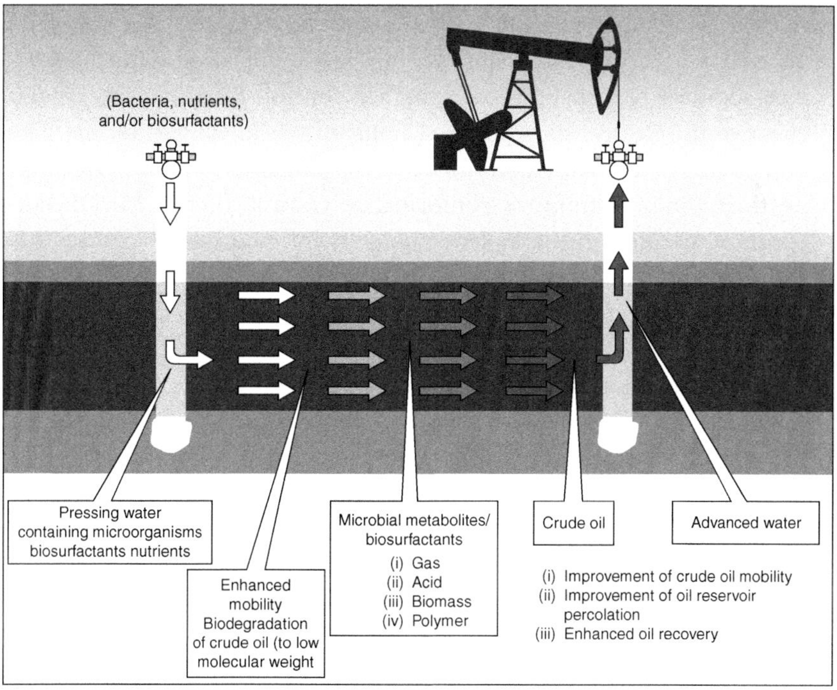

Figure 9.3 ***Schematic for MEOR injection.*** *Modified from She et al., 2019.*

and are nonpathogenic. These bacteria produce the desired EOR chemicals in situ through metabolic reactions, which result from multiple biochemical process steps to generate the EOR chemicals. The required bacteria should be small, spherical and less than 20% of the formation pores' throat size (Nwidee et al., 2016). Small cell sizes (between 0.5 and 5.0 μm) easily penetrate through reservoir porous media (Bryant & Burchfield, 1989). Nutrients (e.g., fermentable carbohydrates) are either injected from the surface into the reservoirs, serving as ex-situ nutrients, or they already exist inside the reservoirs as in-situ nutrients. The main purpose of nutrients is to provide favorable conditions for microbial metabolism, which results in the production of biosurfactants, biopolymers, and gases.

- *Biosurfactants.* Biosurfactants are biodegradable and considered more effective in emulsifying aqueous hydrocarbon mixtures than synthetic surfactants (Banat, 1995). Biosurfactants can be produced in bioreactors ex-situ and injected as an aqueous surfactant formulation into the reservoir (Nwidee et al., 2016). As the surfactant migrates towards the oil–aqueous interface, interfacial tension (IFT) is reduced and the capillary number increases (Banat, 1995). This eventually results in displacing and recovering residual oil.
- *Biopolymers.* Biopolymers can be used to improve sweep efficiency. The viscosity of the injected water can be greatly improved with biopolymer additives, which promote mobility control. Biopolymers are also effective in selective plugging and improve the injectivity profile. The biopolymers are mainly used for selective plugging of high permeability thief zones to improve conformance control. In this MEOR application, bacteria in the aqueous phase is injected into the formation and preferentially flows through high-permeability pathways until biomass growth plugs the pore throats, leading to permeability reduction (Nwidee et al., 2016). Plugging of high permeability thief zones diverts the injected water toward the lower permeability zone, which is still saturated with oil (Sen, 2008), thereby improving oil recovery due to the subsequent improved sweep efficiency.
- *Bacterial consortia.* This type of MEOR produces gases such as N_2, CH_4, CO_2, and H_2. These gases are effective in promoting oil viscosity reduction, IFT reduction, and oil swelling (Nwidee et al., 2016). Bacterial consortia are also used for the selective production of biosolvents (i.e., acetone, ethanol, 1-butanol, butanone, etc.) and bioacids (i.e., acetic, propionic, butyric, etc.) (Nwidee et al., 2016). In pressure-depleted reservoirs, biogases help in pressure build-up; meanwhile, at high pressures, biogases can dissolve within the oil phase and reduce its viscosity

(Sen, 2008). Bioacids dissolve carbonate rocks, which leads to an increase in the porosity and permeability of the target reservoir, which qualifies it as a stimulation technique (Sen, 2008).

9.3.3 Field applications of microbial EOR

The MEOR procedure in the field is either injected as part of the flooding mode or the huff-n-puff mode similar to what happens during the injection of thermal- and gas-based EOR methods. If the microbial reaction products are totally prepared on the surface and then injected into the reservoir, then, there is no difference between microbial EOR and chemical EOR. Successful MEOR field projects have been deployed in several countries, including Malaysia, India, Russia, Australia, Trinidad-Tobago, China, former East Germany, Norway, Saudi Arabia, Poland, Canada, Oman, United States, Romania, Hungary, and former Czechoslovakia (Nwidee et al., 2016). Most of the projects have been applied in conventional reservoirs. However, MEOR is generally applied to formations of low temperature (< 98°C) and high permeability (> 50 mD) (Sheng, 2013). However, one field application was applied in a reservoir, which has properties close to unconventional formations. The pilot test was applied in a low-permeability reservoir (Sheng, 2013). The MEOR pilot was performed in Wangyao, in the Ansai field, China, where the porosity was 14%, the permeability was 5.22 mD, and the reservoir temperature was 45°C. The water cut reached 66.1%, and the average well oil rate was 1.48 tons/d before the test. One injector, Wang 15-5, was chosen with seven producers around it to handle the project. A total of 1,150 m^3 solution was injected. The solution consisted of two slugs, the first slug contained microbes and nutrients, and the second slug contained microbial reaction products. The injection was conducted from June 28 to August 21, 2009. The recommended response time was 40 days. By the end of 2009, the cumulative oil had increased to 297.65 tons. The estimated ultimate incremental oil was 550 tons based on the injection process.

9.3.4 Pros and cons of microbial EOR

MEOR is environmentally friendly since the products of bio-MEOR are biodegradable (Lazar et al., 2007). The MEOR technology has low oil production costs (Nwidee et al., 2016). Therefore, MEOR is not strongly dependent on oil prices like conventional chemical EOR processes. Also, MEOR consumes less energy than thermal EOR processes (Sen, 2008). MEOR is economically attractive for application in marginally producing oil fields because the injected bacteria and nutrients are cost efficient (Lazar et al., 2007).

The MEOR process is complex because of its many variables such as reservoir chemistry for bacteria functionality and isolation (Nwidee et al., 2016). Another disadvantage is its relatively low incremental oil recovery (Sen, 2008).

9.4 Air-foam injection

The air foam EOR technique is a novel enhanced oil recovery (EOR) technique, which combines the reduction of oil/water interfacial tensions (IFTs) to ultralow values with the generation of foam drive for mobility control (Janssen, Mutawa, Pilus, & Zitha, 2019). This is known as foam-assisted flooding. Foam involves a discontinuous gas phase, that is, gas bubbles, within a continuous liquid phase. Foam stability is a strong function of the lamellae thickness. The lamellae are thin aqueous films that separate gas bubbles within the foam texture. The thinner lamellae tend to rupture more easily. The formation of foam is facilitated through the addition of a foaming agent. For example, adding surfactant to the aqueous phase could inhibit coalescence of separated gas bubbles, and thus promote foam stability. For reservoirs or blocks that are under high water cut or extra-high water cut for a period of time, by means of foam injection, injected air into the reservoirs can be controlled effectively to avoid earlier breakthrough (Dong, Liu, Sun, Zheng, Sun, 2012). Meanwhile, the sweep efficiency could be also improved (Gauglitz, Friedmann, Kam, & Rossen, 2002; Kovscek & Bertin, 2002). Many factors affect the performance of air-foam injection such as temperature, pressure, reservoir mineral components, reservoir dip angle, oil viscosity, water saturation, air-injection rate, and the gas/liquid ratio of foam (Dong et al., 2012). Literature on air-foam injection for shale and tight reservoirs has been scarce. However, this technique is viable EOR from shale and tight oil reservoirs.

9.5 Energized fluids for EOR

Energized fluids are defined as fluids with one or more compressible gas components, such as CO_2, N_2, or any combination of gases, dispersed in a small volume of liquid (Karadkar, Bataweel, Bulekbay, Alshaikh, 2018). Such types of fluids can offer attractive stimulation techniques, which are often an alternative to conventional stimulation fluids, such as low reservoir pressure, water-sensitive formations, and/or the need to shorten the flow-back period. Facilitating and speeding up the flow back is very important to improving the oil recovery in shale and tight oil reservoirs because fluid remaining in the formation negatively impacts the physical properties of rock, lowering the conductivity of the reservoir and reducing or impeding the flow of oil and gas (Watts, 2014). Some studies indicate that the hydraulic fracturing process with solutions energized by CO_2 or N_2 can economically achieve more hydrocarbon recovery than non-energized techniques (Watts, 2014). One such study found that using energized fluids could improve the well productivity performance by 1.6 to 2.1 times, compared with non-energized solutions (Watts, 2014). Energized fluids have many obstacles, such as low stability at high temperature, high friction pressure during pumping, corrosion after using CO2, and the need for specialized surface pumping equipment. Cryogenic liquid N_2 (−320 to −322°F) has been used in the hydraulic fracturing process. Carbon dioxide and nitrogen share a long and successful history in hydraulic fracturing and well stimulation, including energizing water-based fluids for unconventional reservoirs (Watts, 2014). Gomaa, Qu, Nelson, and Maharidge (2014) reported that liquid N_2 can reduce breakdown pressure and increase fracture complexity. CO_2 can be pumped with conventional equipment in liquid form. When CO_2 is heated inside the reservoir, it vaporizes to a gas state. A significant change occurs in the phase behavior of CO_2 during the fracturing process, as shown in Fig. 9.4. The gaseous CO_2 can assist in the flow back which can

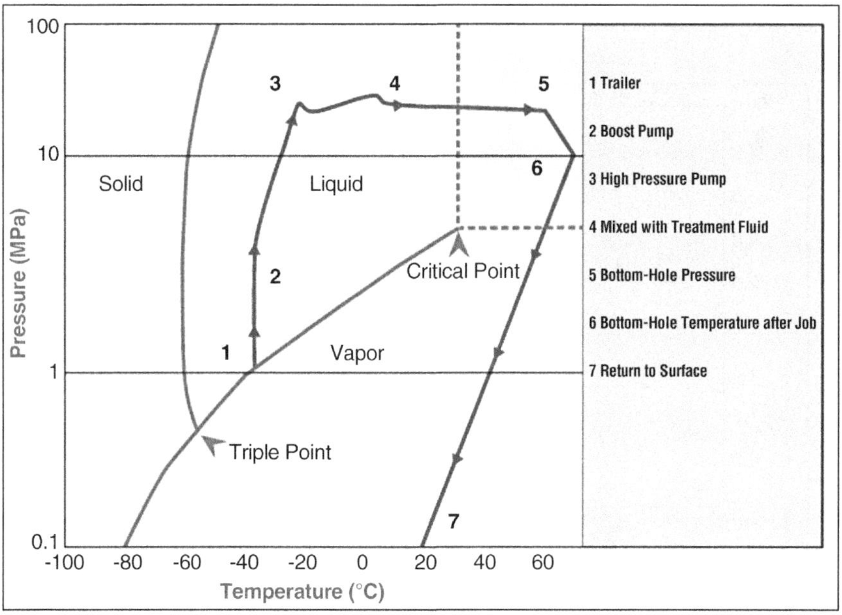

Figure 9.4 ***Phase behavior changes of CO_2 during hydraulic fracturing (Watts, 2014).***

help improve the oil recovery. Liquid CO_2 has a high density, which helps to gain hydrostatic pressure and to carry proppants during the fracturing process. Other benefits to the liquified CO_2 include lower IFT and clay inhibition. The combination of water and CO_2 may lead to the formation of carbonic acid, which can have some other benefits. Energized fluids in different types of foams are often used to improve the oil recovery process (Karadkar et al., 2018).

In unconventional liquid-rich reservoirs, energized solutions such as liquified-CO_2 provide the necessary energy to move hydrocarbons in low-pressure zones or in areas with strong capillary forces, because not as much water invades the formation to create the blocking problems for hydrocarbon flowability (Watts, 2014). The solubility and miscibility properties of CO2 during the hydraulic fracturing process provide a greater chance to energize the flow of higher viscosity hydrocarbons (Watts, 2014). This effect results in higher well productivity than the hydraulic fracturing process using slick water fracturing as shown in Fig. 9.5.

To sum up, energizing the fracturing fluid with CO_2 or N_2 will: (1) reduce the amount of water necessary to pump the frac job, (2) lower the leak-off of the water phase, and (3) improve flow back, since the invaded water saturation is lower (Watts, 2014). In other words, energized fracturing fluids can improve the total flow-back volume and rate from unconventional liquid-rich reservoirs, and when foamed, the energized fluids

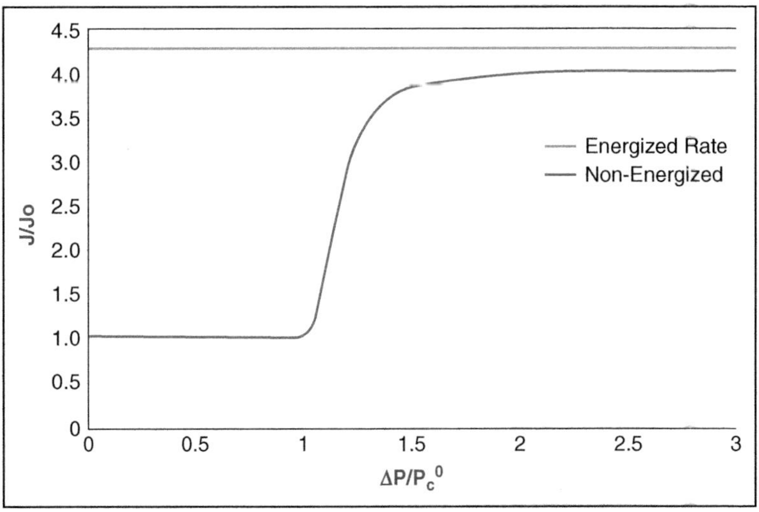

Figure 9.5 ***Comparison of predictive productivity (unit-less measure of J/J0) of energized solutions versus non-energized water fraction stimulations (Watts, 2014).***

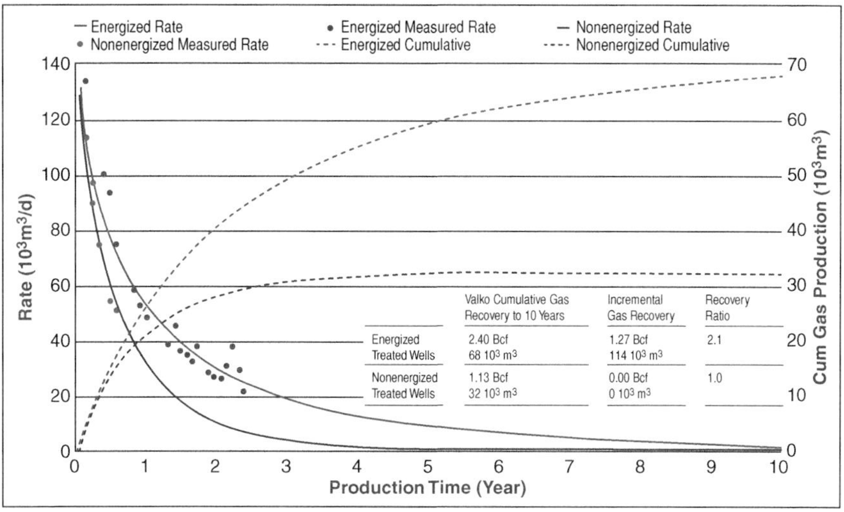

Figure 9.6 ***Averaged 10-year decline analysis for energized versus of non-energized frac fluids (Montney shale). Modified from Watts, 2014.***

can significantly lower the total liquid leak-off to minimize fluid retention (Watts, 2014). This proves that the energized fluids have higher cumulative hydrocarbon production compared with non-energized fluids as shown in Fig. 9.6.

9.6 Other EOR methods

9.6.1 CO_2 huff-n-puff followed by surfactant assist spontaneous imbibition (SASI) technique

One method which has EOR potential in shale and tight oil reservoirs is the sequential method of CO_2 huff-n-puff injection and the surfactant-assist spontaneous imbibition (SASI). Some researchers have conducted lab experiments on CO_2 injection in huff-n-puff mode followed by SASI. They reported that the oil recovery factor could be significantly increased by this technique. However, the incremental oil recovery is a function of the operating parameters such as injection pressure and number of cycles. They recommended that CO_2 huff-n-puff injection should be performed first. Then, the imbibition process is carried out by injecting the surfactant. The oil recovery factor resulting from surfactant imbibition is much lower compared with that of CO_2 huff-n-puff. However, this technique should be complementary to achieve the highest oil recovery from the target reservoir.

9.6.2 Chemical blend-based EOR for tight and shale oil reservoirs

Chemical blends are defined as different chemical components such as an anionic surfactant, an organic solvent, an oxidizing agent, clay stabilizer, scale inhibitor, and other components which can be used to improve oil recovery in unconventional liquid-rich reservoirs. The chemical blends were injected in the same well in a huff-n-puff mode. Since the main player for the chemical blends was the surfactant, some researchers call them "surfactant blends." The main purpose of a surfactant is to reduce the IFT and increase the water-wetness of the shale rocks. This surfactant role is important in facilitating the flow back of fracturing fluids and mitigating the water blocking problems. The organic solvent can be used in solubilizing oil and heavy organic components such as asphaltenes. The oxidizing agent can oxidize some of the kerogen to form a weak acid which dissolves part of the calcite. Since some shale rocks are rich in calcite (>50 wt.%); thus, when a chemical agent reacts with the calcite in the shale rocks, it leads to calcite dissolution. This dissolution resulted in increasing the permeability of shale rocks. Therefore, the performance of surfactant blenders is much better than using surfactant alone. The permeability enhancement, wettability alteration, and oil solubilization are the main mechanisms by which chemical blends can enhance oil recovery in shale and tight oil reservoirs.

9.6.3 Fracturing improved oil recovery (IOR) techniques

Although some researchers may not agree on how to classify these methods as EOR techniques, they can be definitely be classified as IOR methods. Stimulated reservoir volume (SRV) fracturing in tight oil reservoirs usually leads to the formation of complex fracture-network growth, which has a fundamentally different production performance than traditional planar bi-winged fracturing (Su, Ren, Meng, Xu, & Wang, 2015). Branch fracturing is one of the improved oil recovery (IOR) techniques which may be considered for shale and unconventional reservoirs. During the fracturing process, the injection flow rate and viscosity are important parameters which control fracture propagation. Based on some laboratory experiments conducted by Wang, Guo, Zhang, Ren, and Ni, (2019), it was found that the hydraulic fracture would cross the natural fracture under the high flow rate and viscosity, while the hydraulic fracture would deviate to the natural fracture under the low flow rate and viscosity. After the hydraulic fractures are generated, adding fibers with a specific percentage in the fracturing fluid would create a temporary bridge blockage within the hydraulic fractures.

Due to this bridge blockage, the fracturing fluid diverts to different directions forming branched fractures. This technique results in enhancing oil recovery from the same well by using fracturing techniques.

The Zipper frac method is another example of fracturing based on IOR techniques. Zipper fracturing is a completion methodology that is often executed in shale wells. Initially, operators used zipper fracs to enhance operational efficiency by reducing the operating time between frac stages. However, recent reports indicate zipper fracturing can lead to enhanced production and improved recoveries (Patel, Cadwallader, & Wampler, 2016). The zippering technique is used during the horizontal well plug-and-perf completions. Patel et al. (2016) explained the methodology for zipper fracs technique. During the pumping operations of a well's frac stage, wireline is rigged up on an offset well running in a hole to set a plug and perforate the casing. At the end of the stimulation job, the wireline is brought down from the offset well and moved to the next well on the pad to prep for pumping operations. The wells that have a completed stage are isolated by the crew, who redirect the pumps to frac the well that was just prepared with a wireline. This process is further described by Patel et al. (2016) as: "The sequence is reminiscent of a zipper closing to link the protruding teeth along the seams of a jacket –one by one, each stage is completed in an alternating sequence." For conventional fracturing patterns, each well is independently fractured from the toe to the heel. However, during the zipper fracturing configuration, after one stage from well A is fractured, another stage from well B is fractured to the early stage of well A as shown in Fig. 9.7. The sequence is

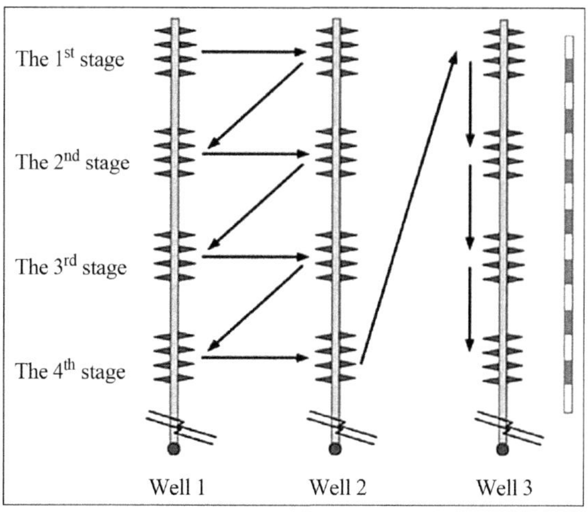

Figure 9.7 ***Schematic diagram of zipper-fracturing operation mode (Watts, 2014).***

continuously repeated until the wells are done. The main purpose of this pattern is to create fracture network complexity close to the tips of the fractures properly network the fractures and maximize the oil form for the best production from the tight oil-rich matrix. This pattern of fracturing has been successful in enhanced oil recovery from shale and tight oil reservoirs.

9.7 SUMMARY

- This chapter summarizes the other EOR methods, which are feasible to be applied in shale and tight oil reservoirs.
- The oil recovery factor could be increased if the CO_2 huff-n-puff injection followed by SASI is performed in shale oil reservoirs.
- The mechanisms of air foam EOR technique are very crucial to enhance oil recovery in shale and tight oil reservoirs. The EOR potential is also significant if there are no injectivity problems on the field scale.
- Chemical blends techniques can be used to improve oil recovery in unconventional liquids-rich reservoirs. The performance of such chemical mixtures in terms of enhancing oil recovery is better than using surfactant alone, as in the synergy process.
- Different mechanisms for each EOR method have been reported in the literature.
- Fracturing IOR techniques could be viable to enhance oil recovery in shale oil reservoirs. However, their benefits for a short time and on limited areas, only the areas which are close to the wellbore of shale plays.

References

Banat, I. M. (1995). Biosurfactants production and possible uses in microbial enhanced oil recovery and oil pollution remediation: A review. *Bioresource Technology*, *51*, 1–12.

Brown, L. R. (2010). Microbial enhanced oil recovery (MEOR). *Current Opinion in Microbiology*, *13*, 1–5.

Bryant, R. S., & Burchfield, T. E. (1989). Review of microbial technology for improving oil recovery. *SPE Reservoir Engineering Journal*, *4*, 151–154.

Dong, X., Liu, H., Sun, P., Zheng, J., & Sun, R. (2012). *Air-foam-injection process: An improved-oil-recovery technique for waterflooded light-oil reservoirs*. Society of Petroleum Engineers. https://doi.org/10.2118/163044-PA.

Egboga, N. U., Mohanty, K. K., & Balhoff, M.T. (2017). A feasibility study of thermal stimulation in unconventional shale reservoirs. *Journal of Petroleum Science and Engineering*, *154*, 576–588.

Gauglitz, P. A., Friedmann, F., Kam, S. I., & Rossen, W. R. (2002). *Foam generation in porous media*. Presented at the SPE/DOE Improved Oil Recovery Symposium, Tulsa, 13–17 April. SPE-75177-MS. https://doi.org/10.2118/75177-MS.

Gomaa, A. M., Qu, Q., Nelson, S., & Maharidge, R. (2014). *New insights into shale fracturing treatment design*. SPE 167754 presented at the SPE/EAGE European Unconventional

Conference and Exhibition held Vienna, Austria, 25–27 February 2014. https://doi.org/10.2118/167754-MS.
Janssen, M. T. G., Mutawa, A. S., Pilus, R. M., & Zitha, P. L. J. (2019). Foam-assisted chemical flooding for enhanced oil recovery: Effects of slug salinity and drive foam strength. *Energy & Fuels*, *33*(6), 4951–4963. https://doi.org/10.1021/acs.energyfuels.9b00645.
Karadkar, P., Bataweel, M., Bulekbay, A., & Alshaikh, A. A. (2018). *Energized fluids for upstream production enhancement: A review*. Society of Petroleum Engineers. https://doi.org/10.2118/192255-MS.
Koottungal, L. (2010). Special report 2010 worldwide EOR survey. *Oil & Gas Journal*, *108*(14), 41–53.
Kovscek, A. R., & Bertin, H. J. (2002). *Estimation of foam mobility in heterogeneous porous media*. Presented at the SPE/DOE Improved Oil Recovery Symposium, Tulsa, 13–17 April. SPE-75181-PA. https://dx.doi.org/10.2118/75181-MS.
Lazar, I., Petrisor, I. G., & Yen, T. F. (2007). Microbial enhanced oil recovery (MEOR). *Petroleum Science and Technology*, *25*, 1353–1366.
Nwidee, L. N., Theophilus, S., Barifcani, A., Sarmadivaleh, S., & Iglauer, S. (2016). EOR processes, opportunities and technological advancements. http://dx.doi.org/10.5772/64828.
Ollivier, B., & Magot, M. (2005). *Petroleum microbiology*. Herndon, VA: ASM Press.
Patel, H., Cadwallader, S., & Wampler, J. (2016). Zipper fracturing: Taking theory to reality in the Eagle Ford Shale. In *Unconventional Resources Technology Conference*. https://doi.org/10.15530/URTEC-2016-2445923.
Sen, R. (2008). Biotechnology in petroleum recovery: The microbial EOR. *Progress in Energy and Combustion Science*, *34*(6), 714–724.
She, H., Kong, D., Li, Y., Hu, Z., & Guo, H. (2019). Recent advance of microbial enhanced oil recovery (MEOR) in China. *Geofluids*. https://www.hindawi.com/journals/geofluids/2019/1871392/.
Sheng, J. (2013). Introduction to MEOR and its field applications in China. *Enhanced oil recovery field case studies* (pp. 543–559).
Sino Australia. (2013). *Introduction to enhanced oil recovery techniques,* online information on company approach to EOR. Available from: http://large.stanford.edu/courses/2015/ph240/zerkalov2/docs/sino.pdf.
Su, Y., Ren, L., Meng, F., Xu, C., & Wang, W. (2015). Theoretical analysis of the mechanism of fracture network propagation with stimulated reservoir volume (SRV) fracturing in tight oil reservoirs. *PLoS ONE*, *10*(5), e0125319. https://doi.org/10.1371/journal.pone.0125319.
Thomas, S. (2008). Enhanced oil recovery—An overview. *Oil & Gas Science and Technology – Revue d'IFP Energies nouvelles*, *63*(1), 9–19.
Wang, J., Guo, Y., Zhang, K., Ren, G., & Ni, J. (2019). Experimental investigation on hydraulic fractures in the layered shale formation. *Geofluids*. https://www.hindawi.com/journals/geofluids/2019/4621038/.
Wang, L., Tian, Y., Yu, X., Wang, C., Yao, B., Wang, S., et al. (2017). Advances in improved/enhanced oil recovery technologies for tight and shale reservoirs. *Fuel*, *210*, 425–445. doi: 10.1016/j.fuel.2017.08.095.
Watts, R. (2014). https://www.aogr.com/web-exclusives/exclusive-story/fluid-systems-energized-nitrogen-carbon-dioxide-optimize-frac-effectiveness.
Zobell, C. (1946). Action of microorganisms on hydrocarbons. *Bacterial Reviews*, *10*, 1–49.

CHAPTER 10

The effects of nanopore confinement on different enhanced oil recovery methods

Contents

10.1 Introduction 201
10.2 Effect of pore confinement on primary production 204
10.3 Effect of pore confinement on enhanced oil recovery methods 206
- 10.3.1 Effect of nanopore confinement on air injection performance 207
- 10.3.2 Effect of nanopore confinement on CO_2-EOR performance 207
- 10.3.3 Effect of nanopore confinement on natural gas-based EOR 210
- 10.3.4 Effect of nanopore confinement on N_2-EOR 213

10.4 Summary 214
References 214

Abstract

This chapter presents the effect of nanopore confinement on phase behavior and pressure-volume-temperature (PVT) properties of unconventional reservoir fluids. The impacts of pore confinement on the production performance of shale wells during the primary production stage are also discussed. Moreover, the positive and negative impacts of nanopore confinement on different enhanced oil recovery methods are extensively explained. This chapter presents how pore confinement should be implemented in the reservoir simulation to accurately mimic the performance of miscible gas injection in unconventional reservoirs.

Keywords: nanopore confinement; bubble point shifting; shale oil phase behavior; capillary pressure; unconventional miscible-gas; enhanced oil recovery (EOR)

10.1 Introduction

Unconventional resources, such as shale gas, tight oil, and shale oil have played a significant role in increasing hydrocarbon production worldwide over the last decade. Understanding recovery mechanisms and the phase behavior associated with that recovery is essential to the successful development of unconventional reservoirs. However, major uncertainties remain that can hinder understanding of the complex phase behavior affecting such reservoirs (Dong, Holditch, McVay, & Ayers, 2011). Thus, extensive studies

http://dx.doi.org/10.1016/B978-0-12-818343-4.00010-3

have been conducted to investigate the phase behavior of reservoir fluids in the confined space of shale reservoirs (Zhang, 2019).

In shale rock reservoirs, the matrix consists of micropores less than 2 nm in diameter to mesopores with diameters in the range of 2–50 nm (Kuila & Prasad, 2011). The confined space in shale nanopores can alter the phase behavior of reservoir fluids by changing the critical properties of the petroleum mixture's constituent components, such as pressure and temperature. Another potential threat to reservoir fluid flowability is the large gas-oil capillary pressure in confined nanopores (Zhang, 2019). Substantially, high capillary pressure can greatly influence the phase behavior of unconventional fluids resulting in different performances by the primary recovery mechanisms. Large capillary pressure due to confined nanopores could explain the change in the pressure–volume–temperature (PVT) properties, inconsistent gas–oil ratio (GOR), reduction in bubble pressure, and the change in dew point pressure. All of these problems have been observed in both unconventional and conventional reservoirs (Nojabaei, 2015) (Fig. 10.1). Honarpour,

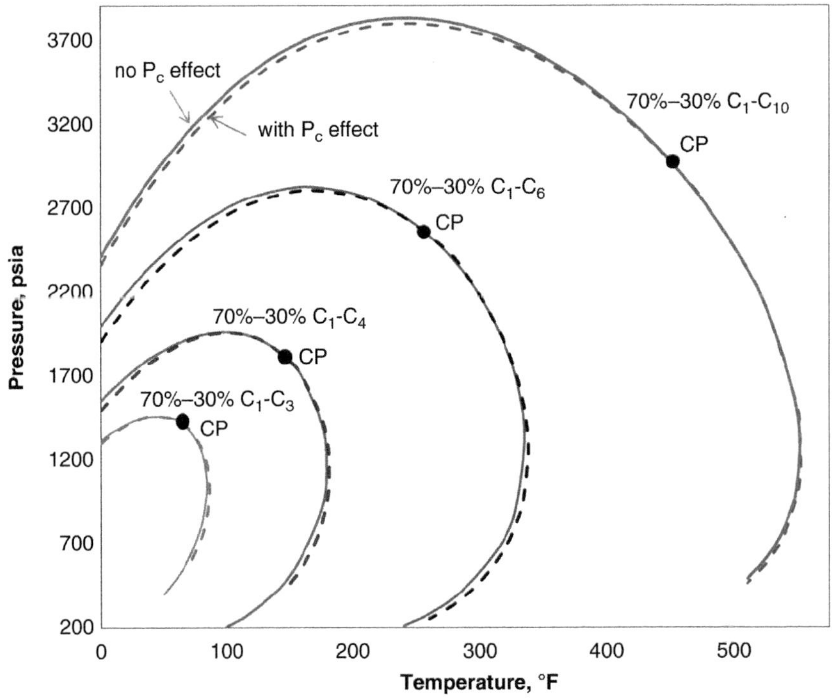

Figure 10.1 ***Phase envelopes with and without capillary pressure for binary mixtures (70-C1-C3, 70-30 C1-C4, 70-30 C1-C6, and 70-30 C1-C10) and a pore radius of 10 nm (Nojabaei, 2015).***

Nagarajan, Orangi, Arasteh, and Yao (2012) explained that the high capillary pressures encountered in unconventional shale reservoirs is responsible for the reduction in bubble point pressure and oil viscosity, as well as the increment in the formation volume factor, resulting in a longer constant GOR time, as shown in Fig. 10.2. The effect of Van der Waals forces on phase behavior have also been investigated, but their contribution was found to be small compared to capillary forces unless pores were very small (<1 nm) and the interfacial tension was low (Zhang, 2019).

In conventional reservoirs, the phase behavior is normal in macropores and oil and gas bubbles form as the unconfined bubble point is achieved, which is favorable for flowing oil droplets in the nanopores and mesopores (Alharthy, Nguyen, Teklu, Kazemi, & Graves, 2013) as shown in Fig. 10.3. However, when pores become smaller, the surface curvature of the phase envelope increased, the bubble-point pressure decreased, and the dew-point pressure increased (Brusllovsky, 1992). Brusllovsky (1992) also concluded that in porous media with varied pore sizes, the bubble point is reached first in the larger pores, while the dew point occurs first in the smaller pores.

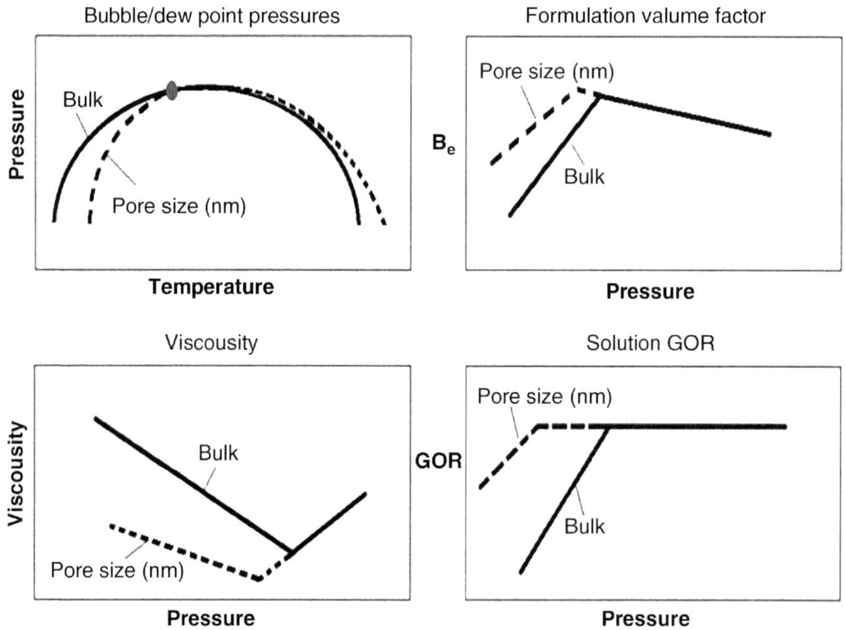

Figure 10.2 *Schematic of the effect of capillary pressure on phase behavior in nanopores (Honarpour et al., 2012).*

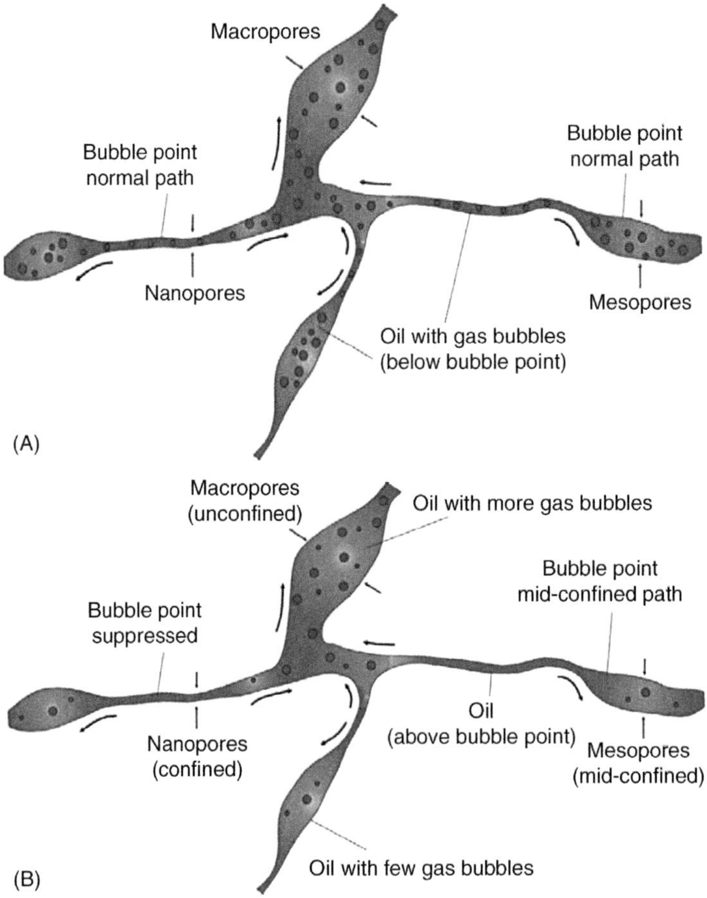

Figure 10.3 ***Phase behavior in different pore sizes for a bubble point system during reservoir depletion (Alharthy et al., 2013).*** (A) Without the phase envelope shift and (B) with the phase envelope shift.

Many other researchers have reported similar findings, which ensure that the pore confinement in unconventional reservoirs plays a significant role in making the phase behavior of unconventional fluids so different from those in conventional reservoirs (Nojabaei, 2015; Yu and Sepehrnoori, 2019).

10.2 Effect of pore confinement on primary production

The effect of pore confinement is not limited to the shifting of the phase behavior for shale fluids. This effect also results in abnormal production performance in the primary depletion stage. Due to the impact of nanopores

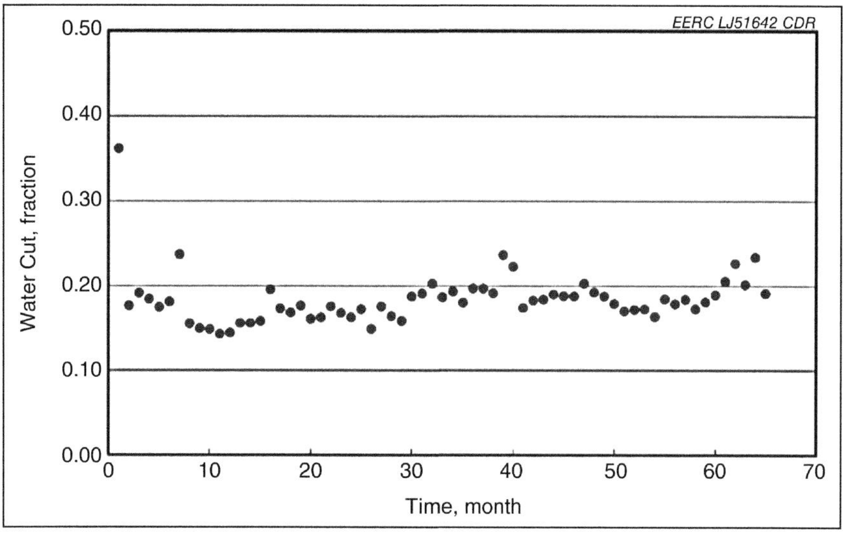

Figure 10.4 ***Trend of water cut development in a typical Bakken well (Jin et al., 2016).***

on fluid/rock properties, several anomalous production phenomena have been observed in many unconventional oil production wells without fully understandable physical reasoning behind them (Nojabaei, 2015). First, the hydraulic fractured wells in the Bakken Petroleum System (BPS) begin their production with a high daily rate due to high initial pressure and fracture connectivity/conductivity; however, a rapid decline in production usually follows (Kurtoglu, Rasdi, Salman, & Kazemi, 2013). Many wells have a decline rate of more than 70% in the first year and an additional 50% in the following two years (Jin et al., 2016).

The production rate gradually stabilizes over time (Baihly, Altman, & Aviles, 2012; Ran & Kelkar, 2015). A fast depletion might happen due to fracture closures and very tight matrices with pore confinement. Second, the water cut seems to be very constant over the production period, as shown in Fig. 10.4, which is notably different from that for conventional wells (Jin & Wojtanowicz, 2010; Salako & MacBeth, 2015). The initial high-water cut might be due to the flow back of fracturing fluid. The water cut is stable with production time indicating the presence of free water in the formation coexisting with oil in the same pore space where the capillary effect is greater than the gravity segregation (Nojabaei, Johns, Chu, 2013; Jin & Wojtanowicz, 2014; Jin et al., 2016). Third, one of the most common observations is the long-lasting flat production GOR with

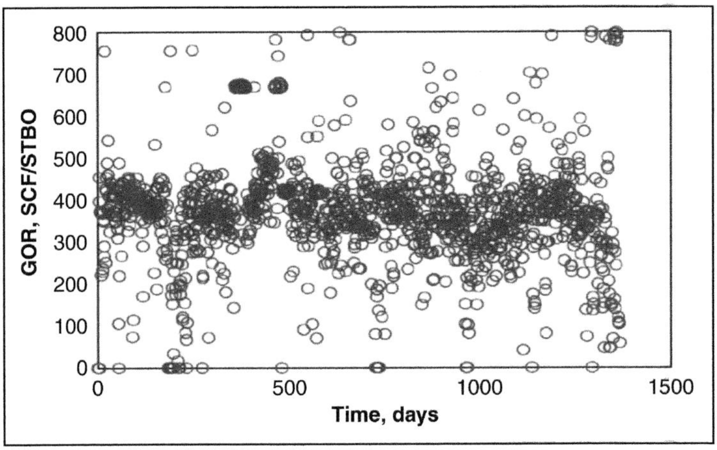

Figure 10.5 ***Production GOR for a well in Bakken reservoir (Nojabaei et al., 2013).***

or without a short step-wise jump observed in many Bakken wells, as shown in Fig. 10.5. Fig. 10.5 shows that the production GOR for a typical multi-stage fractured horizontal well in the Middle Bakken was markedly close to the initial GOR (Rsi) during the 4 years of production. Meanwhile, the buildup data and pressure measurements from the observation wells revealed that both the flowing bottom-hole pressure and the pressure around the wells have been below the bubble point pressure. Such GOR behavior could not be matched by using the PVT properties reported by the lab (Nojabaei, 2015). Nojabaei (2015) also noted that the combination of enhanced critical gas saturation in tight rocks and the suppressed bubble point pressure due to the tiny pore sizes are the only two reasons that might explain such phenomena.

10.3 Effect of pore confinement on enhanced oil recovery methods

The effect of pore confinement is not only significant in the primary depletion stage but also in both the secondary and tertiary recovery stages. Pore confinement significantly affects many enhanced oil recovery (EOR) methods applied in unconventional reservoirs. Pore confinement can have both a positive and negative effect on the performance of EOR methods. The next sub-sections explain how pore confinement affects the performance of unconventional EOR methods in detail.

10.3.1 Effect of nanopore confinement on air injection performance

Research has indicated that the confinement of pore spaces can affect air injection project success in several ways (Huang and Sheng, 2018). The confinement inhibits the distillation endothermic process. It also serves to increase low-temperature oxidation exothermic activity. The confinement effect leads to a better fuel deposition. Smaller pore sizes also result in a lower high-temperature oxidation (HTO) exothermic peak temperature. These factors have really a positive effect on the performance of air injection. However, as indicated they are more beneficial after achieving the HTO stage. The activation energy, as defined, is the energy, which is required for initiating combustion reactions. Pore sizes also affect the reduced activation energy; thus, the environment helps initiate the air injection combustion reactions. Hence, confined pore spaces generally have a positive impact on the success of air injection projects.

Some considerations make the unconventional reservoir unfavorable for air injection. The low air injectivity of shale oil reservoirs might increase the concentration of the oxygen in production wells above 10%, which creates safety hazards (Zhang, 2019). Furthermore, the tight matrix of unconventional reservoirs would make the heat loss bigger compared with conventional reservoirs. The excessive mature shale oil has a low content of unsaturated components, mainly aromatics and asphaltenes. This is unfavorable for a fuel loaded during the air injection process. The very thick zone needs a high amount of air injected to guarantee a stable oxidation reaction as well as a thermal front. The nano-Darcy permeability matrix and limited available fractures are the barriers for achieving a high injection rate. The nanometer-diameter pore can have a drastic effect on the arrangement of molecules inside the pores. This confinement effect can significantly change the phase behavior in shale oil reservoirs, increasing the complexity of crude oil oxidation, evaporation, cracking, and combustion in the nanopores. More hydraulic and thermal-induced fractures are needed to diminish the pore confinement effect as well as improve air injection performance in shale oil reservoirs.

10.3.2 Effect of nanopore confinement on CO_2-EOR performance

Because shale and tight reservoir rocks have pore throats ranging in size from micrometers to nanometers, they are subject to large capillary pressure. When permeability is ultra-low ($k < 200$ nD), diffusion flow can overrule advection, which means the performance of gas injection may no longer be controlled by the minimum miscibility pressure (MMP) (Zhang, 2019). When this happens, diffusion flow due to pore confinement mandates the adoption of several operating parameters, which control the success or failure of CO_2-EOR in shale oil reservoirs (Alfarge et al. 2017a). Therefore, pore confinement can have a significant impact on the CO2-EOR in shale oil reservoirs. In tight reservoirs where $k > 200$ nD and capillary pressure is still large, the advection likely dominates over diffusive transport so that the MMP becomes important (Zhang, 2019). Some studies have determined that the nanopore confinement can alter component critical pressures and temperatures. It can also change the critical point, which in turn will significantly impact the MMP (Zhang, 2019). Even though MMP has a major impact on traditional CO2-EOR performance, applying traditional phase behavior and simulation models of CO_2-EOR operations on unconventional reservoirs could easily provide inaccurate results (Zhang, 2019). Therefore, new algorithms which consider the effect of nanopore confinement on phase behavior and critical properties are needed for more accurate reservoir simulation that can capture the physics of fluid flow through the micro- to nanometer sized pores in such reservoirs. Notably, pore confinement has a clear influence on CO_2-EOR in shale oil reservoirs, regardless of whether the permeability is greater or less than 200 nD.

10.3.2.1 Identification of flow type in shale oil reservoirs

Identification of the flow regimes in shale oil reservoirs is crucial when evaluating the pore confinement effect on CO_2-EOR performance. Alfarge et al. (2017b) extensively investigated the flow regimes in a shale oil reservoir by using numerical simulation methods. The Péclet (*Pe*) number was locally calculated in each grid for both the natural-depletion stage and the EOR-stage. In the shale matrix, results indicated that the *Pe* number is far below 1 for both gas and oil phases, which means that diffusion is the most dominant flow in the matrix, as shown in Fig. 10.6. However, in hydraulic fractures, the viscous flow is clearly dominated as shown by the high *Pe* value, which greatly exceeds 100. Fig. 10.7 shows that the *Pe* number does not significantly change during the natural depletion stage; however, its changes do have a noticeable effect during the EOR stage. Fig. 10.7 also shows two different behaviors in the gas and oil phases of the EOR stage, that is, the *Pe* number increases with time in the gas phase; but decreases with time in the oil phase.

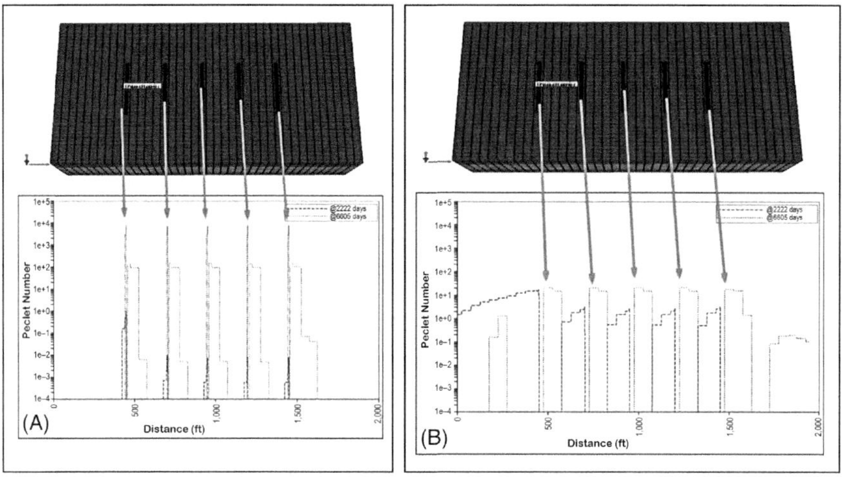

Figure 10.6 ***Péclet number distribution along cross-section in the matrix-model for shale oil reservoirs.*** (A represents the gas phase while B represents the oil phase) (Alfarge et al., 2017b).

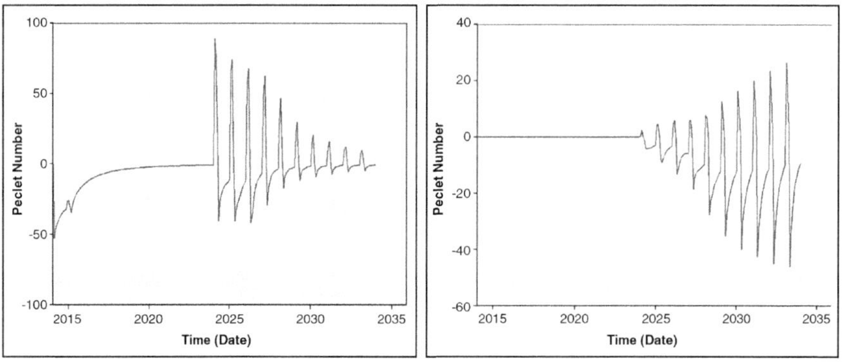

(A) Oil Phase (B) Gas Phase

Figure 10.7 ***Péclet number change with time (at 10 ft from the hydraulic fracture) in unconventional reservoirs (Alfarge et al., 2017b).***

In natural fractures, results indicate that the *Pe* number is significantly changeable where *Pe* can be < 1 in areas which are far away from hydraulic fractures; however, in areas close to hydraulic fractures, the *Pe* can be >100. According to the average value of the *Pe* number in natural fractures, the dispersion flow could be the most dominant flow. Flow regimes for both the natural-depletion stage and the EOR-stage in unconventional reservoirs are summarized in Table 10.1.

Table 10.1 Summary of flow regimes for natural depletion and EOR stage in unconventional reservoirs (Alfarge et al., 2017b).

Primary production			EOR production		
Areas	**Oil phase**	**Gas phase**	**Areas**	**Oil phase**	**Gas phase**
HF	Convection	Convection	HF	Convection	Convection
NF	Diffusion	Diffusion	NF	Dispersion	Convection
Matrix	Diffusion	Diffusion	Matrix	Diffusion	Diffusion

10.3.2.2 Effect of nanopore confinement on minimum miscible pressure

MMP in unconventional reservoirs can differ from one reservoir to another based on the nano-size confinement in shale (Zhang, 2019). Zhang et al. (2016) researched the shift value in critical properties of shale fluids in confined pores and reported that MMP could be reduced significantly for pore diameters below 50 nm. Zhang et al. (2018) simultaneously researched both the pressure inequalities of vapor and liquid phases as well as the shift in fluid critical properties using their flash calculations. They reported that the MMP value under confinement is 5.4% lower than that found in the bulk fluid phase (non-nanopore confinement). They also concluded that molecular diffusion and capillary pressure can improve oil recovery by gas injection. The main reason behind the reduction in MMP value is due to the tie-line length caused by nanoscale pore confinement, which consequently contributes to a lower MMP (Zhang, 2019). Used the perturbed-chain statistical associating fluid theory (PCSAFT) with a parachor model to estimate the CO_2 MMP for bulk phase versus nanopores, and observed a reduction in MMP primarily caused by a shift in critical properties Teklu et al. (2014) reported no change in MMP due to the influence of large capillary pressure on phase equilibrium. Adel et al. (2018) performed experiments for CO_2 injection in shale cores and observed that recovery factors will still increase for pressures above the MMP; therefore, based on this observation, they concluded that MMP is not an important parameter for CO_2-EOR in unconventional reservoirs. Notably, Adel et al. (2018) measured MMPs using slim tube experiments without considering the nanopore confinement effect. They also concluded that MMP would decrease due to the confinement effect based on recent research (Zhang, 2019).

10.3.3 Effect of nanopore confinement on natural gas-based EOR

Natural gas injection is considered one of the most effective enhanced oil recovery methods for unconventional reservoirs. Both lean and rich gases have been used to enhance oil recovery in shale oil reservoirs. Recently,

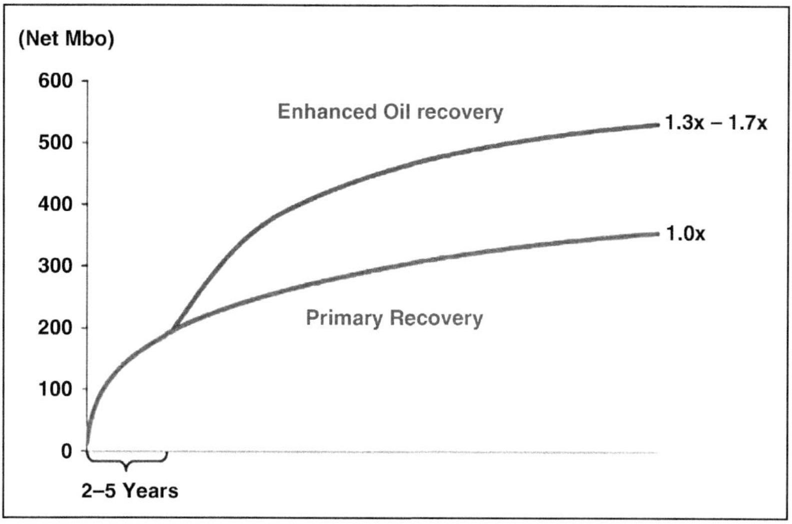

Figure 10.8 ***The reported expected recovery from huff-n-puff gas injection into the Eagle Ford (Hoffman, 2018).***

seven pilot projects were initiated in 49 wells. Here, natural gas was injected via a huff-n-puff mode in Eagle Ford and showed very promising results (Hoffman, 2018). Fig. 10.8 shows that the wells under the operations were predicted to produce 30%–70% more oil after the huff-n-puff gas injection was deployed. The confinement is clearly significant in low permeability formations due to the nanopores.

Based on the experiment data from the Eagle Ford core samples, around 80% of the nanopores were 20 nm or less (Sanaei, Jamili, Callard, & Mathur, 2014). Zhang et al. (2018) calculated the fluid properties under 5 nm, 10 nm, and 15 nm as shown in Table 10.2.

Table 10.2 Critical temperatures and pressures under different pore sizes (Zhang et al., 2018).

	Critical temperatures (K)				Critical pressures (bar)			
Components	**Bulk**	**15 nm**	**10 nm**	**5 nm**	**Bulk**	**15 nm**	**10 nm**	**5 nm**
CO_2	304.2	296.7	293.1	282.2	73.8	72.0	71.1	68.4
N_2	126.2	123.2	121.7	117.4	33.9	33.1	32.7	31.6
CH_4	190.6	185.9	183.6	176.8	46.0	44.9	44.3	42.7
C_2-nC_5	274.7	266.6	262.5	250.6	37.0	35.9	35.3	33.7
C_6-C_{10}	438.7	421.4	413.0	388.1	25.4	24.4	24.0	22.5
C_{11+}	740.3	701.4	682.4	627.0	17.7	16.8	16.3	15.0

Zhang et al.'s (2018) results show that critical temperatures and pressures suppress as the pore sizes reduce. Zhang et al. (2018) also calculated the phase equilibrium and obtained the bubble point pressure under different pore sizes as shown in Fig. 10.9. Their results indicate that the bubble point pressure significantly decreases, especially for the small pores, which further impacts the oil recovery in tight oil reservoirs. The effect of pore confinement in reservoir simulation was, therefore, determined able to increase in well performance during the CH_4 injection process (Zhang et al., 2018).

The effect of nanopores confinement on well production is shown in Fig. 10.10. The estimated incremental oil recovery factor at the end

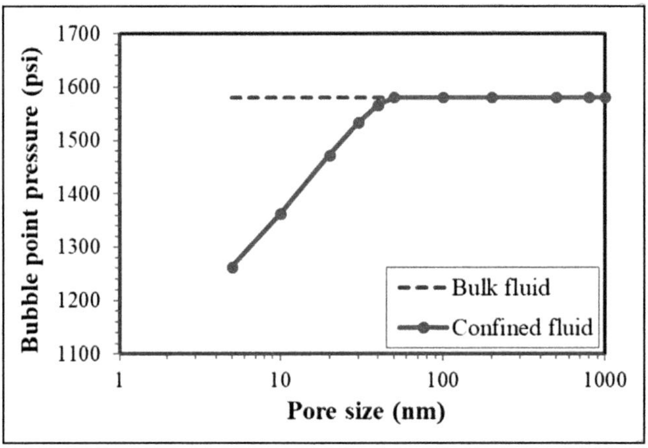

Figure 10.9 ***The reported expected recovery from the huff-n-puff gas injection into the Eagle Ford shale field study site (Zhang et al., 2018).***

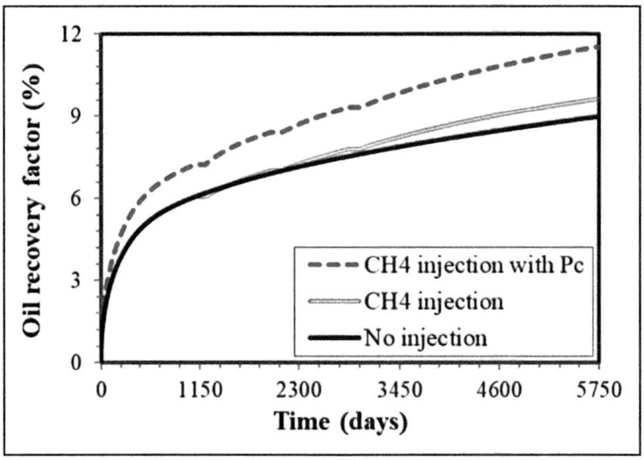

Figure 10.10 ***Effect of nanopore confinement on well performance in a 15-year period (Zhang et al., 2018).***

of 15 years was 2.3% and at a pore size of 10 nm and 0.8% bulk (Zhang et al., 2018).

This clearly indicates that the effect of confinement has a positive influence on the cyclic CH_4 injection effectiveness. Due to the nanopores confinement, the miscibility minimum pressure of the mixture and the oil viscosity decreased and led to the improvement of well performance under miscible gas huff-n-puff operations. Moreover, the bubble point pressure in the confined space was lower than that for the bulk phase, which means that more time was allotted for the single-phase production during production (Zhang et al., 2018). Hence, confinement should be correctly evaluated in the calculations of miscible gas-based EOR in shale reservoirs.

10.3.4 Effect of nanopore confinement on N_2-EOR

N_2 injection has not been widely used because of the disadvantage of nitrogen's physicochemical properties (Wu et al., 2019). Nitrogen gas has a high MMP which makes it difficult to dissolve in crude oil. Consequently, the injected nitrogen does not usually offer the benefits of reduction in oil viscosity or the interfacial tension needed to lower the capillary trapping effect. However, nitrogen injection has many other benefits such as good elastic energy during flooding, which helps maintain reservoir pressure (Lu, Li, Li, Hou, Zhang, 2017; Nguyen, Carey, Viswanathan, Porter, 2018; Yu, Li, Sheng, 2016a; Yu, Meng, Sheng, 2016b; Yu & Sheng, 2015). N_2 injection has other advantages such as its abundance in the atmosphere, relatively low-cost and well-established harnessing and capture of technology, as well as its noncorrosive property (Du & Cheng, 2014). This makes nitrogen an ideal candidate for injection into shale oil reservoirs to improve oil recovery. Siregar et al. (2007) studied oil recovery by injecting nitrogen during laboratory experiments; however, their results did not show N_2 injection leading to higher oil recovery. Other researchers pointed out that oil could be recovered at a high and stable production rate under N_2 flooding (Yu & Sheng, 2016). Joslin, Ghedan, Abraham, & Pathak, 2017 studied various EOR techniques, and their results showed N_2 flooding to be the most effective method via pressure maintenance when the permeability value is lower than 0.03 mD (Wu et al., 2019). Clearly, nitrogen molecules are very small compared with other gases. In the confined nanopores of shale reservoirs, the molecular diffusion is the most dominant flow. Therefore, N_2 diffusivity has an advantage over other gases due to its superior performance in N_2 injection.

10.4 SUMMARY

- The confined space in shale nanopores can significantly alter the phase behavior of reservoir fluids either by changing the critical properties of constituent components, such as critical pressures and temperatures, or by taking advantage of induced high capillary pressure.
- The effect of pore confinement was not limited to the shifting of phase behavior for shale liquids. This effect also results in abnormal production performance in the primary depletion stage.
- The pore confinement significantly affects many EOR methods applied in unconventional reservoirs. The pore confinement impacts the performance of EOR methods positively and sometimes negatively.
- According to the literature, MMP could be significantly reduced due to pore confinement. The reduction in MMP could enhance the performance of many miscible gas-based EOR methods in unconventional reservoirs.
- The effect of nanopore confinement is so significant that it should be implemented in reservoir simulation to capture the real physics during the modeling process for miscible gas-based EOR in unconventional reservoirs.

References

Adel, I. A., Tovar, F. D., Zhang, F., & Schechter, D. S. (2018). *The impact of MMP on recovery factor during CO_2—EOR in unconventional liquid reservoirs*. Society of Petroleum Engineers. https://doi.org/10.2118/191752-MS.

Alfarge, D., Wei, M., & Bai, B. (2017a). Factors affecting CO_2-EOR in shale-oil reservoirs: Numerical simulation study and pilot tests. *Journal of Energy & Fuel*. https://doi.org/10.1021/acs. energyfuels.7b01623.

Alfarge, D., Wei, M., & Bai, B. (2017). Comparative study for CO_2-EOR and natural gases injection-techniques for improving oil recovery in unconventional oil reservoirs. In: Carbon Management Technology Conference. https://doi.org/10.7122/485175-MS.

Alharthy, N. S., Nguyen, T., Teklu, T., Kazemi, H., & Graves, R. (2013). *Multiphase compositional modeling in small-scale pores of unconventional shale reservoirs*. Society of Petroleum Engineers. https://doi.org/10.2118/166306-MS.

Baihly, J. D., Altman, R. M., & Aviles, I. (2012). Has the economic stage count been reached in the Bakken shale? In SPE-159683, Proceedings of the SPE Hydrocarbon Economics and Evaluation Symposium. Calgary, AB, September 24–25.

Brusllovsky, A. L. (1992). Mathematical simulation of phase behavior of natural multicomponent systems at high pressure with an equation of state. *SPE Reservoir Engineering*, 7, 117.

Dong, Z., Holditch, S. A., McVay, D., & Ayers, W. B. (2011). *Global unconventional gas resource assessments*. Society of Petroleum Engineers. https://doi.org/10.2118/148365-MS.

Du, B., & Cheng, L. (2014). Experimental study of enhanced oil recovery with CO_2 slug+ N_2 flood in low permeability reservoir. *Geosystem Engineering*, *17*(5), 279–286.

Hoffman, B. T. (2018). Huff-n-puff gas injection pilot projects in the Eagle Ford. SPE-189816-MS, Proceedings of the 18th SPE improved oil recovery symposium, Tulsa, OK, USA, 14–18 April 2012, https://doi.org/10.2118/189816-MS.

Honarpour, M. M., Nagarajan, N. R., Orangi, A., Arasteh, F., & Yao, Z. (2012). *Characterization of critical fluid PVT, rock, and rock-fluid properties—Impact on reservoir performance of liquid rich shales*. Society of Petroleum Engineers. https://doi.org/10.2118/158042-MS.

Huang, S., & Sheng, J. (2018). Effect of nanopore confinement on crude oil thermal-oxidative behavior. *Energy Fuels*. https://doi.org/10.1021/acs.energyfuels.8b02177.

Jin, L., Sorensen, J. A., Hawthorne, S. B., Smith, S. A., Bosshart, N. W., Burton Kelly, M. E., et al. (2016). *Improving oil recovery by use of carbon dioxide in the Bakken unconventional system: A laboratory investigation*. Society of Petroleum Engineers. https://doi.org/10.2118/178948-PA.

Jin, L. & Wojtanowicz, A. K. (2010). Coning control and recovery improvement using in situ water drainage/injection in bottom-water-drive reservoir. SPE 129663, Proceedings of SPE improved oil recovery symposium, Tulsa, OK, April 24–28.

Jin, L., & Wojtanowicz, A. K. (2014). Progression of injectivity damage with oily wastewater in linear flow. *Petroleum Science, 11*, 550–562.

Joslin, K., Ghedan, S., Abraham, A., Pathak, V. (2017). *EOR in tight reservoirs, technical and economical feasibility*. In: SPE Unconventional Resources Conference. Society of Petroleum Engineers.

Kuila, U., & Prasad, M. (2011). *Surface area and pore-size distribution in clays and shales*. Presented at the SPE Annual Technical Conference and Exhibition, Denver, Colorado.

Kurtoglu, B., Rasdi, F., Salman, A., Kazemi, H. (2013). *Evaluating long term flow regimes in unconventional oil reservoirs with diverse completion technology*. SPE 167145, Proceedings of the SPE Unconventional Resources Conference-Canada, Calgary, AB, November 5–7, 2013.

Lu, T., Li, Z., Li, J., Hou, D., & Zhang, D. (2017). Flow behavior of N_2 huff and puff process for enhanced oil recovery in tight oil reservoirs. *Scientific Reports*, 7(1), 15695.

Nguyen, P., Carey, J. W., Viswanathan, H. S., & Porter, M. (2018). Effectiveness of supercritical-CO_2 and N_2 huff-and-puff methods of enhanced oil recovery in shale fracture networks using microfluidic experiments. *Applied Energy, 230*, 160–174.

Nojabaei, B. (2015). Phase behavior and flow analysis of shale reservoirs using compositionally-extended black-oil approach (PhD dissertation). https://etda.libraries.psu.edu/catalog/27487.

Nojabaei, B., Johns, R. T., & Chu, L. (2013). Effect of capillary pressure on phase behavior in tight rocks and shales. *SPEREE, 16*(3), 281–289.

Ran, B., & Kelkar, M. (2015). *Fracture stages optimization in Bakken shale formation*. URTeC 2154796, Proceedings of the Unconventional Resources Technology Conference, San Antonio, TX, July 20–22.

Salako, O. & MacBeth, C. (2015). *Effective imaging of reservoir fluid changes*. Proceedings of the 77th EAGE Conference and Exhibition, Madrid, June 2015.

Sanaei, A., Jamili, A., Callard, J., Mathur, A. (2014). Production modeling in the Eagle Ford gas condensate window: Integrating new relationships between core permeability, pore size, and confined PVT properties. In Proceedings of the SPE Western North American and Rocky Mountain Joint Meeting, Denver, CO, USA, 16–18 April.

Siregar, S., Hidayaturobbi, A., Wijaya, B., Listiani, S. N., Adiningrum, T., Irwan, I., et al. (2007). Laboratory experiments on enhanced oil recovery with nitrogen injection. *Journal of Engineering Science and Technology, 39*(1), 20–27.

Teklu, T. W., Alharthy, N., Kazemi, H., Yin, X., Graves, R. M., & AlSumaiti, A. M. (2014). *Phase behavior and minimum miscibility pressure in nanopores*. Society of Petroleum Engineers. https://doi.org/10.2118/168865-PA.

Wu, S., Li, Z., Wang, Z., Sarma, H. K., Zhang, C., & Wu, M. (2019). *Investigation of CO_2/N_2 injection in tight oil reservoirs with confinement effect.* Energy Science & Engineering. https://doi.org/10.1002/ese3.578.

Yu, Y., Li, L., & Sheng, J. J. (2016a). *Further discuss the roles of soaking time and pressure depletion rate in gas huff-n-puff process in fractured liquid-rich shale reservoirs.* In: SPE Annual Technical Conference and Exhibition. Society of Petroleum Engineers.

Yu, Y., Meng, X., & Sheng, J. J. (2016b). Experimental and numerical evaluation of the potential of improving oil recovery from shale plugs by nitrogen gas flooding. *Journal of Unconventional Oil and Gas Resources., 15*, 56–65.

Yu, Y., & Sheng, J. J. (2015). *An experimental investigation of the effect of pressure depletion rate on oil recovery from shale cores by cyclic N2 injection.* In: Unconventional Resources Technology Conference, San Antonio, Texas, 20–22 July 2015. Society of Exploration Geophysicists, American Association of Petroleum, pp. 548–557.

Yu, Y., & Sheng, J. J. (2016). *Experimental evaluation of shale oil recovery from Eagle Ford core samples by nitrogen gas flooding.* In SPE Improved Oil Recovery Conference. Society of Petroleum Engineers.

Yu, W., & Sepehrnoori, K. (2019). *Shale gas and tight oil reservoir simulation.* Elsevier Book. https://doi.org/10.1016/C2017-0-00263-X.

Zhang, K. (2019). CO_2 minimum miscibility pressure and recovery mechanisms in heterogeneous low permeability reservoirs (Master thesis). http://hdl.handle.net/10919/93728.

Zhang, Y., Di, Y., Shi, Y., & Hu, J. (2018). Cyclic CH_4 injection for enhanced oil recovery in the Eagle Ford shale reservoirs. *Energies, 11*(11), 3094. https://doi.org/10.3390/en11113094.

Zhang, K., Liu, Q., Wang, S., Feng, D., Wu, K., Dong, X., Dong, X., Chen, S., Chen, Z. (2016). *Effects of nanoscale pore confinement on CO_2 displacement.* SPE/AAPG/ SEG Unconventional Resources Technology Conference. San Antonio, Texas, USA.

CHAPTER 11

The impacts of geomechanics coupling on CO_2-EOR

Contents

11.1 Theory 218
  11.1.1 Flow type in unconventional liquid-rich reservoirs 218
  11.1.2 Molecular diffusion 218
  11.1.3 Dissolution 221
  11.1.4 Multicomponent adsorption 221
  11.1.5 Coupling between the geomechanics module and the reservoir flow model 222
11.2 Compositional models for the formation fluids 227
11.3 Reservoir modeling 228
11.4 Results and discussion 231
  11.4.1 Molecular diffusion effect 231
  11.4.2 Adsorption effect 233
  11.4.3 Dissolution effect (water solubility effect) 236
  11.4.4 Geomechanic coupling effects 237
11.5 Summary 241
References 242

Abstract

This chapter discusses numerical simulation methods of compositional models incorporated with LS–LR–DK (logarithmically spaced, locally refined, and dual permeability) models and local grid refinements of hydraulic fracture conditions to simulate the feasibility of CO_2 injection in shale oil reservoirs. Different mechanisms for CO_2 interactions with organic surfaces, shale brine, and shale oil were presented with different scenarios in numerical models. Molecular diffusion mechanisms, adsorption effects, and aqueous solubility effects were discussed in this chapter. In addition, linear elastic models and stress-dependent correlations were discussed in relation to geomechanic coupling effects on production and injection processes of CO_2-EOR in shale oil reservoirs. Some of the results presented in this chapter were validated by matching the performance of some CO_2 field pilot projects performed in the Bakken formation, which extends over portions of North Dakota and Montana.

Keywords: geomechanic coupling; CO_2-EOR mechanisms in shale and tight oil reservoirs; CO_2-EOR huff-n-puff; EOR history matching; reservoir simulation/field production data

http://dx.doi.org/10.1016/B978-0-12-818343-4.00011-5

11.1 Theory

11.1.1 Flow type in unconventional liquid-rich reservoirs

Gravity drainage, physical diffusion, viscous flow, and capillary forces are common forces, which control the fluid flow in the porous media. However, one force might overshadow the contributions of other forces depending on the reservoir properties and operating conditions. Molecular diffusion is defined as the movement of molecules caused by Brownian motion or composition gradients in a mixture of fluids (Mohebbinia et al., 2017). This type of flow would be the most dominant flow in the fractured reservoirs with a low-permeability matrix when gravitational drainage is inefficient. Gas injection is the most common EOR process affected by molecular diffusion considerations (Hoteit and Firoozabadi, 2009; Mohebbinia et al., 2017). Ignoring or specifying incorrect diffusion rates during the simulation process can lead to an overestimate or underestimate of oil recovery production by injected gas. Such inaccuracies are not only due to the variance in the miscibility process between the injected gas and the formation oil but also due to the path changes for the injected gas species from the fractures to the formation-matrix. The Péclet number (*Pe*) is a class of dimensionless numbers, which has been used to measure the relative importance of molecular diffusion flow to convection flow. This number can be calculated as shown in Eq. (11.1). If the *Pe* number is less than 1, the molecular diffusion is the dominant flow; however, if the *Pe* is greater than 50, convection is the dominant flow. The dispersion flow is dominant when the *Pe* ranges from 1–50 (Hoteit and Firoozabadi, 2009).

$$Pe = \frac{\text{Diffusion time}}{\text{Convection time}} = \frac{L^2/D}{L/v} = \frac{Lv}{D} \tag{11.1}$$

where v is the bulk velocity, L is a characteristic length, and D is the molecular diffusion coefficient.

11.1.2 Molecular diffusion

The main transportation process through the shale reservoirs with nano-Darcy permeability is subjected to Fick's law. The mechanism responsible for this process is called the molecular diffusion mechanism. The molecular diffusion process is more dominant in tight reservoirs with a significant heterogeneity (Hoteit and Firoozabadi, 2009). As the heterogeneity increases, the molecular diffusion mechanism will be more pronounced. This is because the more heterogeneous a reservoir is, the greater the area of lower permeability and higher residual oil saturation, which are two of the most important parameters contributing to the molecular diffusion mechanism's dominance (Yu and Sepehrnoori, 2019). The effectiveness of this mechanism is also dependent on the type of injected gases. For example, shale oil has a high concentration of light components, which favor the production of natural gases such as methane. At the same time, shale oil has a low concentration of CO_2. Therefore, the molecular diffusion mechanism in the simulation process for injected CO_2 has a significant effect on the obtained oil recovery. However, the molecular diffusion mechanism in the simulation process for injected methane has a minor effect on obtained oil recovery. Thus, the binary molecular diffusion effect of injected CO_2 on the production of shale oil is simulated in this work by using two different experimental correlations conducted by Sigmund (1976a, 1976b) and Wilke and Chang (1955). The following polynomial equation was fitted with Sigmund's (1976a, 1976b) experimental results.

$$D_{ij} = \frac{\rho_k^0 D_{ij}^0}{\rho_k}(0.99589 + 0.096016\rho_{kr} - 0.22035\rho_{kr}^2 + 0.032874\rho_{kr}^3) \tag{11.2}$$

where D_{ij} is the binary diffusion coefficient in cm^2/s between components i and j in the mixture, $\rho^0 D_{ij}^0$ is the zero-pressure limit of the density-diffusivity product, ρ_k is the density of the diffusion mixture in kg/m^3, and ρ_{kr} is the reduced density, which can be calculated by Eq. (11.3). The subscript k denotes the phase which could be either water, oil, or gas.

In the simulator, the product of mixture density and diffusion coefficient can be calculated by Eq. (11.4), and the diffusion coefficient of the component i in the mixture can be calculated by Eq. (11.5).

$$\rho_{kr} = \rho_k \frac{\sum_i^{nc} y_{ik} v_{ci}^{5/3}}{\sum_i^{nc} y_{ik} v_{ci}^{2/3}} \tag{11.3}$$

where y_{ik} is the mole fraction of the i species in phase k, and v_{ci} is the critical volume of the i species.

$$\rho_k^0 D_{ij}^0 = \frac{0.18583\, T^{0.5}}{\sigma_{ij}^2 \mho_{ij} R} \cdot \left\{ \frac{1}{M_i} + \frac{1}{M_j} \right\}^{0.5} \tag{11.4}$$

where R is the universal gas constant, T is the temperature, M is the molecular weight, σ_{ij}^2 is the collision diameter between i and j, and $\mho_{ij}$ is the collision integral of the Lennard–Jones potential.

$$D_i = \frac{1-y_i}{\sum_{i\neq j} y_i D_{ij}^{-1}} \tag{11.5}$$

where D_i is the diffusion coefficient of component i in the mixture, and y_i is the mole fraction of component i. In the above equations, the collision diameter and the collision integral of the Lennard–Jones potential are related to the component critical properties through the following equations:

$$\sigma_i = \left(2.3551 - 0.087\,\omega_i \cdot \left\{\frac{T_{ci}}{P_{ci}}\right\}^{1/3}\right) \tag{11.6}$$

$$\varepsilon_i = K_B(0.7915 + 0.1963\,\omega_i)T_{ci} \tag{11.7}$$

$$\sigma_{ij} = \frac{\sigma_i + \sigma_j}{2} \tag{11.8}$$

$$\varepsilon_{ij} = \sqrt{\varepsilon_i \varepsilon_j} \tag{11.9}$$

$$T_{ij}^{*} = \frac{K_B}{\varepsilon_{ij}} \tag{11.10}$$

$$\mho_{ij} = 1.06306\left\{T_{ij}^{*}\right\}^{-0.1561} + 0.193\exp\left(-0.47635\,T_{ij}^{*}\right) + 1.03587\exp\left(-1.52996T_{ij}^{*}\right) + 1.7647\exp\left(-3.89411T_{ij}^{*}\right) \tag{11.11}$$

where K_B is the Boltzmann's constant (=1.3805E–16 ergs/K).

The second method is Wilke and Chang's (1955) correlation, in which the diffusion factor, $D_i\mu / T$, is correlated with the molar volume of solutes and the molecular volume of solvents based on different experimental data. The following equation fitted their experimental data:

$$D_i = \frac{7.4E-8\left(M_i^{'}\right)^{0.5} T}{M v_{bi}^{0.6}} \tag{11.12}$$

where, $M_i^{'} = \sum_{j\neq i} x_j M_j / \left(1-x_i\right)$, $v_{bi} = 0.285\,v_c^{1.048}$, D_i is the diffusion of component I in the mixture, $M_i^{'}$ is the molecular weight of solvent, T is the temperature, μ is the viscosity, and v_{bi} is the partial molar volume of component i at the boiling point, x_i is the mole fraction of component i, M_j is the molecular weight of component j, and v_c is the critical volume.

11.1.3 Dissolution

The general trapping mechanisms have not been clearly investigated in previous studies. They are responsible for the immobilization of CO_2 in geological media, and their functions include dissolution, as well as residual and mineral trapping. In this study, the aqueous solubility for CO_2 has been considered. Henry's Law has been used to simulate the dissolution effects on CO_2-EOR performance. Pressure, temperature, and salinity effects have been considered on solubility of CO_2 in shale brine. Harvey's correlation has been used to calculate Henry's law constant to estimate the real CO_2 solubility in shale brine.

According to Li and Nghiem (1986), calculating the dissolution of the component *i* in the reservoir fluid can be done by Henry's law as follows:

$$f_{iw} = y_{iw} H_i \tag{11.13}$$

where f_{iw} is the fugacity of component *i* in aqueous phase, y_{iw} is the mole fraction of component *i* in the aqueous phase, and H_i is Henry's constant of component *i*. To make f_{iw} equal to the fugacity of component *i* in the gas phase, f_{ig}, both the aqueous phase and the gaseous phase are assumed to be in thermodynamic equilibrium (Kim, Park, & Lee, 2015). f_{ig} can be computed from Peng and Robinson's (1976) equation of state (EOS). Henry's constant H_i at any pressure and temperature is calculated by the following equation (Stumm and Morgan, 1996).

$$\text{In}\, H_i = \text{In}\, H_i^* + \frac{\overline{v}_i \left(p - p^*\right)}{RT} \tag{11.14}$$

where H_i^* is Henry's constant for component *i* at reference pressure, p^*, v_i^- is the partial molar volume of component *i*, p^* is the reference pressure, *R* is the universal gas constant, and *T* is the temperature.

11.1.4 Multicomponent adsorption

Kim et al. (2015) reported that CO_2 is preferably adsorbed in organic shale in contrast to methane; thus, CO_2 is 5 times stronger than CH_4. Since shale oil has a high methane mole fraction, injecting CO_2 can enhance shale oil recovery by adsorption effects. Therefore, CO_2 adsorption has been simulated in this research. A multicomponent extended isothermal Langmuir coefficient has been used to simulate the adsorption effects on CO_2-EOR in shale oil reservoirs. The following equation has been used to calculate CH_4 and CO_2 binary gas sorption:

$$\omega_i = \frac{\omega_{i,\max} B_i y_{ig} P}{1 + P\sum_i B_j y_{ig}} \tag{11.15}$$

where ω_i is the moles of the adsorbed component i per unit mass of rock. $\omega_{i,\max}$ is the maximum moles of adsorbed component i per unit mass of rock, B_i is the parameter for the Langmuir isotherm relation, y_{ig} is the molar fraction of the adsorbed component i in the gas phase, and P is the pressure.

11.1.5 Coupling between the geomechanics module and the reservoir flow model

To simulate the geomechanic effects in shale reservoirs, this chapter combines stress-dependent correlations with a linear-elastic constitutive model. To mimic the reduction in the porosity and permeability of the formation matrix, as well as the natural fractures and hydraulic fractures during the production and injection process, different stress-dependent porosity and permeability correlations have been combined with linear-elastic constitutive models in this study. First, exponential correlation is applied to compute stress-dependent properties as shown in Eqs. (11.16) and (11.17).

$$\phi = \phi_0 \exp^{-a(\sigma' - \sigma_0')} \tag{11.16}$$

$$K - K_0 \exp^{-a(\sigma' - \sigma_0')} \tag{11.17}$$

where σ' is the effective stress and a and b are experimental coefficients. The subscript 0 indicates the initial state. Experimental coefficients were gathered from literature dealing with the Bakken formation. Cho, Ozkan, and Apaydin (2013) present more details about these coefficients, including the porosity and permeability multiplier for both the matrix and fractures. This enabled the coupling of the porosity and permeability multipliers to the reservoir fluid flow simulator by inserting tables.

Second, two linear-elastic constitutive models wee simulated to consider different aspects of geomechanic effects. These models are described as:

1. **3-D two-way coupling (the iterative coupled approach)**: In this approach, the geomechanics calculations are performed one step behind the reservoir flow calculations. In this type of coupling, the data calculated in the reservoir simulator and in the geomechanic module would

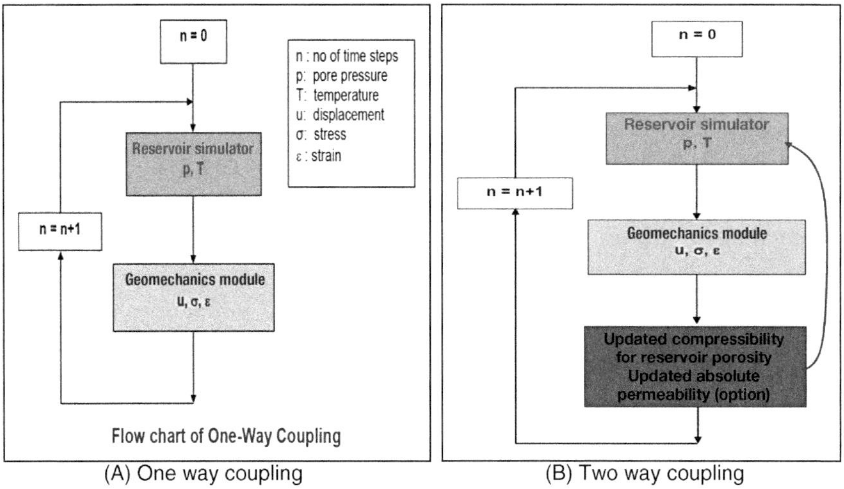

Figure 11.1 ***The systematic procedure for both types of geomechanic coupling by using linear-elastic constitutive models.***

exchange back and forth (CMG, 2018). As a result, the reservoir flow simulation process would be affected by the geomechanic responses. The systematic procedure for this type of coupling is shown in Fig. 11.1.

2. **3-D one-way coupling (explicit coupled approach)**: This method is considered a special case of the iterative coupled approach where the information from a reservoir flow simulator is sent to a geomechanic module, but the calculations in the geomechanic module are not sent back to the reservoir simulator (Tran et al., 2005). As a result, the reservoir flow is not affected by the geomechanical responses calculated in the geomechanic module. This method is useful when an empirical model for the compaction/dilation process is used in the reservoir flow simulator to match the observed pressure in the field. Then, the geomechanic module is used to calculate the actual deformation in the field (Tran, Nghiem, & Buchanan, 2005). The systematic procedure for this type of coupling is shown in Fig. 11.1.

Most numerical simulation studies in this area assume that the porous media, and thereafter the numerical grids, do not deform or move on a bulk basis (Tran et al., 2005). Each grid cell bulk volume, V_b, will remain constant at its original value. V_b^0. Accordingly, the changes happening in the pore volume are merely due to the changers in pore diameters, without any changes in the pore centers themselves moving in space (Tran et al., 2005). In contrast to previous numerical simulation studies which ignore any bulk medium movement accompanying rock expansion and contraction, this study has accounted for the fact that porous media will deform and move to some extent. Commonly, local deformation of porous media is expressed as a change in bulk volume V_b. As a result, the volumetric strain would be $\varepsilon_v = \left(V_b^0 - V_b\right)/V_b^0$ The basis for coupling these linear-elastic constitutive models with our flow model starts with a continuity equation in a deformable medium as shown in Eq. (11.18):

$$\frac{\partial}{\partial t}\left(\rho_f \phi\left(1-\varepsilon_v\right)\right) - \nabla.\left\{\sum_{j=1}^{np} \rho_f \frac{k}{\mu}\left[\nabla p - \rho_f b\right]\right\} = Q_f \tag{11.18}$$

where ϕ is the true porosity, which is defined as follows:

$$\phi = \text{True porosity} = \frac{\text{Current pore volume}}{\text{Current bulk volume}} = \frac{V_p}{V_b} \tag{11.19}$$

The volumetric strain in Eq. (11.18) reflects the changes in the bulk volume of the porous media. These changes in the bulk volumes of porous media are ignored in conventional simulation methods. In a conventional simulator, different porosities (called reservoir porosities) are defined as:

$$\phi^* = \text{Reservoir porosity} = \frac{\text{Current pore volume}}{\text{Current bulk volume}} = \frac{V_p}{V_b} \tag{11.20}$$

By comparing Eqs. (11.19) and (11.20), we can produce Eq. (11.21) as follows:

$$\phi^* - (1-\varepsilon_v)\phi \tag{11.21}$$

Substituting Eq. (11.21) into Eq. (11.18) gives:

$$\frac{\partial}{\partial t}\left(\rho_f \phi^*\right) - \nabla.\left\{\sum_{j=1}^{np} \rho_f \frac{k}{\mu}\left[\nabla p - \rho_f b\right]\right\} = Q_f \tag{11.22}$$

Strain displacement relations can be defined depending on the definition of strain and rotation caused by moving a rigid body (Jabbari, Ostadhassan, & Rabeie, 2015). To explain how the geomechanic effects are reflected in this task, we need to start with the force equilibrium equation:

$$\nabla.\sigma - \rho_r B = 0 \tag{11.23}$$

where σ is the stress tensor and B is the force per unit mass that accounts for gravity. Let u be the displacement vector which is the shortest distance from the initial to the final position of a point on a particular body with a symmetric matrix. Hence, the gradient of the displacement vector u can be decomposed as:

$$\nabla u = \frac{1}{2}\left[\nabla u + \nabla u^T\right] + \frac{1}{2}\left[\nabla u + \nabla u^T\right] \tag{11.24}$$

where the superscript T denotes the matrix transpose. On the right side of Eq. (11.24), the first term is a symmetric matrix equivalent to the strain tensor, which is a result of changing the length or shape within a body. The second term is a skew-symmetric matrix equivalent to the rotation tensor R, which is the result of moving a rigid body (Tran et al., 2005). Therefore,

$$\varepsilon = \frac{1}{2}\left[\nabla u + \nabla u^T\right] \tag{11.25}$$

and

$$R = \frac{1}{2}\left[\nabla u + \nabla u^T\right] \tag{11.26}$$

In Eq. (11.23), the stress is the total stress tensor. In reality, only the effective stress can affect the strength of solid grains in the porous media (Tran et al., 2005). The following equation shows the total stress relationship to the effective stress

$$\sigma = \sigma' + \alpha pI \tag{11.27}$$

where α is Biot's coefficient, P is the pore pressure, and I is the identity matrix.

The constitutive relationship between stress, strain, and temperature in the geomechanic process in one dimension can be written in the following equation:

$$\sigma' = E\left(\varepsilon - \beta_r \Delta t\right) \tag{11.28}$$

where E is Young's modulus and β_r is the linear thermal expansion coefficient of the solid rock.

For multiple dimensions, the general constitutive relation is:

$$\sigma' = C : \varepsilon - \eta \Delta TI \tag{11.29}$$

where C is the tangential stiffness tensor (equivalent to Young's modulus in a 1-D linear case), and $\eta = E\beta_r / (1 - 2v)$ for 3-D and plane strain while $\eta = E\beta_r / (1 - v)$ for plane stress.

By substituting Eqs. (11.22), (11.24), and (11.26) into Eq. (11.23), the displacement equation results are as follows:

$$\nabla.\left[C:\frac{1}{2}\left(\nabla u+(\nabla u)^{T}\right)\right]+\nabla.\left[(\alpha p-\eta\Delta T)I\right]=\rho_{r}B \tag{11.30}$$

The displacement equation (Eq. (11.30)) depends on primary unknowns, which are P, T, and vector u, along with intermediate unknowns, which are the reservoir porosity ϕ^{*} and permeability tensor (K) (Tran et al., 2005).

According to the study's model, the reservoir porosity is a function of three variables (pore pressure, temperature, and mean stress) while the previous conventional simulation studies present reservoir porosity as the only function of the first two parameters. Therefore, in the simulation process of the cases which have been simulated in this work, the porosity is calculated as:

$$\phi_{n+1}^{*}=\phi_{n}^{*}+C_{n}^{0}\Delta p+C_{n}^{1}\nabla T \tag{11.31}$$

The parameters of Eq. (11.31) are:

$$C_{n}^{0}=\left(C_{0}+C_{2}a_{1}\right)n \tag{11.32}$$

$$C_{n}^{1}=\left(C_{1}+C_{2}a_{2}\right)n \tag{11.33}$$

$$C_{0}=\frac{1}{V_{b}^{0}}\left[\frac{dV_{p}}{dp}+V_{b}\alpha c_{b}\frac{d\sigma_{m}}{dp}+V_{p}\beta_{p}\frac{dT}{dp}\right] \tag{11.34}$$

$$C_{1}=-\frac{V_{b}}{V_{b}^{0}}\beta_{p} \tag{11.35}$$

$$C_{2}=-\frac{V_{b}}{V_{b}^{0}}\alpha c_{b} \tag{11.36}$$

$$a_{1}=\Gamma\left\{\frac{2}{9}\frac{E}{1-\nu}\left(C_{b}-C_{r}\right)\right\} \tag{11.37}$$

$$a_{2}=\Gamma\left\{\frac{2}{9}\frac{E}{1-\nu}\beta\right\} \tag{11.38}$$

The coefficients C_{n}^{0} and C_{n}^{1} are analogous in form to the formation compressibility and thermal expansion coefficient, respectively

(Tran et al., 2005). These coefficients do contain this information but also reflect the geomechanic response corresponding to pressure and temperature changes.

Nevertheless, the permeability tensor (K) is indirectly related to porosity as follows:

$$\phi^* = \phi^*(\phi, \varepsilon_v) \tag{11.39}$$

$$K = K(\phi^*) \tag{11.40}$$

However, there are some other approaches to link the permeability to the volumetric strain, such as:

$$\text{In}\left(\frac{k}{k_0}\right) = C_{n1}\varepsilon_v \tag{11.41}$$

where k and k_0 are the permeability at current and initial conditions, and C_{n1} is a parameter determined from laboratory experiments.

11.2 Compositional models for the formation fluids

In this study, two different compositional models were constructed to represent Bakken oil compositions. The oil used in this model has 42 API°, 725 SCF/STB, and 1850 psi as oil gravity, gas oil ratio, and bubble point pressure, respectively. Notably, compositional models are the most time-consuming models due to the large number of components in a typical formation oil. Our model has 34 components. With this many components, it would take a long time for the simulator to complete running one scenario. The common practice in numerical simulations for such situations is to carefully lump the reservoir oil components into a short representative list of pseudo-components. These pseudo-components must match the laboratory measured phase behavior data. The supplied data for such compositional models must have a description of the associated single carbon numbers and their fractions, saturation pressure test results, separator results, constant composition expansion test results, differential liberation test results, and swelling test results (CMG, 2018). These steps can be used for tuning the EOS to match the fluid behavior. WinProp-CMG (2018) has been used to lump the original 34 components into seven pseudo-components as presented in Table 11.1. WinProp offers an EOS based on fluid behavior and the PVT modeling package. In WinProp, the laboratory data for fluids

Table 11.1 Compositional data for the Peng–Robinson EOS in the model oil.

Component	Mole fraction	Critical pressure (atm)	Critical tempera-ture (K)	Acentric factor	Molar weight (g/gmole)
CO_2	0	7.28E + 01	3.04E + 02	0.225	4.40E + 01
N_2–CH_4	0.2704	4.52E + 01	1.90E + 02	0.0084	1.62E + 01
C_2H–NC_4	0.2563	4.35E + 01	4.12E + 02	0.1481	4.48E + 01
IC_5–CO_7	0.127	3.77E + 01	5.57E + 02	0.2486	8.35E + 01
CO_8–C_{12}	0.2215	3.10E + 01	6.68E + 02	0.3279	1.21E + 02
C_{13}–C_{19}	0.074	1.93E + 01	6.74E + 02	0.5672	2.20E + 02
C_{20}–C_{30}	0.0508	1.54E + 01	7.92E + 02	0.9422	3.22E + 02

can be imported and an EOS can be tuned to match the physical behavior for the lab data. Fluid interactions can be predicted; then, a fluid model can be created for use in CMG software. Table 11.2 presents the Peng–Robinson EOS fluid description and binary interaction coefficients of Bakken crude oil with different injected gases. Fig. 11.3 represents the two-phase envelope for the Bakken oil which was generated by using WinProp-CMG.

11.3 Reservoir modeling

The LS–LR–DK (logarithmically spaced, locally refined, and dual permeability) model constructed in this paper simulates the Bakken formation and was chosen for it well known ability to accurately simulate the fluid flow in fractured shale-oil reservoirs (Wan and Sheng, 2015). In this study, we built a numerical model which has the typical fluid and rock properties of the Bakken formation areas which have handled some of the CO_2-EOR pilot tests. Then, both the LS–LR–DK model and fluid model were combined to simulate compositional interactions of the reservoir fluid and the injected CO_2 during the enhanced oil recovery processes. In the models presented in this chapter, we injected CO_2 as a huff-n-puff process with the same operating parameters used in the pilot test to work in a hydraulically fractured well. Most of the potential EOR mechanisms for CO_2 were considered in the construction of this model. In this field case study, the production well was stimulated with five hydraulic fractures. The spacing between the hydraulic fractures is 200 ft. The simulated model includes two regions which are stimulated reservoir volume (SRV) and unstimulated reservoir volume (USRV) as shown in Fig. 11.2.

Table 11.2 Binary interaction coefficients for Bakken oil.

Component	CO_2	N_2–CH_4	C_2H–NC_4	IC_5–CO_7	CO_8–C_{12}	C_{13}–C_{19}	C_{20}–C_{30}
CO_2							
N_2–CH_4	1.01E − 01						
C_2H–NC_4	1.32E − 01	1.30E − 02					
IC_5–CO_7	1.42E − 01	3.58E − 02	5.90E − 03				
CO_8–C_{12}	1.50E − 01	5.61E − 02	1.60E − 02	2.50E − 03			
C_{13}–C_{19}	1.50E − 01	9.76E − 02	4.24E − 02	1.72E − 02	6.70E − 03		
C_{20}–C_{30}	1.50E − 01	1.45E − 01	7.79E − 02	4.27E − 02	2.51E − 02	6.00E − 03	

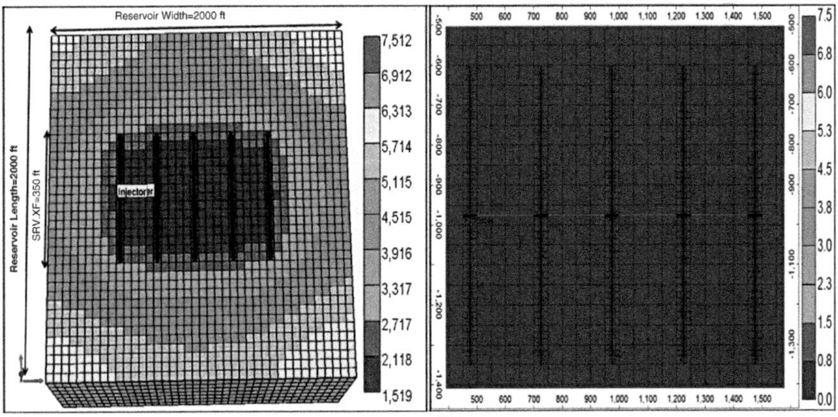

(A) (B)

Figure 11.2 (A) Average pressure in a depleted well in Bakken. (B) A closed view for SRV of production well.

The dimensions of the reservoir model are 2000 ft × 2000 ft × 42 ft, which corresponds to length, width, and thickness, respectively. The dimensions of the fractured region represent five fractures with a half-length of 350 ft in the J direction, a width of 0.001 ft in the I direction, and a fracture height of 42 ft in the K direction. Fracture conductivity is 15 mD ft. The other model input parameters are presented in Table 11.3 (Fig. 11.3).

Table 11.3 Model input parameters for the base case.

Parameter	Value	Unit
Model dimensions	2000 × 2000 × 42	ft
Production time	20	year
Reservoir top	8000	ft
Reservoir temperature	240	°F
Reservoir pressure	7500	psi
Initial water saturation	0.3	value
Total compressibility	1×10^{-6}	psi^{-1}
Matrix permeability	0.005	mD
Matrix porosity	0.085	value
Horizontal well length	1000	ft
Total number of fractures	5	value
Fracture conductivity	15	mD ft
Fracture half-length	250	ft
Fracture height	42	ft

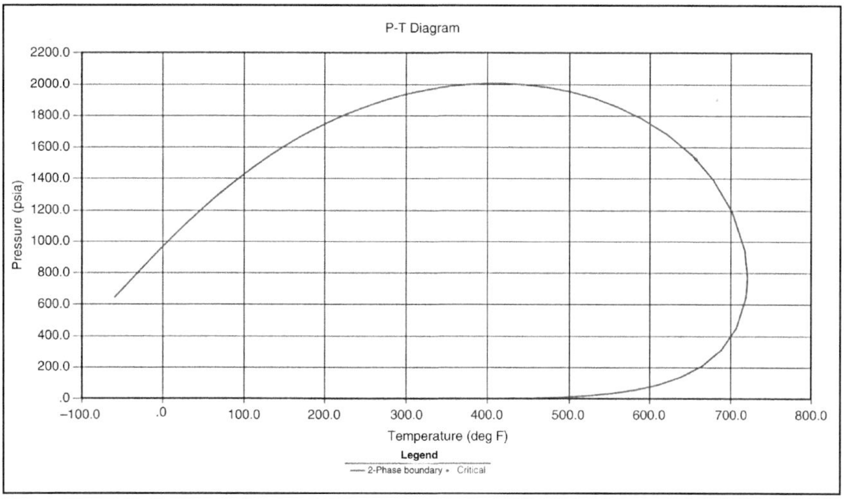

Figure 11.3 ***The two-phase envelope for Bakken oil which was generated by WinProp-CMG.***

11.4 Results and discussion

11.4.1 Molecular diffusion effect

Different scenarios have been simulated in both huff-n-puff scenarios and flooding processes. The results indicate that the molecular diffusion mechanism has a positive effect on CO_2-EOR performance in the huff-n-puff process as shown in Fig. 11.4A. This occurs when molecular diffusion mechanisms enable CO_2 to penetrate the shale matrix instead of solely concentrating on natural fractures as shown in Figs. 11.5 and 11.6. However, Fig. 11.4B shows the molecular diffusion mechanism having a negative effect on the CO_2-EOR performance due to the continuous injection process during flooding. The reason for this negative effect in the flooding process is due to two main issues: (1) In the flooding process, the reservoir pressure derived from the injection process is usually much higher than the minimum miscibility pressure (MMP) required for CO_2 as shown Fig. 11.9. This leads to a downgrade of the CO_2-EOR performance in naturally fractured reservoirs because the best CO_2-EOR performance can only be achieved at an injection pressure, which is below or near miscible conditions. This conclusion is consistent with the conclusions of Hoteit and Firoozabadi (2009). (2) The second reason is that the molecular diffusion mechanism cannot redistribute the injected CO_2 between natural fractures and the shale matrix as quickly as the injection process provided

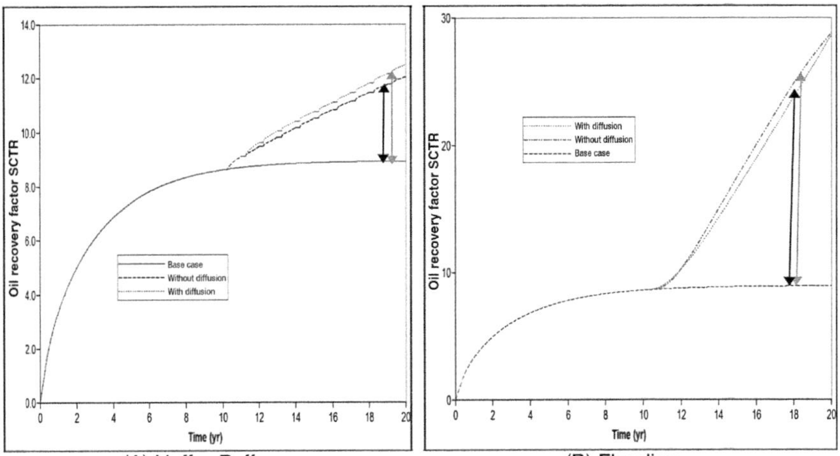

(A) Huff-n-Puff (B) Flooding

Figure 11.4 ***Effect of molecular diffusion mechanism on CO_2-EOR performance in ULR.*** (A) Huff-n-puff process and (B) during flooding.

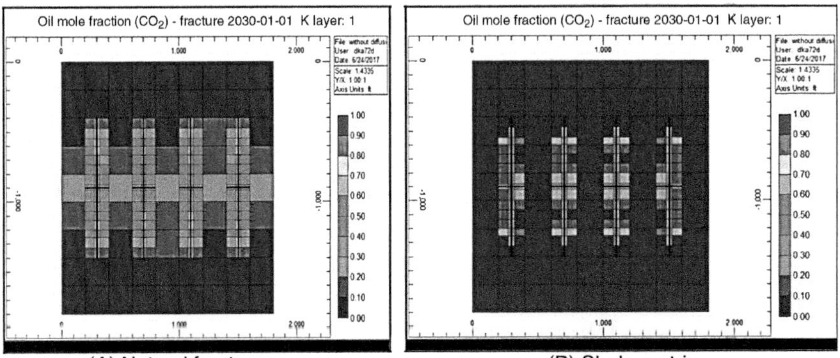

(A) Natural fractures (B) Shale matrix

Figure 11.5 ***The distribution of the injected CO_2 in huff-n-puff process (without diffusion).***

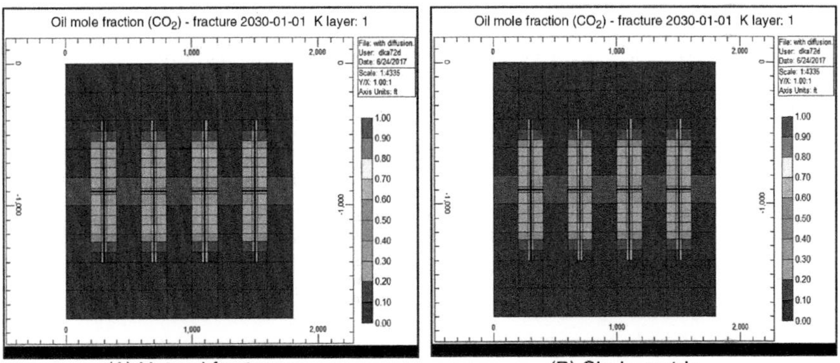

(A) Natural fractures (B) Shale matrix

Figure 11.6 ***The distribution of the injected CO_2 in huff-n-puff process (with diffusion).***

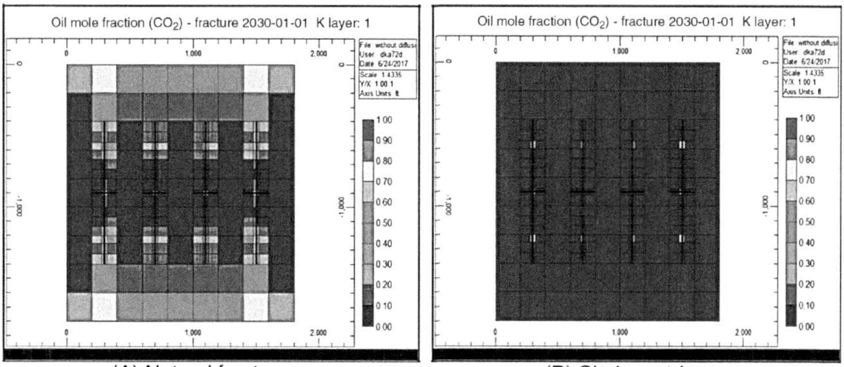

(A) Natural fractures (B) Shale matrix

Figure 11.7 ***Distribution of injected CO_2 in flooding process (without diffusion).***

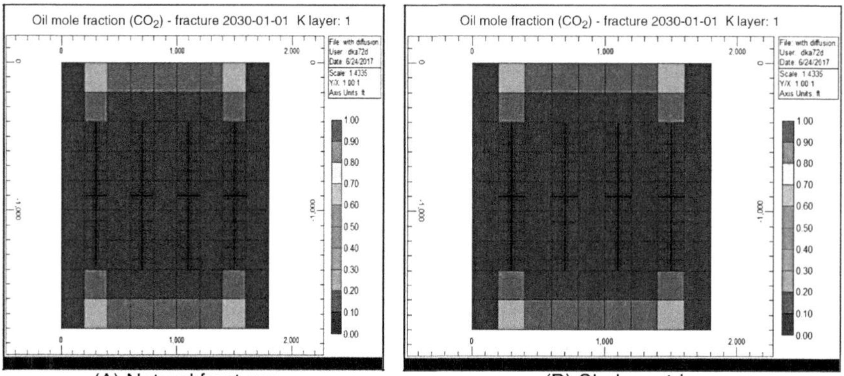

(A) Natural fractures (B) Shale matrix

Figure 11.8 ***Distribution of injected CO_2 in flooding process (with diffusion).***

by the flooding mode as shown in Figs. 11.7 and 11.8. The CO_2 molecular diffusion mechanism in the continuous injection process could improve oil recovery, but it takes too much time to do so (Fig. 11.9).

11.4.2 Adsorption effect

The sorption process has a clear impact on understanding fluid flow mechanisms of gas production from shale gas reservoirs. The adsorption rate of hydrocarbon components increases strongly with molecular weight (Jiang and Younis, 2016). The adsorption experimental data shown in Fig. 11.10 provides clear evidence of this relationship. Fig. 11.10 shows CO_2 possessing a greater adsorption affinity than the lighter components. Since shale oils have a higher light hydrocarbon content than most other minerals, injecting CO_2 can help extract these light components by facilitating their adsorption into the organic materials on the shale surfaces.

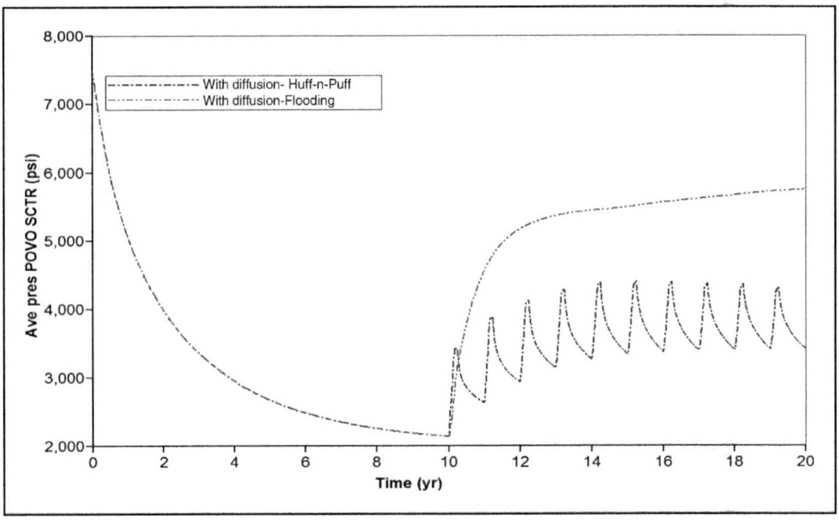

Figure 11.9 ***Average reservoir pressure in primary depletion and EOR stage in both the huff-n-puff model and flooding scenarios.***

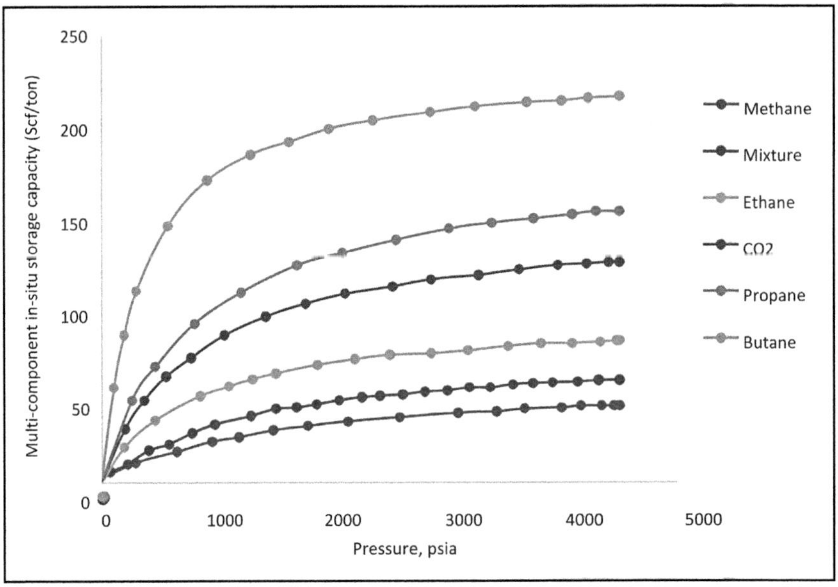

Figure 11.10 ***Laboratory-measured single-component and extended Langmuir isotherms for typical naturally occurring gases (Ambrose, Hartman, Akkutlu, 2011).***

Meanwhile, the light hydrocarbon components which were desorbed from the shale surface increased oil recovery by the adsorption mechanism. However, our simulation results indicated that adsorption has a negative effect on CO_2-EOR performance as shown in Fig. 11.11. The cause of the negative role attributed to the CO_2-EOR adsorption mechanism was clear to us. Since adsorption in the modeling process decreases the injected volume of CO_2 diffusion into shale oil because some portion of that injected CO_2 was adsorbed into shale surfaces as shown in Fig. 11.12. Therefore, the

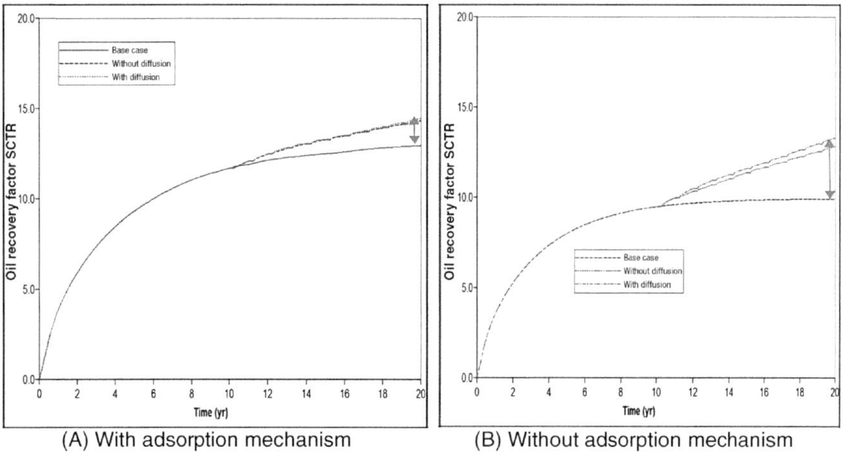

Figure 11.11 ***Effect of adsorption mechanism on CO_2-EOR performance in ULR.***

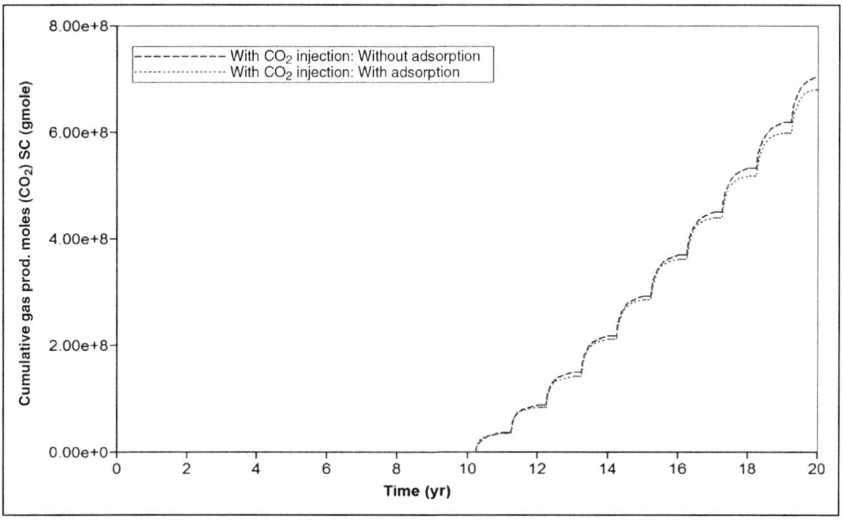

Figure 11.12 ***Volume of CO_2 produced using-adsorption models vs. non-adsorption models.***

negative effect of adsorption on the CO_2-EOR performance was not based on the adsorption itself but rather the interactions with molecular diffusion. The simulation results proved that the impact of CO_2 diffusion into shale oil has a greater potential to increase oil recovery compared to the adsorption effect.

11.4.3 Dissolution effect (water solubility effect)

Different scenarios have been simulated to define the effect of CO_2 solubility in shale brine on CO_2-EOR performance in shale oil reservoirs. Our simulation results indicated that water solubility has a negative effect on CO_2-EOR performance as shown in Fig. 11.13. Much like the adsorption mechanism performance in the last section, the water solubility model led to a reduced injected volume of CO_2 diffusing into shale oil because some portion of that injected CO_2 would dissolve into the shale brine. Therefore, the negative dissolution effect on the CO_2-EOR performance was not connected to the main effect of the dissolution itself but resulted from the interaction with molecular diffusion. The simulation results also proved that the impact of the CO_2 diffusion into shale oil has a greater potential to increase oil recovery than the dissolution effect.

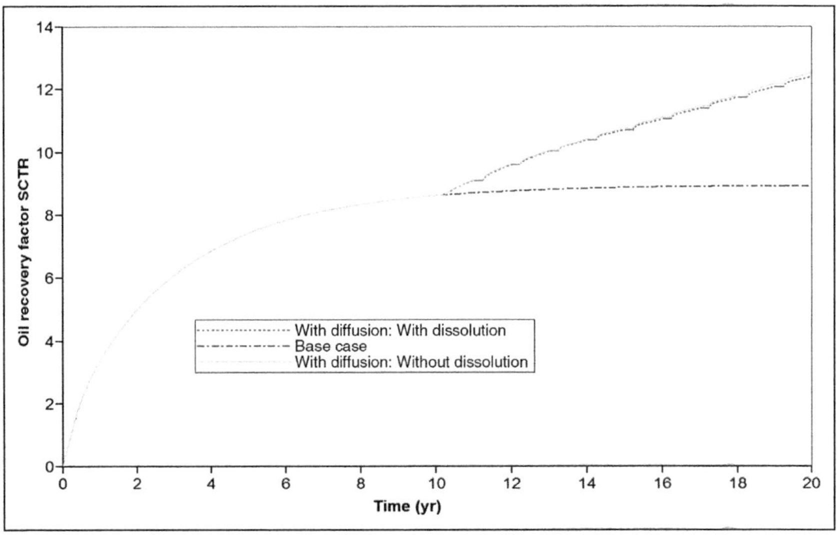

Figure 11.13 *Effect of water solubility on CO_2-EOR performance in ULR.*

11.4.4 Geomechanic coupling effects

Three different types of geomechanic models were constructed—compressibility models, linear elastic models, and stress-dependent correlations. We used a geomechanic module in a GEM simulator with input data from the Bakken formation to simulate linear elastic models. Stress-dependent porosity and permeability were also considered by applying exponential correlations. We also used Cho et al.'s (2013) experimental coefficients in the exponential correlations for the Bakken formation. Fig. 11.14 presents the porosity and permeability multiplier for the Bakken formation with different pressures. These agree well with Kim, Cho, and Lee (2017) results. Our simulation focused on the geomechanic coupling effects of CO_2-EOR as they apply to flooding and the huff-n-puff method. Our simulation results indicate that both the fracture and matrix properties of tight rocks are sensitive to stress and strain variations resulting from the injection and production processes shown in Fig. 11.15, which shows how the natural fracture permeability decreased rapidly in the primary production process due to pressure depletion. However, natural fracture permeability increased during the CO_2 injection process due to pressure maintenance. The same scenario is shown in Fig. 11.15B during the huff-n-puff process.

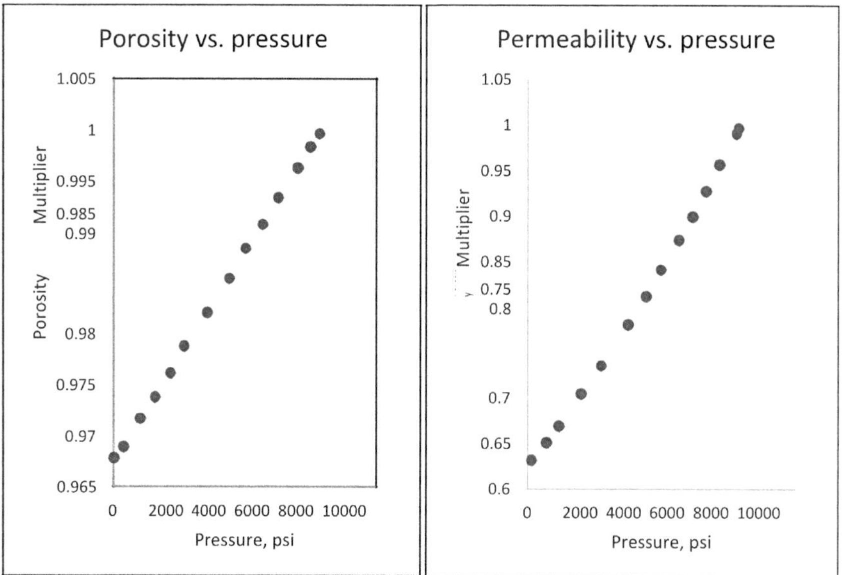

(A) Porosity multiplier (B) Permeability multiplier

Figure 11.14 ***Reduction in porosity and permeability of Bakken formation with reservoir pressure depletion.***

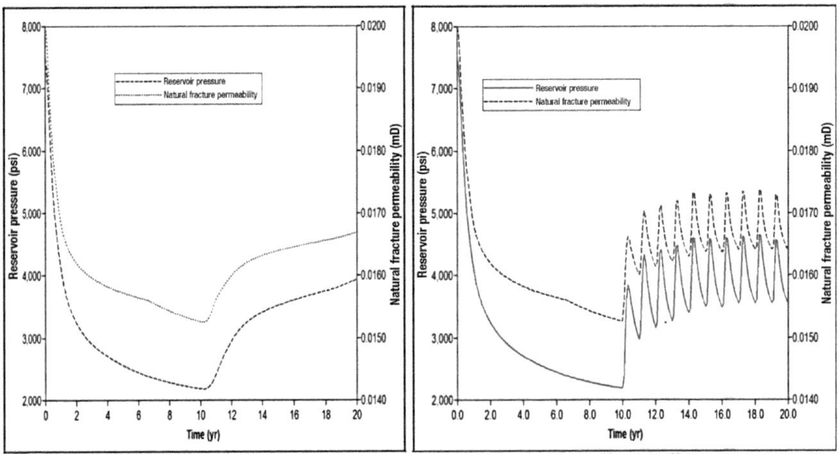

(A) Flooding process (B) Huff -n-Puff

Figure 11.15 ***Natural fracture permeability of the Bakken shale model with CO_2 flooding and huff-n-puff processes.***

The simulated geomechanic coupling effects of the CO_2-EOR process have two blades of effect; one is negative and the other is positive. The negative effect is a result of the compaction effect, which reduced the porosity and permeability of the target formation during the depletion process. However, the positive effect was a result of reservoir deformation due to its elasticity, which created additional pressure maintenance during the depletion effect. Overall, the geomechanic coupling had a negative effect on the CO_2-EOR performance in both the huff-n-puff and flooding process as shown in Figs. 11.16 and 11.17, respectively. This was mainly due to the pseudo dilation phenomenon as shown in Fig. 11.18.

During the CO_2-injection process, the target reservoir had no way to restore its original shape or its original properties even after injecting CO_2 at a pressure equal to the initial reservoir pressure. Fig. 11.18 clearly shows the amplitude size difference during the first and last cycle of the CO_2 injection huff-n-puff process.

Different models of geomechanics were incorporated into the reservoir flow model. Then, each model was fitted with a CO_2-EOR pilot test conducted in the Bakken formation's Burning Tree field, Well 36-2H, in Montana (2009). The results indicated that among the three different correlations of geomechanic coupling, stress-dependent correlations agreed well with the pilot test performance as shown in Fig. 11.19. However, linear-elastic constitutive models were not a good match with that pilot test

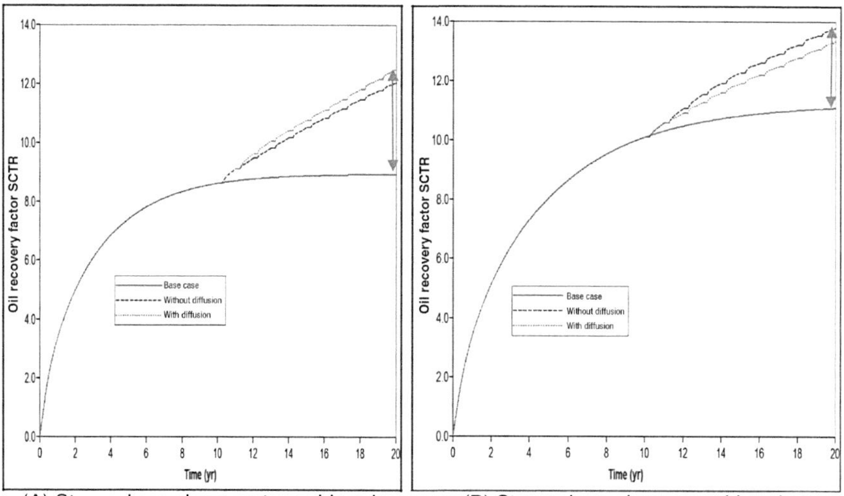

(A) Stress dependency not considered (B) Stress dependency considered

Figure 11.16 ***CO_2-EOR performance with and without stress dependency in huff-n-puff operations.***

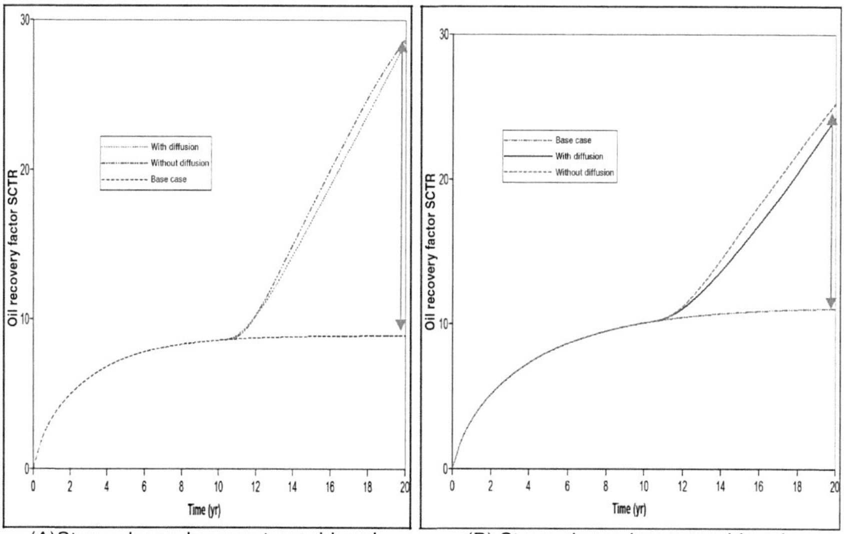

(A)Stress dependency not considered (B) Stress dependency considered

Figure 11.17 ***CO_2-EOR performance with and without stress dependency in continuous flooding operations.***

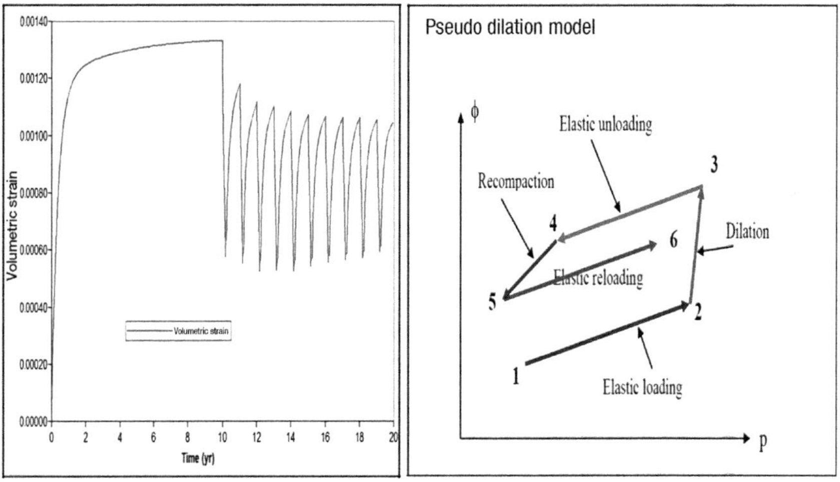

(A)Volumetric strain change with time (B) Pseudo dilation model

Figure 11.18 ***Explanation of geomechanical coupling's negative effect on the CO_2-EOR performance.***

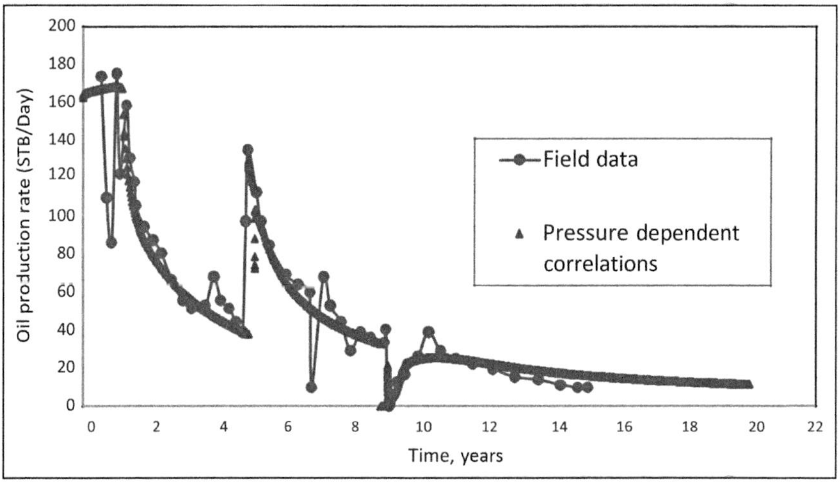

Figure 11.19 ***Coupling geomechanics effects with CO_2-EOR modeling for CO_2 pilot test conducted in the burning tree 36-2H well of the Bakken formation in Montana (2009).***

performance. Bakken formation results indicate that linear-elastic constitutive models over-predict the sensitivity of shale formations to stress change when the shale formations are not notably stress-sensitive, which describes the Bakken formation. Linear-elastic constitutive models can give a good representation of formations which are very soft such as the Eagle Ford

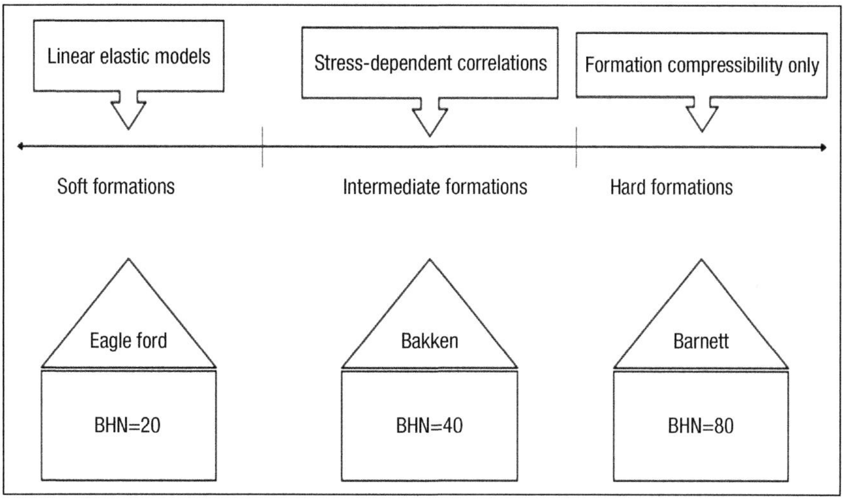

Figure 11.20 ***Selection criteria for geomechanic models according to formation hardness and Young's modulus.***

formation shown in Fig. 11.20. However, the compressibility model is not set up to represent the geomechanic effect in hard formations found in Barnett play. However, our stress-dependent correlations are an excellent match for formations with mid-hardness, such as the Bakken formation.

11.5 SUMMARY

1. This chapter presents different mechanisms for CO_2 interactions with an organic surface, shale brine, and shale oil. The investigated mechanisms have been implemented in different scenarios of numerical models.
2. This chapter shows the CO_2 molecular diffusion mechanism to have a clearly positive effect on CO_2-EOR injection in the huff-n-puff protocol; however, this mechanism had a relatively negative effect on the continuous flooding mode with CO_2-EOR injection.
3. Both dissolution and adsorption mechanisms have a negative effect on CO_2 performance in terms of enhancing oil recovery in unconventional formations.
4. The CO_2 molecular diffusion mechanism plays a dominant role over other CO_2 mechanisms, which helps control the success or failure of CO_2-EOR injections in an unconventional liquid reservoir (ULR).
5. Geomechanic coupling has a clear effect on CO_2-EOR performance; however, different geomechanics approaches have different measures of success in these shale plays. Stress-dependent correlations are the best

match for CO_2-EOR pilot tests in the Bakken formation while linear elastic models are the best match for CO_2-EOR pilot tests in the Eagle Ford formation.

References

Ambrose, R. J., Hartman, R. C., & Akkutlu, I. Y. (2011). Multi-component sorbed phase considerations for shale gas-in-place calculations. *Society of Petroleum Engineers*. doi: 10.2118/141416-MS (January 1).

Cho, Y., Ozkan, E., & Apaydin, O. G. (2013). Pressure-dependent natural-fracture permeability in shale and its effect on shale-gas well production. *Society of Petroleum Engineers*. doi: 10.2118/159801-PA (May 1).

CMG. 2018. Computer modeling group, GEM manual. https: www.CMG.Ca/ accessed 2018.

Hoteit, H., & Firoozabadi, A. (2009). Numerical modeling of diffusion in fractured media for gas-injection and recycling schemes. *SPE Journal*, *14*(02), 323–337. doi: 10.2118/103292-PA SPE-103292-PA.

Jabbari, H., Ostadhassan, M., & Rabeie, M. (2015). Geomechanics modeling in CO_2-EOR: Case study. *Society of Petroleum Engineers*. doi: 10.2118/175908-MS.

Jiang, J., & Younis, R. M. (2016). Compositional modeling of enhanced hydrocarbons recovery for fractured shale gas-condensate reservoirs with the effects of capillary pressure and multicomponent mechanisms. *Society of Petroleum Engineers*. doi: 10.2118/179704-MS.

Kim, T. H., Cho, J., & Lee, K. S. (2017). Modeling of CO_2 flooding and huff and puff considering molecular diffusion and stress-dependent deformation in tight oil reservoir. *Society of Petroleum Engineers*. doi: 10.2118/185783-MS.

Kim, T. H., Park, S. S., & Lee, K. S. (2015). Modeling of CO_2 injection considering multicomponent transport and geomechanical effect in shale gas reservoirs. *Society of Petroleum Engineers*. doi: 10.2118/176174-MS.

Li, Y. K., & Nghiem, L. X. (1986). Phase equilibria of oil, gas and water/brine mixtures from a cubic equation of state and Henry's law. *The Canadian Journal of Chemical Engineering*, *64*(3), 486–496.

Mohebbinia et al. (2017). Molecular diffusion calculations in simulation of gasfloods in fractured reservoirs. SPE-182594-MS paper Presented at the SPE Reservoir Simulation Conference held in Montgomery, TX, USA, February 20–22.

Peng, D. Y., & Robinson, D. B. (1976). A new two-constant equation of state. *Industrial & Engineering Chemistry Fundamental*, *15*(1), 59–64.

Sheng, J. J. (2015). Enhanced oil recovery in shale reservoirs by gas injection. *Journal of Natural Gas Science and Engineering*, *22*, 252–259. doi: 10.1016/j.jngse.2014.12.002.

Sigmund, P. M. (1976a). Prediction of molecular diffusion at reservoir conditions. Part 1 – Measurement and prediction of binary dense gas diffusion coefficients. *Petroleum Society of Canada*. doi: 10.2118/76-02-05.

Sigmund, P. M. (1976b). Prediction of molecular diffusion at the reservoir conditions. Part II – Estimating the effects of molecular diffusion and convective mixing in multicomponent systems. *Journal of Canadian Petroleum Technology*, *15*(3), 53–62. doi: 10.2118/76-03-07 PETSOC-76-03-07.

Stumm, W., & Morgan, J. J. (1996). *Aquatic chemistry: Chemical equilibria and rates in natural waters* (third edition). New York, NY: Wiley-Interscience.

Tran, D., Nghiem, L., & Buchanan, L. (2005). An overview of iterative coupling between geomechanical deformation and reservoir flow. *Society of Petroleum Engineers*. doi: 10.2118/97879-MS.

Wan, T., & Sheng, J. (2015). Compositional modelling of the diffusion effect on EOR process in fractured shale-oil reservoirs by gasflooding. *Society of Petroleum Engineers*. doi: 10.2118/2014-1891403-PA.
Wilke, C. R., & Chang, P. 1955. Correlation of diffusion coefficients in dilute solutions. AIChE J1 (2): 264–270. http://dx.doi.org/10.1002/aic.690010222.
Yu, W., & Sepehrnoori, K. (2019). Shale Gas and Tight Oil Reservoir Simulation. Elsevier Book. ISBN: 978-0-12-813868-7. https://doi.org/10.1016/C2017-0-00263-X.

CHAPTER 12

Comparative and optimization of CO_2 and natural gas EOR methods

Contents

12.1 Reservoir modeling 245
12.2 Sensitivity analysis 249
12.3 Results and discussion 251
12.3.1 Parameters which control CO_2-EOR 251
12.3.2 Parameters control NG-EOR 257
12.3.3 Comparison between CO_2-EOR and NGs-EOR 261
12.4 Summary 264
References 265

Abstract

In this chapter, numerical simulation methods of compositional models are incorporated with LS–LR–DK (logarithmically spaced, locally refined, and dual permeability) models to mimic the performance of CO_2 as well as natural gases (lean gas and rich gas) in terms of enhanced oil recovery (EOR). A numerical simulation investigation is presented in this chapter covering different operational strategies implemented by unconventional EOR reservoirs, which explore and optimize the most applicable gases with the most effective parameters designed to manage EOR production. Two design of experiment (DOE) methods were used to conduct the sensitivity analysis including the response surface methodology (RSM) and the one-parameter-at-a-time (OPAAT) approach. This chapter generally presents the guidelines on how to select and optimize the most feasible EOR technique based on the rock and fluid properties of the candidate reservoir. It also serves as a guidance on how to use numerical simulation to investigate the most important factors controlling EOR methods.

Keywords: sensitivity analysis; shale and tight oil reservoirs; natural gas EOR; CO_2 EOR; design of experiment (DOE); response surface methodology (RSM); one-parameter-at-a-time (OPAAT) method

12.1 Reservoir modeling

The LS-LR-DK (logarithmically spaced, locally refined, and dual permeability) model was initially constructed to simulate Bakken formation characteristics, which established its ability to accurately simulate the fluid

Fundamentals of Enhanced Oil Recovery Methods for Unconventional Oil Reservoirs
http://dx.doi.org/10.1016/B978-0-12-818343-4.00012-7

flow in fractured shale oil reservoirs (Wan & Sheng, 2015). Then, an advanced general equation-of-state (EOS) compositional simulator was used to build the formation fluid model. Then, both the LS–LR–DK model and EOS-based formation fluid model were combined to simulate compositional interactions of the reservoir fluid and the injected gases during the enhanced oil recovery (EOR) processes. The molecular diffusion model through the constructed LS–LR–DK model, that has the typical fluid and rock properties of the Bakken formation, and fluid compositional model was conducted.

Throughout the constructed models presented in this chapter, different miscible gases are injected into shale formations, using the huff-n-puff mode through a hydraulically fractured well. Most of the EOR mechanisms reported for such miscible gases are considered in the presented models. In the base case, the production well is stimulated with five hydraulic fractures. The spacing between the hydraulic fractures is 200 ft. The simulated model includes two regions: (1) the stimulated reservoir volume (SRV) and (2) the unstimulated reservoir volume (USRV) as shown in Fig. 12.1. The dimensions of the reservoir model are 2000 ft × 2000 ft × 42 ft, which corresponds to the reservoir length, width, and thickness. The dimensions of the fractured region are five fractures with a half-length of 350 ft in the J direction, a width 0.001 ft in I direction, and a fracture height of 42 ft in the K direction. Fracture conductivity is 15 mD ft. The other model's input parameters are given in Table 12.1. The simulated oil used in the presented models has 42 API°, 725 SCF/STB, and 1850 psi as oil gravity, gas oil ratio, and bubble point pressure, respectively. It is known that compositional models are the most time-consuming models due to the large number of components in the typical formation oil. In our model, we have 34 components, which would take a long time for the simulator to complete running one scenario.

The common practice in numerical simulations for such a situation is to carefully lump the reservoir oil components into a short representative list of pseudo-components. These pseudo components are acceptable if they match the laboratory measured phase behavior data. The supplied data for such compositional models must have a description of associated single carbon numbers and their fractions, saturation pressure test results, separator results, constant composition expansion test results, differential liberation test results, and swelling test results (CMG, 2018). These steps can be used for tuning the EOS to match the fluid behavior. WinProp-CMG was used

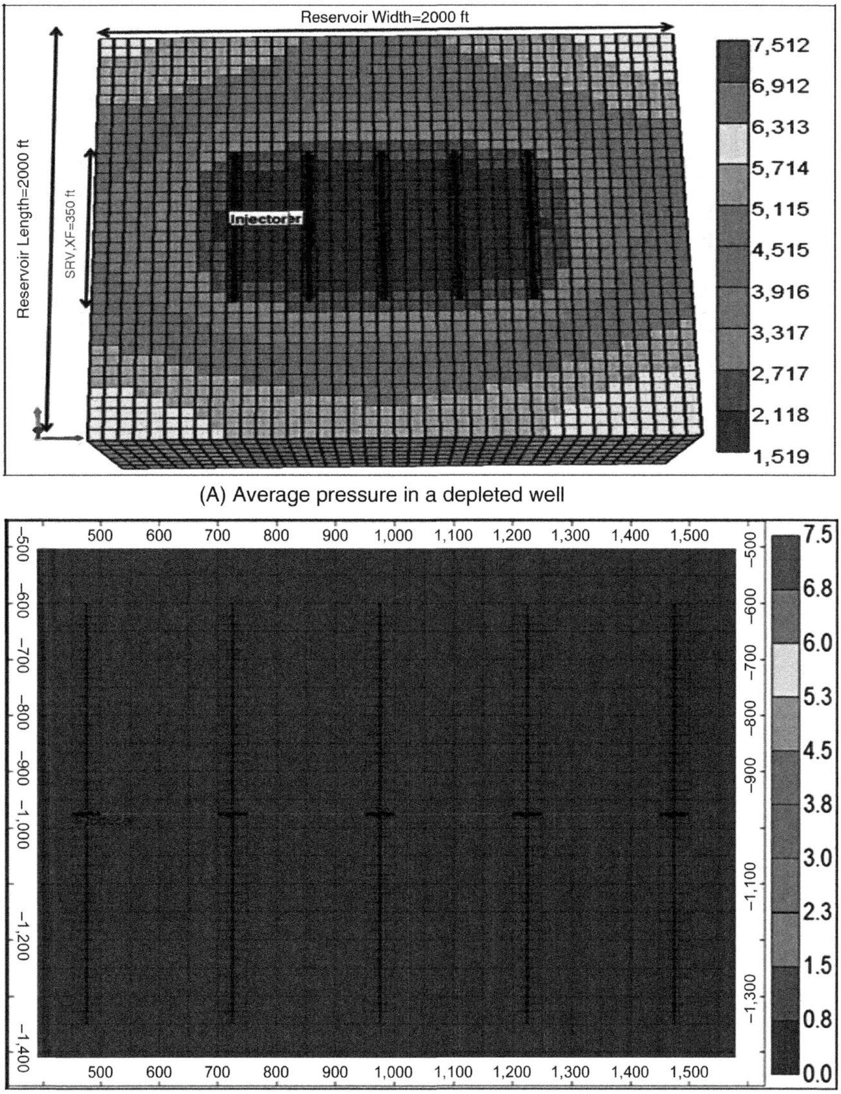

Figure 12.1 (A) Average pressure in a depleted well in Brakken and (B) a closed view for the stimulated reservoir volume (SRV) of the production well.

to lump the original 34 components into seven pseudo components as presented in Table 12.2.

WinProp offers an EOS based on fluid behavior and the PVT modeling package. In WinProp, the laboratory data for fluids can be

Table 12.1 Model input parameters for the base case.

Parameter	Value	Unit
Model dimensions	2000 × 2000 × 42	ft
Production time	20	year
Reservoir top	8000	ft
Reservoir temperature	240	°F
Reservoir pressure	7500	psi
Initial water saturation	0.3	value
Total compressibility	1×10^{-6}	psi^{-1}
Matrix permeability	0.005	mD
Matrix porosity	0.085	value
Horizontal well length	1000	ft
Total number of fractures	5	value
Fracture conductivity	15	mD-ft
Fracture half-length	250	ft
Fracture height	42	ft

Table 12.2 Compositional data for the Peng–Robinson EOS in the model oil.

Component	Mole fraction	Critical pressure (atm)	Critical temperature (K)	Acentric factor	Molar weight (g/gmol)
CO_2	0	7.28E + 01	3.04E + 02	0.225	4.40E + 01
N_2–CH_4	0.2704	4.52E + 01	1.90E + 02	0.0084	1.62E + 01
C_2H–NC_4	0.2563	4.35E + 01	4.12E + 02	0.1481	4.48E + 01
IC_5–CO_7	0.127	3.77E + 01	5.57E + 02	0.2486	8.35E + 01
CO_8–C_{12}	0.2215	3.10E + 01	6.68E + 02	0.3279	1.21E + 02
C_{13}–C_{19}	0.074	1.93E + 01	6.74E + 02	0.5672	2.20E + 02
C_{20}–C_{30}	0.0508	1.54E + 01	7.92E + 02	0.9422	3.22E + 02

imported, and an EOS can be tuned to match the physical behavior for the lab data. Fluid interactions can be predicted, and a fluid model can then be created for use in CMG software. Table 12.3 presents the Peng–Robinson EOS fluid description and binary interaction coefficients of the Bakken crude oil with different injected gases. Fig. 12.2 represents the two-phase envelope for the Bakken oil, which was generated using WinProp-CMG.

Table 12.3 Binary interaction coefficients for Bakken oil.

Component	CO_2	N_2–CH_4	C_2H–NC_4	IC_5–CO_7	CO_8–C_{12}	C_{13}–C_{19}
CO_2						
N_2–CH_4	1.01E – 01					
C_2H–NC_4	1.32E – 01	1.30E – 02				
IC_5–CO_7	1.42E – 01	3.58E – 02	5.90E – 03			
CO8–C_{12}	1.50E – 01	5.61E – 02	1.60E – 02	2.50E – 03		
C_{13}–C_{19}	1.50E – 01	9.76E – 02	4.24E – 02	1.72E – 02	6.70E – 03	
C_{20}–C_{30}	1.50E – 01	1.45E – 01	7.79E – 02	4.27E – 02	2.51E – 02	6.00E – 03

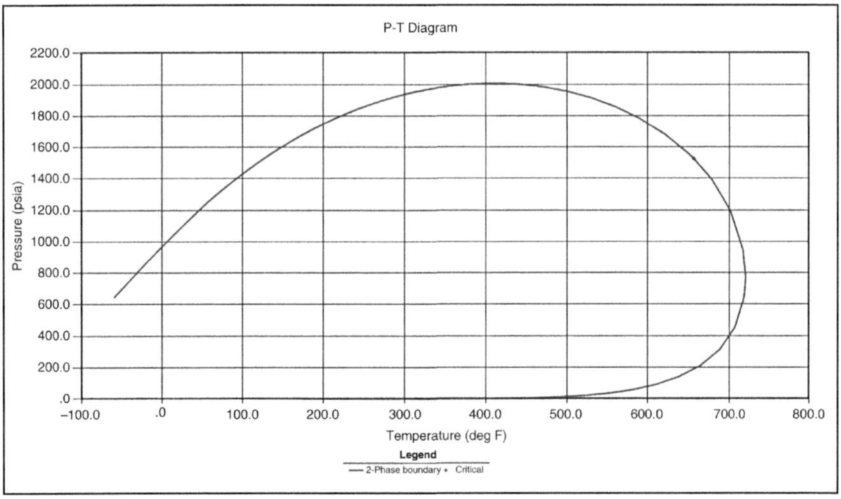

Figure 12.2 *Two-phase envelope for Bakken oil, generated by WinProp-CMG.*

12.2 Sensitivity analysis

The purpose of the sensitivity analysis is to determine how sensitive an objective function is to different parameters qualitatively and quantitatively. Identifying the parameters which have a high impact on CO_2-EOR as well as natural-gas EOR huff-n-puff performance can be used to predict where and when these EOR techniques will succeed or fail, depending on the reservoir characteristics prior to the field application. In this study, two methods were used to conduct sensitivity analyses for the parameters influencing CO_2-EOR performance and natural gas EOR performance in shale reservoirs. The first method is the one-parameter-at-a-time (OPAAT) approach, which analyzes each parameter individually. The output results are then produced in a tornado chart. The second method is the response surface method (RSM), which analyzes the parameters at the same time (simultaneously). These output results have been produced in three types of plots, which are discussed in the results section. The main benefit of the second method over the first method is that the interaction between parameters and influential parameters are reported. Both approaches are used in the basic calculations of RSM as follows:

Sobol method: The Sobol method is one of the variance-based sensitivity analysis methods used to quantify the amount of variance that each input factor, Xi, contributes to the unconditional variance of output, V(Y) (CMG, 2017). For example: In a given case with three inputs and one output, if 50% of the output change can happen by changing the first input, 30% by changing the second input, 10% by changing the third one, and 10% due to interactions between the first two input parameters, these percentages are clearly reflected in measures of sensitivity.

Morris method: The Morris method (also referred to as the elementary effects (EE) method) is one of the screening methods used to determine the effect of the input parameters on the model outputs (CMG, 2017). The Morris approach has two measures, the mean and standard deviation, which are used together. The mean reflects the linear influence of an input factor on the output function while the standard deviation reflects the nonlinear or interaction functionality.

Tornado plot: This visual tool provides a qualitative and quantitative effect for the input parameters on the output ones, with a higher value meaning the objective function is more sensitive to that parameter and vice versa (CMG, 2017). For more information about the basics and principles of this method, the CMG manual (CMG, 2017) can be reviewed.

Sensitivity parameters: The investigated parameters and their range of values are the same for the three types of gases (CO_2, lean gas, and rich gas) listed in Table 12.4. The selected parameters have been analyzed within a range and the simulation results have been compared with the values used for the base reservoir model. The objective functions used for sensitivity analysis purposes are (1) the oil recovery

Table 12.4 Parameters for sensitivity analysis.

Parameter	Symbol	Minimum	Base case	Maximum
Fracture spacing in I- direction (ft)	DIFRAC	0	0.5	0.625
Fracture spacing in J- direction (ft)	DJFRAC	0	0.5	0.625
Molecular diffusion rate (cm^2/s)	Diff_Rate	6E–10	8E–06	0.000008
Injection rate (scf/day)	Inj_Rate	500,000	1,875,000	1,500,000
Matrix permeability in I- direction	MA_PERMI	1E–05	0.009	0.1
Matrix permeability in J-direction	MA_PERMJ	1E–05	0.009	0.1
Matrix permeability in K-direction	MA_PERMK	1E–06	0.009	0.1
Natural permeability in I-direction	NF_PERMI	0.0001	0.02	1
Natural permeability in J-direction	NF_PERMJ	0.0001	0.02	1
Natural permeability in K-direction	NF_PERMK	0.0001	0.04	1
Matrix porosity	POR	0.03	0.03	0.1

factor, (2) the oil production rate, and (3) the cumulative oil production. The time frame of the objective functions covered 10 years of the CO_2-EOR and NG-EOR huff-n-puff process following 10 years of primary depletion.

12.3 Results and discussion

12.3.1 Parameters which control CO_2-EOR

OPAAT methodology. Test results from this study indicate that the matrix porosity is the most significant parameter affecting the three objective functions: (1) the oil recovery factor, (2) cumulative production, and (3) the oil production rate as shown in Figs. 12.6–12.8, respectively. The oil recovery factor increased from 12.6% to 19.3% by increasing the matrix porosity from 3% to 10%. The cumulative oil production increased from 1.5 E + 05 STB to 2.7 E + 05 STB by increasing the matrix porosity from 3% to 10%. The daily oil production rate per well increased from 13.2 STB/day to 25.8 STB/day by increasing the matrix porosity from 3% to 10%.

Since the diffusion flow is the most dominant flow in shale oil reservoirs, increasing the contact area between the injected CO_2 and the target reservoir oil would result in enhancing the CO_2 diffused into the shale oil. If the injected CO_2 penetrates deeper into the shale reservoir, it will swell more oil and make it less viscous. This explains why CO_2 needs a high porosity in the matrix to increase the oil recovery factor in the shale reservoir. The second important factor, which controls the CO_2-EOR performance, is the formation average permeability (the average permeability value, which includes the hydraulic fracture (HF), natural fracture (NF), and matrix permeability). Figs. 12.3–12.5 show that the second factor that controls the success of CO_2-EOR is the average conductivity of shale formation. Our tests showed the conductivity of oil pathways to have a significant impact on CO_2-EOR performance in all the scenarios, which were simulated. We determined that increasing the conductivity of oil pathways in shale increases the kinetics of the oil recovery process in the productive areas. Conversely, a decreased (limited) CO_2 diffusion rate in the formation oil causes the CO_2 to experience less exposure time with the formation oil causing its production to be pushed back (decreased). Hence, the CO_2-EOR performance is downgraded as the oil pathway conductivity increases. The verification process was achieved by injecting CO_2 into the low conductivity hydraulic fractures

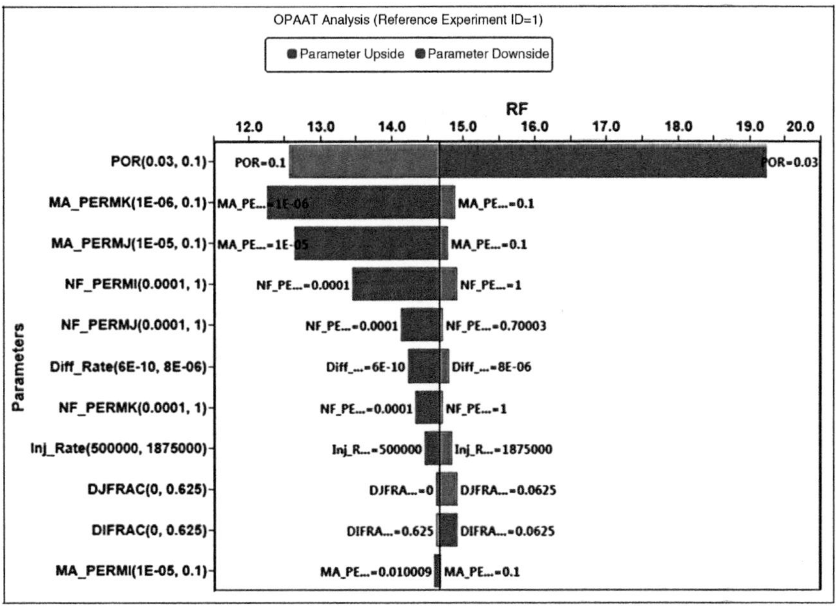

Figure 12.3 *Tornado chart for the factors affecting RF with CO_2-EOR (OPAAT).*

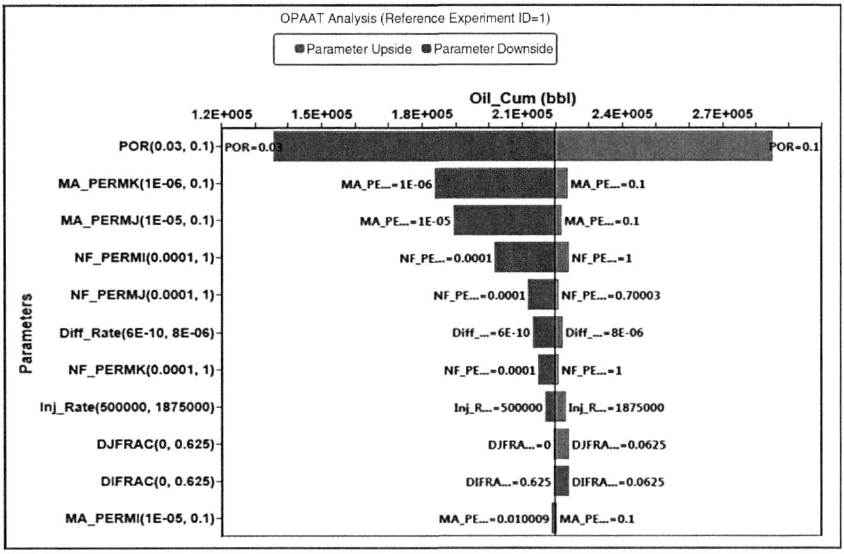

Figure 12.4 *Tornado chart for factors affecting cumulative oil with CO_2-EOR.*

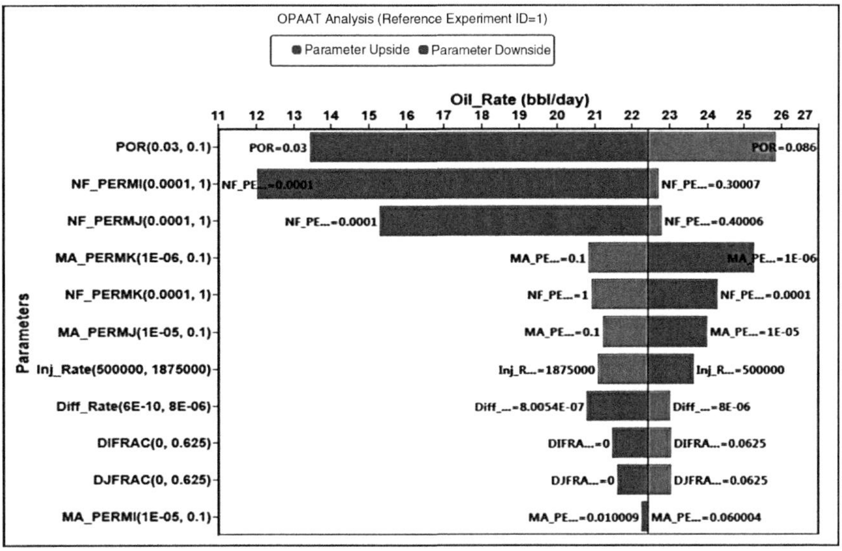

Figure 12.5 *Tornado chart for factors affecting oil rate with CO_2-EOR.*

instead of injecting the CO_2 in the high conductivity hydraulic fractures. The results indicated that the injected CO_2 will enhance oil recovery in the reservoirs with low conductivity fractures more efficiently than in the reservoirs with high conductivity fractures. To sum up, as long as the kinetics of the oil recovery process in the productive areas do not exceed the CO_2 diffusion rate, the injected CO_2 will experience more exposure time with the formation oil before its production is pushed back (decreases).

The following are results based on the response surface methodology (RSM):

1. **Formation total porosity (including natural fracture porosity):** Both the Sobol approach and the Morris method indicate that the most important factor affecting obtained oil recovery by CO_2-EOR is the total porosity of the shale formation as shown in Fig. 12.6A and B. We concluded that as long as the formation porosity and fracture intensity increased, oil recovery obtained by CO_2-EOR would also increase, which means that the total porosity of the shale formation has a positive effect on CO_2-EOR performance as shown in Fig. 12.6B. The lesson learned from this behavior is that increasing the total porosity of a formation will increase the contact area between the injected CO_2 and the formation oil; thus the CO_2-EOR performance will be enhanced.
2. **Formation average permeability (average permeability with attention to HF, NF, and matrix permeability)**: Both the Sobol approach and Morris method indicate that the second parameter which controls the success of CO_2-EOR is the average conductivity of the shale formation, as shown in Fig. 12.6A&B. We found that as long as the conductivity of the oil pathways increased, oil recovery obtained by CO_2-EOR decreased, which means that the conductivity of oil pathways had a negative effect on the CO_2-EOR performance as shown in Fig. 12.8. The interpretation used to explain this behavior is that increasing the conductivity of oil pathways in shale formations will result in increasing the kinetics of the oil recovery process in the productive areas. As a result, for reservoirs with oil pathways of high conductivity, the injected CO_2 will experience less exposure time with the formation oil before the injected CO_2 is being produced back. This explains why CO_2 only penetrates in the limited areas around the hydraulic fractures in shale reservoirs with high conductivity fractures. Therefore, CO_2-EOR performance will be downgraded with the increasing conductivity of oil pathways. However, if the conductivity of oil pathways in shale formations decreased, it would make CO_2 penetrate deeper into the tight matrix, far away from the hydraulic fractures because the injected CO_2 would experience longer exposure time resulting in deeper CO_2 penetration into the tight matrix. As a result, CO_2 would swell more volumes of oil, reduce oil viscosity, and finally produce larger quantities of oil by the counter-current mechanism (Figs. 12.7 and 12.8).

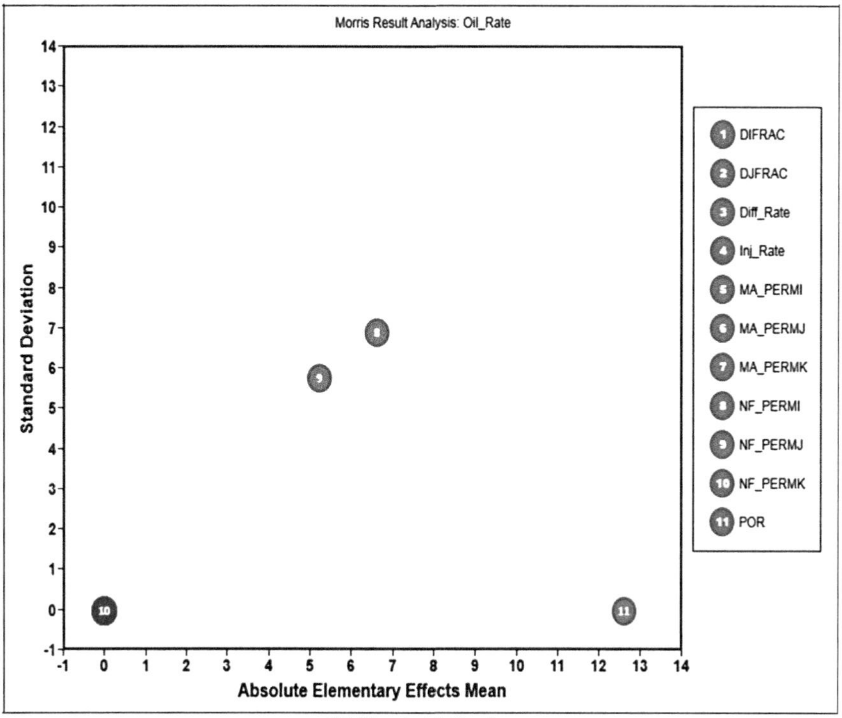

(A) Morris Method

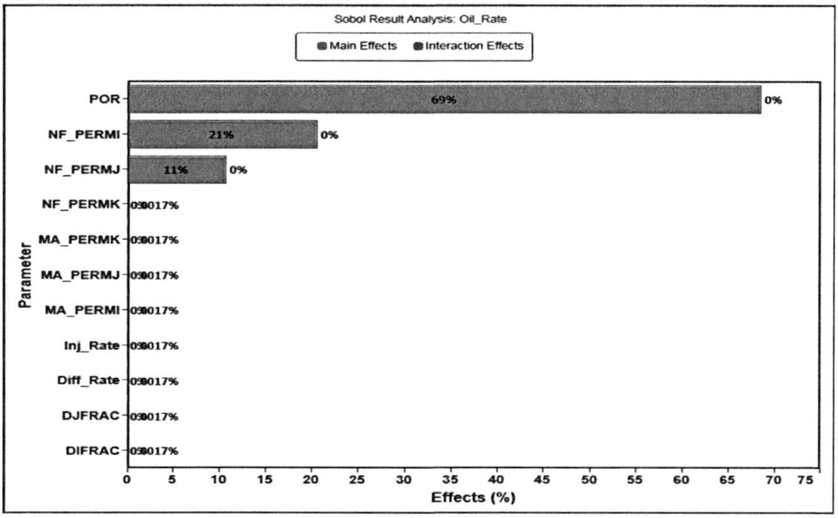

(B) Sobol Method

Figure 12.6 ***RSM results for CO_2-EOR.*** (A) Morris method and (B) Sobol method.

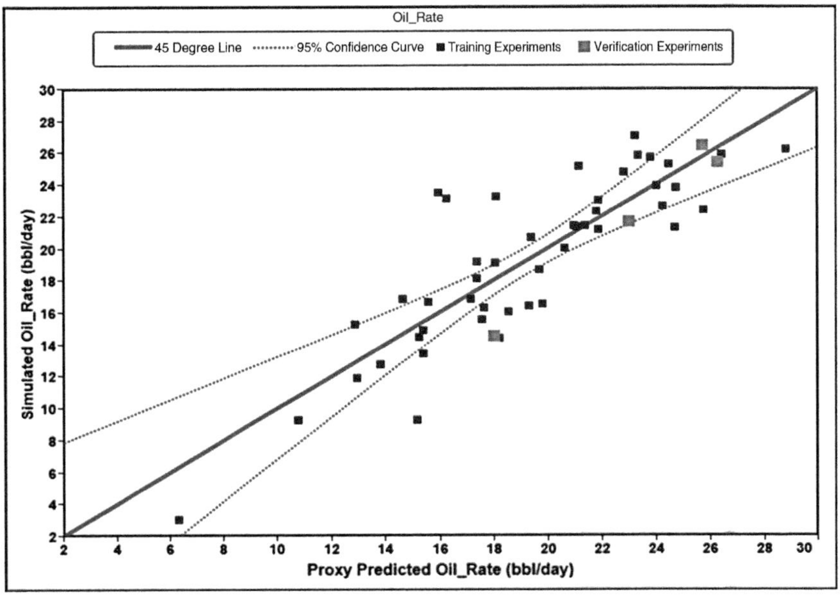

Figure 12.7 *RSM results for CO_2-EOR: proxy versus simulated model.*

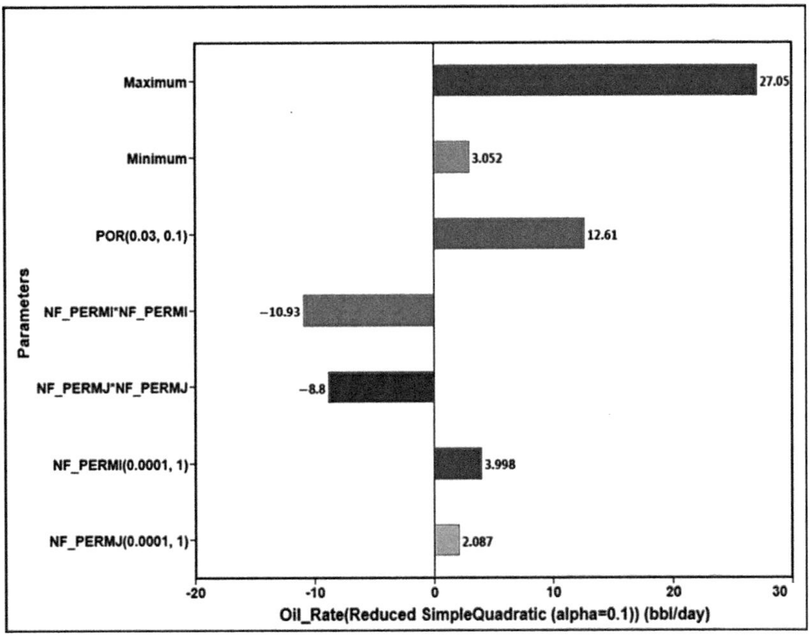

Figure 12.8 *RSM results for CO_2-EOR huff-n-puff process: Tornado chart.*

12.3.2 Parameters control NG-EOR

OPAAT methodology. The results indicate that the natural fracture permeability is the most important parameter, which affects the oil recovery factor as shown in Fig. 12.9. The oil recovery factor increased from 16.8% to 21% by increasing the natural fracture permeability from 0.0001 to 0.6 mD. However, the molecular diffusion rate is the most important parameter controlling the oil production rate as shown in Fig. 12.10. The oil production rate per well increased from 21.7 STB/day to 35.5 STB/day by increasing the molecular diffusion rate from 6E-10 to 8E-07 cm^2/s. The results also show that the matrix porosity is the main factor impacting cumulative oil production as shown in Fig. 12.11. Cumulative oil production increased from 1.6E + 05 STB to 2.8E + 05 STB by increasing the matrix porosity from 3% to 10%.

The main reasons behind the differences in the main factors, which control CO_2-EOR success versus NG-EOR success comes from two main factors. The first factor is the difference in molecular weight for natural gases versus CO_2. The molecular weight of natural gases is much smaller than the molecular weight of CO_2. Therefore, the diffusivity of natural gases is much easier than the diffusivity of. CO_2. This explains why porosity is not that important for NG-EOR. The second factor is due to the difference in the concentration gradient between NGs and shale oil on one side versus CO_2

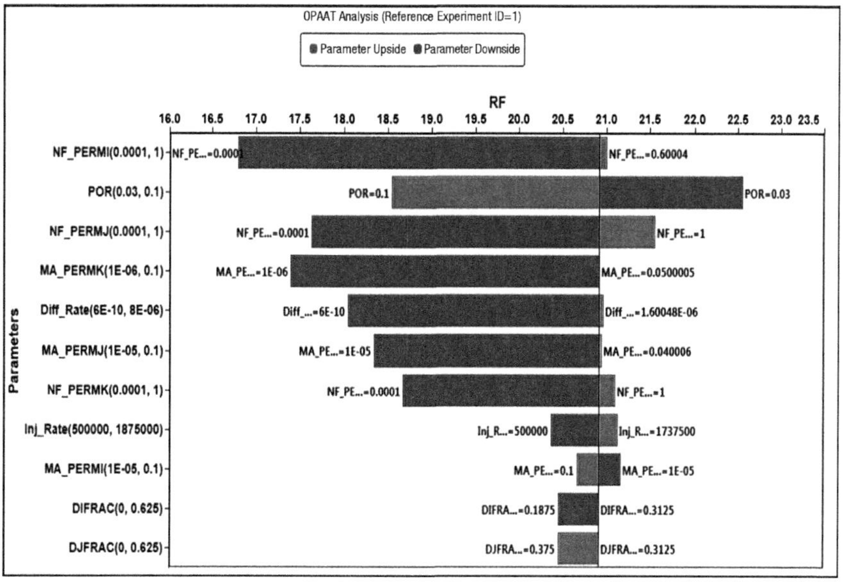

Figure 12.9 *Tornado chart for factors affecting RF with NGs-EOR (OPAAT).*

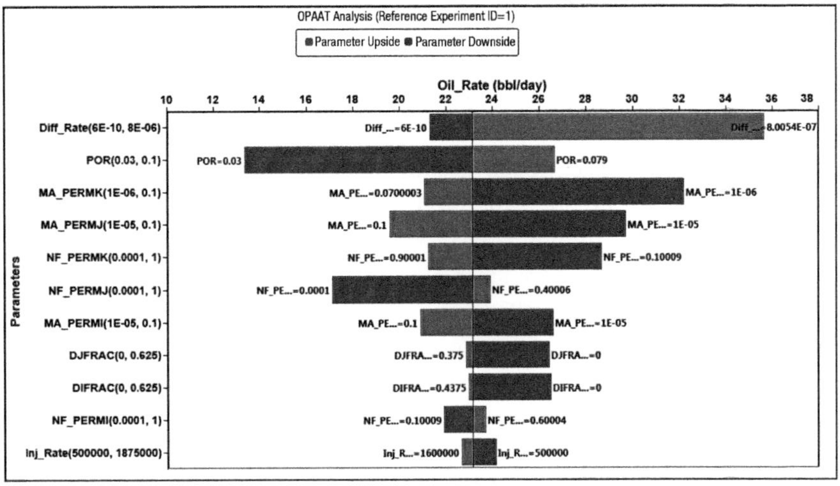

Figure 12.10 *Tornado chart for factors affecting oil rate obtained with NG-EOR.*

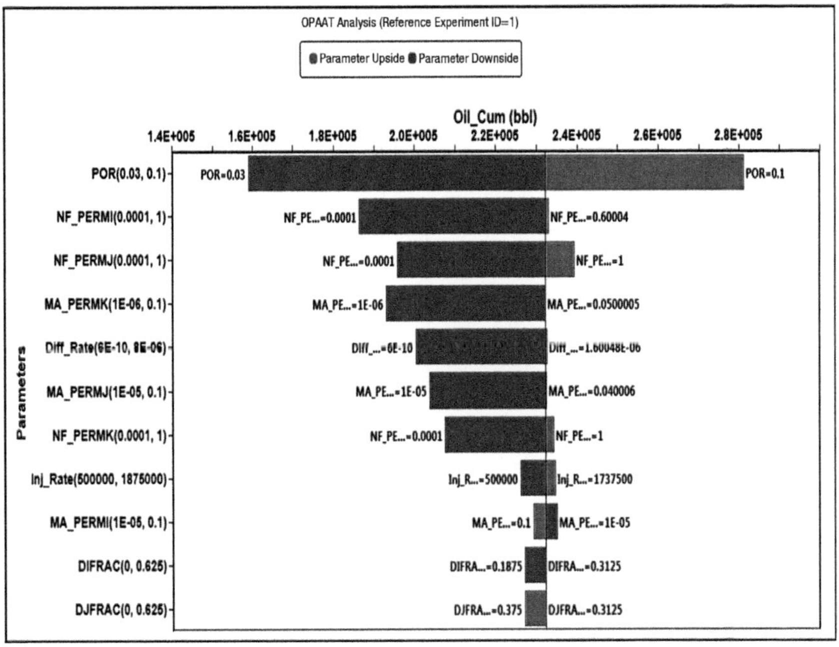

Figure 12.11 *Tornado chart for factors affecting cumulative oil with NG-EOR.*

and shale oil on the other side. As shale oil contains a high concentration of natural gases, the concentration gradient between the injected natural gases and target formation oil is lower than the concentration gradient between the injected CO_2 and target formation oil.

The following results pertain to the response surface methodology (RSM):

1. **Formation total porosity (including natural fracture porosity):** Both the Sobol approach and Morris method point to the total porosity of shale formation as most important factor affecting the obtained cumulative oil (as shown in Figs. 12.12 and 12.13). We determined that as long as the formation porosity and fracture intensity increase, the oil cumulative production obtained by NGs-EOR will also increase. This

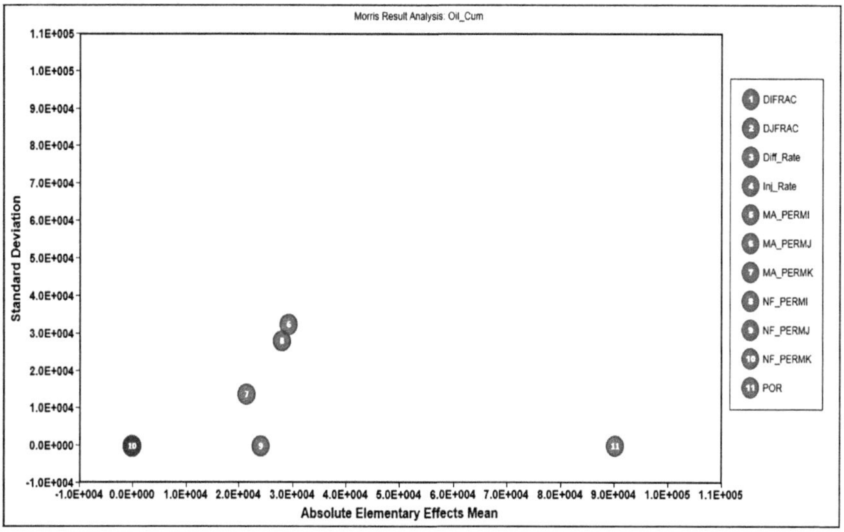

Figure 12.12 *Factors impacting the NGs-EOR huff-n-puff process (Morris method).*

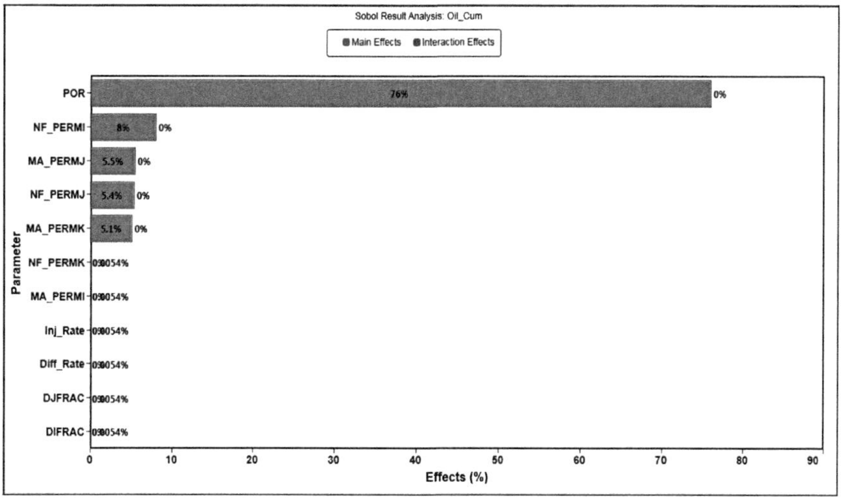

Figure 12.13 *Factors impacting NGs-EOR Huff-n-Puff process (Sobol method).*

means that total porosity of the shale formation has a positive effect on NG-EOR performance as shown in Fig. 12.13. Our interpretation of what is behind this behavior pivots around increasing the total porosity of formation to increase the contact area between the injected NGs and the formation oil, so NG-EOR performance can be enhanced.

2. **Formation average permeability (counting for HF, NF, and matrix permeability):** Both the Sobol approach and the Morris method point to the average conductivity of shale formation as the second most important parameter, which controls the success of NGs-EOR as shown in Figs. 12.12 and 12.13. We found that as the conductivity of the oil pathways increased, the oil recovery obtained by NGs-EOR also decreased, which means that the conductivity of oil-pathways has a negative effect on the NGs-EOR performance as shown in Fig. 12.18. Our interpretation of this behavior is that increasing the conductivity of oil pathways in shale formations would result in increasing the kinetics of the oil recovery process in the productive areas. Thus, to obtain a limited effective diffusion rate for the injected-NGs into formation oil, NGs would experience less exposure time with the formation oil before they are produced back. Therefore, the NG-EOR performance will be downgraded with the increase in oil pathways conductivity (Figs. 12.14 and 12.15).

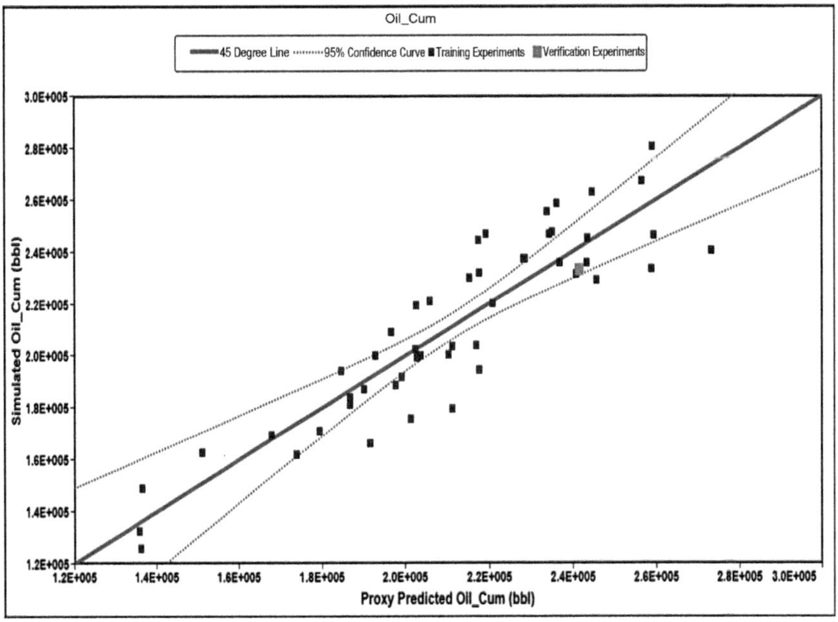

Figure 12.14 ***RSM results for the NGs-EOR: proxy versus simulated model.***

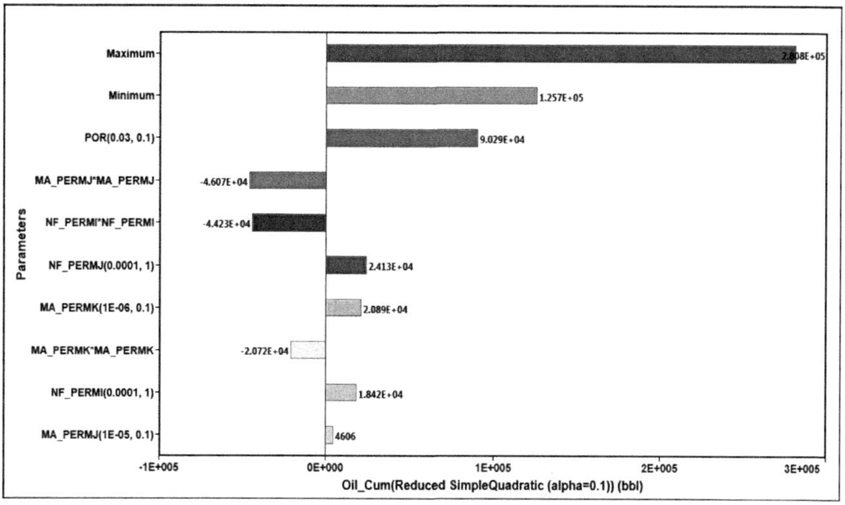

Figure 12.15 *RSM results for NG-EOR huff-n-puff process: Tornado chart.*

12.3.3 Comparison between CO_2-EOR and NGs-EOR

In the same models, the authors conducted a parametric sensitivity for both CO_2-EOR and NGs-EOR for four parameters: (1) formation permeability, (2) formation porosity, (3) natural fracture spacing, and (4) molecular diffusion rate. From Fig. 12.16, we notice that the performance of NGs-EOR exceeds the performance of CO_2-EOR in the permeability

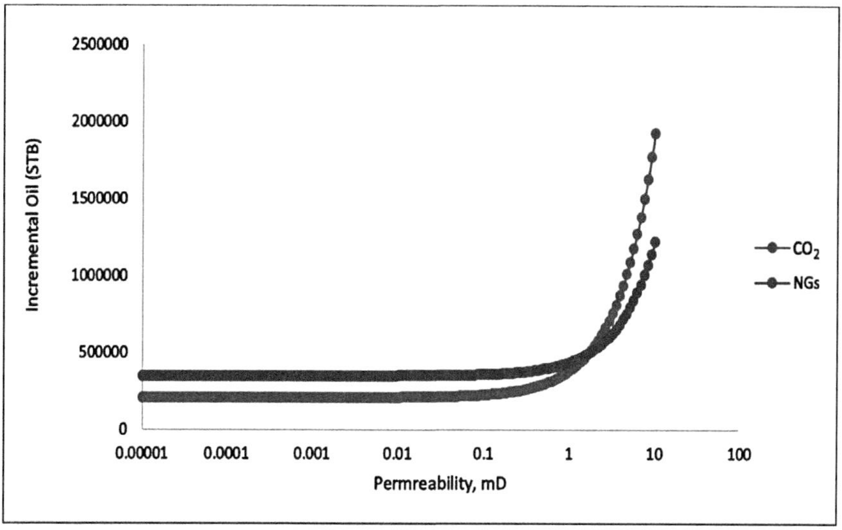

Figure 12.16 *Sensitivity of CO_2-EOR versus NG-EOR with formation permeability.*

range of (0.00001–1 mD). This happens due to the differences in both molecular weight and concentration gradient between CO_2 versus natural gas. Because CO_2 has a higher molecular weight than lean gas, CO_2 needs larger pore throats to invade the matrix of shale oil. However, the performance of CO_2-EOR exceeds the performance of NGs-EOR in the permeability range of more than 1 mD. This happens due to the differences in the concentration gradient between the injected fluid and the formation fluid. The concentration gradient is especially important for CO_2; however, it is low for both lean gas and rich gas. This is particularly true for shale oil because the composition of shale oil usually contains high concentrations of light components (i.e., C1 and C2). However, shale oil contains low concentration of CO_2; therefore, the concentration gradient of CO_2 is much higher than the concentration gradient of NGs under the same conditions.

From Fig. 12.17, both CO_2-EOR and NGs-EOR have the same behavior trend with porosity change. As long as the porosity increased, the performance of both CO_2-EOR and NGs-EOR would be enhanced. Since the diffusion flow is the most dominant flow in the shale oil reservoirs, increasing the contact area between the injected CO_2 and NGs with the target reservoir oil would result in enhancing CO_2 or NGs diffused into shale oil. As a result, if the injected CO_2 or NGs penetrate deeper into the shale reservoir, it would swell more oil and make it less viscous. This explains

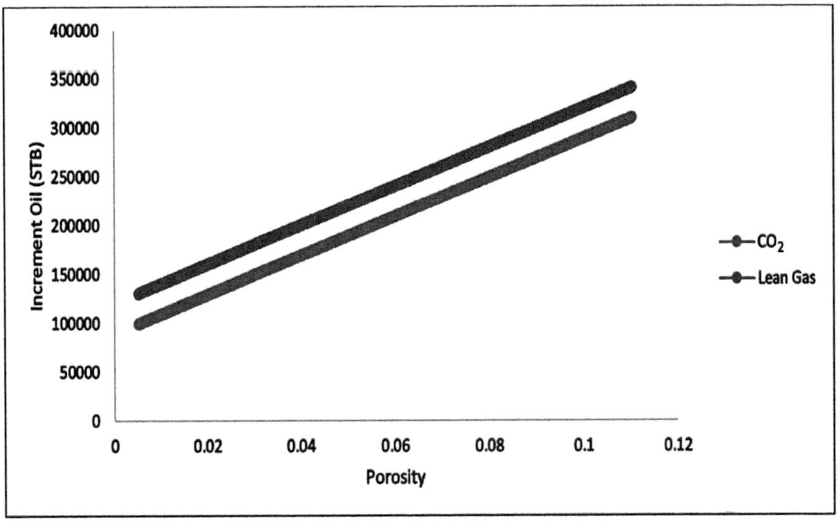

Figure 12.17 ***Sensitivity of CO_2-EOR versus NGs-EOR with formation porosity.***

why CO_2 and NGs need a high porosity to enhance oil production in shale reservoirs.

From Fig. 12.18, CO_2-EOR and NGs-EOR have different behavior trends with changing natural fracture intensity. As long as the natural fracture intensity increased, the performance for both CO_2-EOR and NGs-EOR would be enhanced. However, CO_2-EOR is highly dependent on the fracture intensity compared with NGs-EOR. This difference in the performance between CO_2-EOR and NGs-EOR can be explained by the difference in the molecular weight between natural gases and CO_2 similar to what happened in the permeability sensitivity tests. Fig. 12.19 shows how increasing the molecular diffusion rate results in enhancing the performance of both CO_2-EOR and NGs-EOR. However, the molar diffusion rate has a significant effect on NG performance compared with the CO_2 performance. This is mainly due to the larger CO_2 molecular weight (as evidenced by their larger molecules) compared to the molecular weight of lean gas.

As a result, CO_2 will not penetrate deeper into shale matrix, far away from hydraulic fractures, if the molecular diffusion rate is relatively low according to the diffusivity equation. However, the penetration of lean gas and rich gas into shale matrix is very sensitive to the diffusion rate because there is so much trapping force inside the small pore throats of the shale formations due to their low molecular weight. Therefore, any increase in the diffusion rate would result in penetrating natural gases deeper into shale matrix.

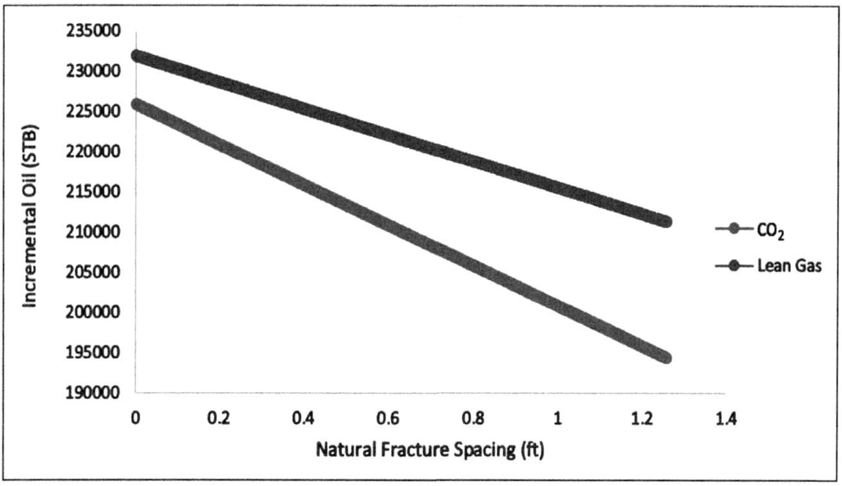

Figure 12.18 *Sensitivity of CO_2-EOR versus NGs-EOR with natural fractures intensity.*

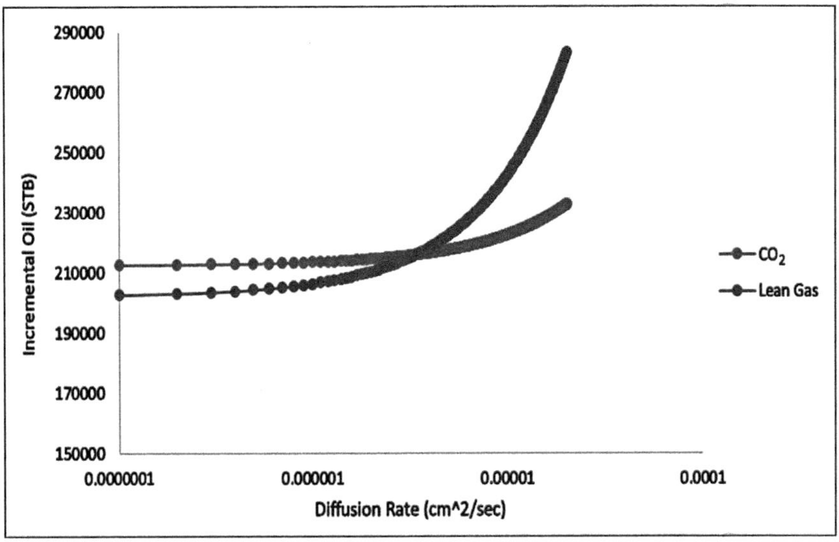

Figure 12.19 ***Sensitivity of CO_2-EOR versus NGs-EOR with molecular diffusion rate.***

12.4 SUMMARY

A sensitivity analysis on the critical parameters affecting CO_2-EOR and NGs-EOR was conducted in this chapter. The RSM and OPAAT approach gave a clear understanding of how CO_2-EOR and NGs-EOR performances vary under different conditions. The main conclusions from this chapter are as follows:

1. Formation porosity is the most important factor when predicting the success or failure of CO_2-EOR in shale formations. Shale formation with a high fracture intensity has an EOR advantage when using CO_2 injection.
2. Formation permeability has a negative effect on both CO_2-EOR and NGs-EOR.
3. The molecular diffusion mechanism has the dominant role over the other mechanisms in controlling the success or failure of both CO_2-EOR and NGs-EOR in ULR. However, NGs have different properties than CO_2, which make the molecular diffusivity behavior much different in the NG performance versus the CO_2 performance.
4. The results of this study indicate that NGs-EOR performance exceeds the performance of CO_2-EOR in formations with very small pore throats (i.e., permeability in range of 0.00001–1 mD). However, injecting CO_2 is highly recommended over injecting NGs in the reservoirs with large pore throats.

5. It has been observed that NGs do not function as strongly as CO_2 under natural fracture intensity. Due to the small molecular weight of NGs, they do not require a contact area as large as that required by CO_2.

References

CMG's GEM 2017. *User's Guide, Version 2017.10*. Calgary, Alberta, AB, Canada: Computer Modeling Group, Ltd.

CMG. 2018. Computer modeling group, GEM manual. https: www.CMG.Ca/ accessed 2018.

Sheng, J. J. (2015). Enhanced oil recovery in shale reservoirs by gas injection. *Journal of Natural Gas Science and Engineering, 22*, 252–259. https://doi.org/10.1016/j.jngse.2014.12.002.

Wan, T., & Sheng, J. (2015). *Compositional modelling of the diffusion effect on EOR process in fractured Shale-Oil Reservoirs by gasflooding*. Society of Petroleum Engineers. https://doi.org/10.2118/2014-1891403-PA.

Index

Note: Page numbers followed by "f" indicate figures and "t" indicate tables.

A

Activation energy, 102–104, 107t, 207
Adsorption effect, on geomechanics coupling, 233, 235
Air combustion
- dry forward combustion, 90
- reverse combustion, 94
- wet forward combustion, 93

Air-foam injection, 192
Air injection
- advantages and disadvantages of, 100
- in conventional reservoirs, 98
- EOR modelling, 97
- pilots in tight oil reservoirs, 103
- recovery mechanisms of, 88
- in unconventional reservoirs, 102

Amphiphilic surfactant molecules, 148
Areal sweep efficiency, 118f

B

Bacterial consortia, 190
Bakken formations, 20, 25f, 26
- CO_2-EOR huff-n-puff process, 40
- rock and fluid properties of, 19

Bakken model, EOR stage for, 176
Biopolymers, 190
Biosurfactants, 190
Bond scission, 95

C

Cap rock, 5
Carbonated water injection (CWI), 121
Chemical enhanced oil recovery methods, for unconventional reservoirs, 141
- surfactant injection effect of
  - during fracturing process, 151
  - mechanisms and types of, 145, 146
  - selection criteria, 154

surfactant selection criteria data compilation, 155
- wettability and interfacial tension, 142t, 155, 158f

CNF. *See* Complex nanofluid (CNF)
CO_2-enhanced oil recovery
- geomechanics coupling on
  - adsorption effect, 233
  - compositional models for formation fluids, 171
  - dissolution, 236
  - dissolution effect (water solubility effect), 236
  - geomechanics coupling effects, 237
  - molecular diffusion, 169
  - molecular diffusion effect, 231
  - multicomponent adsorption, 221
  - reservoir modelling, 228
  - unconventional liquid-rich reservoirs, flow type in, 218
- huff-n-puff pilot test, in Bakken formation, 56
- huff-n-puff projects, in conventional reservoirs, 49
- OPAAT methodology, 251
- Performance
  - nanopores confinement on, 212, 213
- in pilot projects, 77
- rock and fluid properties of unconventional plays, 19

CO_2-molecular-diffusion mechanism, 18
Cold-water injection, 124, 125f
Combustion reactions, modes of, 95
- bond scission or low-temperature oxidation, 95
- oxygen addition, 95
- spontaneous combustion or high-temperature oxidation, 96

Complex nanofluid (CNF), 151
Conventional reservoirs
- air injection pilots in, 97, 103
- CO_2-EOR huff-n-puff projects in, 49
- history matching for CO_2-EOR in, 49
- surfactant selection criteria in, 154
- versus unconventional reservoirs, 160

Core-sample bulk volume, on oil recovery obtained by CO_2 injection, 37
Counter-current mechanism, 18, 31, 254
CSS. *See* Cyclic steam stimulation (CSS)
CWI. *See* Carbonated water injection (CWI)
Cyclic steam stimulation (CSS), 186

D

Data compilation, surfactant selection criteria, 155
Decline zone, 51, 60
Displacement efficiency, 119, 120, 144t
Dissolution, on geomechanics coupling, 237
Dry forward combustion, 90

E

EE method. *See* Elementary effects (EE) method
Elementary effects (EE) method, 250
Energized fluids, for enhanced oil recovery methods, 192
Energy Information Administration (EIA), 20
Enhanced oil recovery (EOR), 100, 121, 141, 168, 185, 186, 206
  air-foam injection, 192
  chemical blends based EOR for tight and shale oil reservoirs, 196
  CO_2 huff-n-puff followed by surfactant assist spontaneous imbibitions (SASI) technique, 195
  energized fluids for, 192
  fracturing increased oil recovery (IOR) techniques, 196
  mechanisms by thermal techniques, 185
  microbial EOR, 188
  nanopores confinement on natural gases based, 210
  OPAAT methodology, 251
  CO_2-EOR, parameters which control, 251
  NGs-EOR, parameters which control, 257
  pore confinement, 206
  reservoir modelling, 228
  sensitivity analysis, 249
  thermal techniques for shale oil reservoirs, 187
EOR stage, for Bakken model, 176
Exothermic intensity of LTO reactions, 95
Explicit coupled approach, 223
Exposure time, on oil recovery obtained by CO_2 injection, effect of, 37

F

Flow-type determination, 175
Formation permeability on oil recovery obtained by CO_2 injection, 28
Formation total porosity, 254, 259
Fracturing improved oil recovery (IOR) techniques, 196

G

Gas cyclic versus continuous flooding, 6
Gaseous CO_2, 192
Geomechanics coupling effects, on geomechanics coupling, 237
Geomechanics coupling, on CO_2 -enhanced oil recovery, 221
  adsorption effect, 233
  compositional models for formation fluids, 227
  dissolution, 221
  dissolution effect (water solubility effect), 236
  geomechanics coupling effects, 237
  molecular diffusion, 168, 177, 218
  molecular diffusion effect, 231
  multicomponent adsorption, 221
  reservoir modeling, 228
  unconventional liquid-rich reservoirs, flow type in, 218

H

High-temperature oxidation (HTO), 92, 96
HTO. *See* High-temperature oxidation (HTO)
Hydrocarbons storage, 2

I

IFT reduction. *See* Interfacial tension (IFT) reduction
IIFs. *See* Injection induced fractures (IIFs)

Impermeable rock Improved oil recovery (IOR) in unconventional reservoirs chemical method, 149t
Infill drilling, 6
Injection induced fractures (IIFs), 81
Injectivity problems, 10, 148
Intensive natural fractures, 150
Interfacial tension, 88, 126, 145, 158f, 159f, 190, 192, 213
Interfacial tension (IFT) reduction, 128t
IOR. *See* Improved oil recovery (IOR)
Iterative coupled approach, 222

L

Laboratory CO_2 EOR projects, dataset for, 27
Liquid CO_2, 192
Liquids-rich shale, 127, 151, 153, 161
Low-salinity water (LSW) injection, 121
Low-temperature oxidation (LTO), 88, 95
LS-LR-DK (logarithmically spaced, locally refined, and dual permeability) method, 169, 245
LSW injection. *See* Low-salinity water (LSW) injection
LTO. *See* Low-temperature oxidation (LTO)

M

Macroscopic efficiency (EV), 142t
Matrix permeability, 18, 26, 102, 104, 171t, 207, 230t, 248t, 251t, 252, 254
Matrix porosity on, oil recovery obtained by CO_2, 31
Mean pore throat radius, on oil recovery obtained by CO_2 injection, 33
MEOR. *See* Microbial enhanced oil recovery (MEOR)
Microbial enhanced oil recovery (MEOR), 189
Microscopic displacement efficiency, definition, 122
Minimum miscible pressure nanopores confinement on, 210
Miscible gases
  based EOR, experimental and simulation studies on injection pilots, 67, 70
Molar diffusion mechanism, 55, 178
Molecular diffusion, 18, 55, 60f, 61, 168, 218
  definition, 168
  effect on geomechanics coupling, 237
  on geomechanics coupling, 218
Morris method, 254, 255f
Multicomponent adsorption, on geomechanics coupling, 221

N

N_2 -enhanced oil recovery nanopores confinement, 213
Nanopores confinement
  on air injection performance, 207
  on CO_2 -EOR performance, 207
  N_2 –EOR, 213
Natural depletion, 174
Natural fracture porosity, 259
Natural gases based enhanced oil recovery in pilot projects, 72
NGs-EOR, OPAAT methodology, 251
Numerical simulation, 49, 54, 57, 67, 70, 105, 147, 169, 173, 208, 223, 246

O

Oil saturation, on oil recovery obtained by CO_2 injection effect, 34
Oil swelling, 9t, 10t, 18t, 47, 49, 51, 62, 88, 89, 122, 171, 190, 227, 246
Oil viscosity reduction, 10t, 23t, 49, 51, 62, 89, 131, 166, 190, 220, 254
One parameter at a time (OPAAT) methodology, 257
  parameters which control, 251
    CO_2 -EOR, 251
    NGs-EOR, 257
  3-D two-way coupling, 222
OOIP. *See* Original oil in place (OOIP)
Original oil in place (OOIP), 6, 99, 141
Oxygen addition, 88, 95, 105

P

PCSAFT. *See* Perturbed-chain statistical associating fluid theory (PCSAFT)
Peak zone, 59
Péclet (Pe) number, 168, 169f, 175f, 176f, 208, 209f

Peng–Robinson equation-of-state fluid description/binary interaction coefficients, 248t
Perturbed-chain statistical associating fluid theory (PCSAFT), 210
Pore confinement
on enhanced oil recovery methods, 206
on primary production, 204
Pressure and temperature, on oil recovery obtained by CO_2 injection, 39
Pseudo components, 227, 246
Pseudo response zone, 59

R

Real response zone, 50, 59, 60
Reservoir modeling, 58, 228, 245
Residual oil saturation, 116, 119f, 130
Reverse combustion, 94
Rock and fluid properties of unconventional plays, 19

S

SAGD. *See* Steam-assisted gravity drainage (SAGD)
SASI technique. *See* Surfactant assist spontaneous imbibitions (SASI) technique
Sensitivity analysis (SA), 40, 249, 251t
Sensitivity parameters, 250
Shale core samples, 49, 51
Shale oil reservoirs
air injection in, applicability of, 104
CO_2 -molecular-diffusion mechanism, 18
core-sample bulk volume on oil recovery obtained by CO_2 injection, effect of, 37
dataset for laboratory CO_2 EOR projects, 27
exposure time on oil recovery obtained by CO_2 injection, effect of, 37
formation permeability on oil recovery obtained by CO_2 injection, effect of, 28
identification of flow type in, 208
matrix porosity on oil recovery obtained by CO_2 injection, effect of, 31
mean pore throat radius on oil recovery obtained by CO_2 injection, effect of, 33
oil saturation on oil recovery obtained by CO_2 injection, effect of, 34
pressure and temperature on oil recovery obtained by CO_2 injection, effect of, 39
rock and fluid properties of unconventional plays, 18
sensitivity analysis, 40
thermal techniques for, 185
total organic carbon content on oil recovery obtained by CO_2 injection, effect of., 30
water injection in, 129
water saturation on oil recovery obtained by CO_2 injection, effect of, 36
Sobol method, 250
Sorption process, 233
SRV. *See* Stimulated reservoir volume (SRV)
Steam-assisted gravity drainage (SAGD), 186
Stimulated reservoir volume (SRV), 196, 228, 246, 247f
Strain displacement relations, definition, 224
Surfactant assist spontaneous imbibitions (SASI) technique, 195

T

Thermal enhanced oil recovery techniques, 185
Thermal techniques
mechanisms by, 189
for shale oil reservoirs, 187
Tight oil reservoirs chemical blends based EOR for definition, 196
TOC. *See* Total organic carbon content (TOC)
Toe-Heel pattern, 72, 75
Total organic carbon content (TOC), 27, 30, 68
Total organic carbon content, on oil recovery obtained by CO_2 injection, effect of, 31
3-D two-way coupling, 222

U

ULR reservoirs. *See* Unconventional liquid-rich (ULR) reservoirs
Ultra-low permeability reservoirs, 7
Ultralow porosity, 131
Unconventional liquid-rich (ULR) reservoirs, 16, 218
Unconventional reservoirs
  air injection in numerical simulation for, 105
  chemical enhanced oil recovery methods for
    surfactant injection, 151
      effect of, 142t, 147, 154
      mechanisms and types of, 146
      selection criteria, 154
      wettability and interfacial tension, 155, 158f
  CO_2-enhanced oil recovery huff-n-puff operation in, 51
  enhanced oil recovery methods for
    air-foam injection, 192
    chemical blends based EOR for tight and shale oil reservoirs, 196
    CO_2 huff-n-puff followed by surfactant assist spontaneous imbibitions (SASI) technique, 195
    energized fluids for fracturing increased oil recovery (IOR) technique, 192
    mechanisms by thermal techniques, 185
    microbial EOR, 188
    thermal enhanced oil recovery techniques, 185
  thermal techniques for shale oil reservoirs, 187
  water injection in
    field applications, 132
    huff-n-puff mode versus flooding mode in shale oil reservoir, 129
    oil recovery due to water flooding, 117
    performance of, 120, 130
    shale oil reservoirs, 126
    in unconventional reservoirs, 124
    waterflooding-derived techniques, 121
    waterflooding, performance of, 120
Unconventional resources, 2
  definition, 2
  IOR methods in, 6
Unstimulated reservoir volume (USRV), 169
USRV. *See* Unstimulated reservoir volume (USRV)

V

Vertical sweep efficiency, 118, 118f, 119

W

WAG injection, 121
Water-assisted combustion, 93
Water clay-bearing rocks, 130
Water CO_2, 126
Water flooding
  derived techniques, 121
  oil recovery due to, 117, 117f
Water huff-n-puff protocol mechanism, 126, 130
Water injection
  field applications, unconventional reservoirs, 132
  huff-n-puff mode versus flooding mode in shale oil reservoirs, 130
  oil recovery due to water flooding, 117
  shale oil reservoirs, 126
  unconventional reservoirs, 124
  waterflooding
    derived techniques, 121
    performance of, 120
Water saturation, on oil recovery obtained by CO_2 injection, effect of, 36
Water solubility effect, on geomechanics coupling, 236
Wet forward combustion, 93
Wettability, 8t, 10t, 24t, 116, 122, 124, 126, 127, 146, 147, 149t, 150t, 154
WinProp-CMG, 171, 227, 231f

Y

Young-Laplace capillary pressure equation, 154

Z

Zipper fracs method, 197

9780128183434